EXPLORERS OF
the Maritime Pacific Northwest

Recent Title in
Mapping the World through Primary Documents

Explorers of the American West: Mapping the World through Primary Documents
Jay H. Buckley and Jeffery D. Nokes

EXPLORERS OF the Maritime Pacific Northwest

Mapping the World through Primary Documents

William L. Lang and James V. Walker

Mapping the World through Primary Documents

ABC-CLIO™

An Imprint of ABC-CLIO, LLC
Santa Barbara, California • Denver, Colorado

Library of Congress Cataloging-in-Publication Data

Names: Lang, William L., editor. | Walker, James V., 1944- editor.
Title: Explorers of the maritime Pacific Northwest : mapping the world through primary documents / [compiled by] William L. Lang and James V. Walker.
Description: Santa Barbara, California : ABC-CLIO, 2016. | Series: Mapping the world through primary documents | Includes bibliographical references and index.
Identifiers: LCCN 2015042134 (print) | LCCN 2016002875 (ebook) | ISBN 9781610699259 (hardback) | ISBN 9781610699266 (ebook)
Subjects: LCSH: Northwest, Pacific—Discovery and exploration—Sources. | Cartography—Northwest, Pacific—History—Sources. | Northwest, Pacific—History—Sources. | BISAC: HISTORY / North America. | HISTORY / Americas (North, Central, South, West Indies).
Classification: LCC F851.5 .E976 2016 (print) | LCC F851.5 (ebook) | DDC 979.5—dc23
LC record available at http://lccn.loc.gov/2015042134

ISBN: 978-1-61069-925-9
EISBN: 978-1-61069-926-6

20 19 18 17 16 1 2 3 4 5

This book is also available on the World Wide Web as an eBook.
Visit www.abc-clio.com for details.

ABC-CLIO
An Imprint of ABC-CLIO, LLC

ABC-CLIO, LLC
130 Cremona Drive, P.O. Box 1911
Santa Barbara, California 93116-1911

This book is printed on acid-free paper ∞

Manufactured in the United States of America

To all students of exploration, maps, and the history of Native peoples
of the Northwest Coast

Contents

Preface

This book is the result of a few chance meetings between a historian and a cartography expert and collector, discussions about our mutual interests, and the discovery that we both were fascinated with the history of exploration and cartography, especially maritime expeditions to the Northwest Coast in the 17th and 18th centuries. For armchair exploration and map enthusiasts, we asked ourselves: What could be more vicariously thrilling than sailing with James Cook or Juan Francisco de la Bodega y Quadra into the rough seas and complex archipelagoes of a wild coastline? And what could be more intellectually stimulating than peeling away the layered maps that cartographers created to represent what explorers saw along the northern edges of North America and Asia? We knew that understanding that history required digging into two distinct subjects, one at sea on pitching decks and encountering Native peoples and the other in the minds of the makers of charts and maps who created representations of the intersections of ocean and land in straits, bays, harbors, and dangerous reefs and shoals. We knew that the subjects were too immense to wrap together in one volume and that the work of thousands of sailors and cartographers were too numerous and varied to meld into one picture, but we also knew that probing even a few of the stories of exploration and mapmaking could reveal some of the truly revolutionary additions to human knowledge that fired our imaginations and interest.

Extraordinary people and exceptional events are at the center of this history. We began by identifying 10 major exploratory experiences that greatly expanded cartographic knowledge about the North Pacific. Each expedition or series of voyages, as in the case of fur trade excursions, is a chapter in the book. Because a main commander or captain led each effort, we begin each chapter with a biography. We also include a simplified map that approximates the route of each exploration except for the multiyear voyages of George Vancouver. But the primary windows on the explorations are 10 or so primary documents that take readers to sea level, from on-the-scene descriptions of the mariners' experiences at sea to their often tenuous engagement with Natives to their return home with information about a remote part of the world.

The documents are arranged to set the background for each expedition, the instructions that the captains carried with them, a few principal events in the voyages and their human costs, the main results of the explorations, and, importantly, how the expeditions expanded the world's cartographic understanding of the North Pacific. Each document includes multiple topics, so we added a brief introduction and longer analysis in each document section to provide context and explain important relationships. Many of the documents reveal personal reflections of the principal actors because they come directly from the commanders' journals or the official logs of the voyage. We also included maps in each chapter to document the ways mariners charted islands and continental coastlines and how those surveys became part of larger maps. European mapmakers sometimes incorporated preconceived ideas with information from direct observation to create maps that they believed were realistic and accurate. The maps also included geopolitical and cultural information, such as place-names, that reflected competition among imperial maritime nations in the Pacific Northwest.

We selected documents that reflect experiences common to most expeditions, such as the continual threat of scurvy on long voyages and the unpredictability of engagements with Native peoples. The unusual and even exceptional events, of course, are also important, such as the violent death of James Cook, the loss of an entire French expedition in 1788, and Robert Gray's sailing into the Columbia River in 1792. It is our hope that readers will be captured as we are by these stories of exploration and mapmaking. There is a lot here, and we found at nearly every turn something new to us, which is one of the main reasons why plunging into a subject such as this never disappoints.

The making of this book benefited from the aid of many people and institutions. The libraries we used, including the University of Oregon, the University of Washington, Portland State University, and the Oregon Historical Society, were crucial to finding and acquiring documents and supporting materials. Online resources, which continue to expand, offered digital primary sources, maps, and background materials. In particular, we want to thank the interlibrary loan staff at Portland State University's Millar Library, Scott Daniels of the Oregon Historical Society Research Library in Portland, and Alethea Steingisser of InfoGraphics at the University of Oregon for the preparation of historic map images and June Black and Lindsay Keast for assistance in obtaining these images.

We owe the readability of the manuscript to an exceptional editor, Marianne Keddington-Lang, who kept us from writing vague or confusing prose and asked us to answer difficult questions. Her experience as a classroom teacher, in addition to a career in scholarly publishing, made her a perfect editor for this project. We also got help from regional writer and exploration specialist Jack Nisbet, who critiqued sections of the manuscript, and from Barbara Walker, who contributed valued editorial advice. Amy Platt created the index. At ABC-CLIO, Kaitlin Ciarmiello encouraged us to pursue the project with some latitude and gave us sound advice on what to include and exclude. Finally, we could not have even considered working up this volume were it not for the astonishing edited journals of the major expeditions, many published by the Hakluyt Society, and the hundreds of books and academic articles on cartography, maritime history, the ethnology of the Northwest Coast, and the pelagic fur trade.

Introduction

We are a migrating species. For thousands of years, humans have spread across the globe from our origins in Africa. By the time of the agricultural revolution some 10,000 years ago, human beings had spread to every habitable continent and major archipelago on the planet. The diversity of human culture over time is the result of migration, the development of communities, and the expansion of some groups at the expense of others. From ancient times, people have used economic and military power to expand their control over land and sea resources. During thousands of years of expansion by ambitious polities in Asia, Europe, South America, and Africa, empires developed, flourished, and declined. The movement of people and their resettlement in new lands, marked by the conquest of other groups and the consequential spread of culture, is the most powerful force of social change in human history. This book is about a particularly dynamic part of that process that took place in the modern era, the period since the Renaissance. In the 15th century, European nations unleashed revolutionary change on the world when they invented deepwater sailing, which relied on a set of technological advances that allowed ships to safely travel across oceans between continents. Up to that time, nearly all vessels navigated seas and along continental ocean coastlines. Only rarely did they cross oceans. The Norse seafaring to North America about 1,000 years ago, for example, was a notable exception, as were the seafaring migrations of groups from Asia to South Pacific islands thousands of years ago. But there was nothing like the Age of Sail, which prevailed from the 15th to the late 19th centuries. It was something new and consequential, and it shaped global change.

The Age of Sail and great explorations of the Pacific Ocean were the result of several coincident and momentous developments. To understand its dimensions and consequences requires us to investigate the histories of technology and science, geopolitical change, and the social context of European expansion. The focus in this book is on a relatively brief period in the Age of Sail: exploration and discovery in the North Pacific Ocean during the 17th and 18th centuries. The North Pacific was one of the last maritime regions to be mapped using modern technologies and graphic depiction. The Pacific had challenged mariners for more than two centuries before the maritime expeditions covered in this book sorted out the relationship of continents and major island archipelagoes between North America and North Asia. It took more than a century of sailing the Pacific, for example, before mariners grasped the ocean's size, more than 45 percent of Earth's water surface. For decades after the first circumnavigation, the Ferdinand Magellan–Juan Elcano voyage in

1519–1522, sailors had crossed the Pacific, finding routes between the coasts of the Americas and Asia. They also searched for two mythic locations, a great continent in the South Pacific and a Northwest Passage through North America that would have shortened the sailing distance between Europe and Asia. By the time mariners shifted their attention to the North Pacific, Spanish, British, and Dutch ship captains had crossed the 10,000-mile Pacific to circumnavigate the world. Between 1603 and 1795, the years of maritime exploration covered in this book, French and American mariners also completed circumnavigations. All of these expeditions in the Pacific resulted in significant changes in patterns of economic exchange around the globe and historic contacts between indigenous peoples and Europeans and Americans. Most important, these explorations generated maps and maritime charts that visually communicated an immense expansion of knowledge about the world, its diversity, and new opportunities for ambitious nations.

This book covers explorations of the North Pacific region, especially the Northwest Coast of North America, beginning in the early 17th century and concluding in the late 18th century. The most concentrated activity, though, lasted for less than six decades, from the 1740s to the 1790s, when Europeans and Americans launched multiple maritime expeditions into a huge area that reached from latitudes 42° north to 70° north and from longitudes 124° west to 160° east. The maps drawn by the mariners discussed in this book covered an enormous geography that spanned the arc of two nearly contiguous continents and millions of square miles of seascape, from Cape Mendocino on the California coast to the Kamchatka Peninsula in eastern Siberia. Their maps were integrated into published geographies that became fundamental to European and American colonial development in the Pacific.

It is difficult to exaggerate the importance of these explorations and the power of the maps they produced. Thinking about the voyages and their results, though, requires that we make an important distinction between exploration and discovery. The two terms are not synonymous. An exploration is based on geographical knowledge, which includes documented relationships among known places and a range of assumptions about what an explorer might find or should seek. The expeditions documented here are explorations, because they were predicated on known and conceptualized geographies that the mariners expected to observe. The official instructions to mariners from expedition sponsors, some of them included in the document sections of the book, generally list places that the explorers are expected to visit, locate, or survey. For the commanders, those particulars were often understood as potential measurements of an expedition's success. In geographical explorations, however, discovery is a narrower term. It refers to unanticipated findings, true additions to knowledge, or confirmations of speculated places. The documents included in this book are filled with perceptions and observations noted by mariners that shocked them or caused them to question their assumptions and even goals. Many discoveries, of course, are relative because they refer to what the discoverer knew and did not know. Christopher Columbus discovered islands in the Caribbean in 1492, but it was not news to the Native inhabitants. Absolute discovery, however, is less frequent and applies to knowledge that is new to humans. Most of the discoveries noted by the explorers in this book are relative, but some were absolute, including accurate geographical place documentation, inclusion of plants and animals to the scientific catalog, and seafaring conditions.

The voyages featured in this book document the accretion of knowledge about an enormous area of the globe, a stunning range of space, environments, Native communities, and ocean conditions. In their Pacific exploratory voyages, the mariners encountered semitropical conditions in Hawaii and other Pacific islands, icebergs and ice floes in the Arctic and

Antarctic regions, and dangerous sailing impediments in narrow inlets and shoaled bays that characterized North American and Asian coastlines in the North Pacific. The region is part of what has been called the Pacific Ring of Fire, which refers to plate tectonic intersections that have created volcanic islands, deep ocean trenches, and mountains that rise directly from coastlines.

The expeditions also had distinct purposes. Most were sponsored by governments that desired geographical information to benefit them in international competitive relationships, find new economic resources, claim territories for colonization, and expand scientific knowledge. To those ends, the expeditions often included scientists to study flora and fauna and make detailed observations. Astronomers and cartographers were also part of most crews. Their technical observations and documentation constitute one of the great intellectual contributions of these voyages. Private and entrepreneurial expeditions that pursued the pelagic fur trade also expanded cartographic understanding of the North Pacific. All expeditions recorded encounters with Native people in an amazing panoply of social and economic conditions, and the documents describing those encounters constitute one of the voyages' other great intellectual contributions. Regardless of the purpose and status of the expedition, all required exceptional planning, technical expertise, and cutting-edge technology. They were their era's equivalent of the space explorations in the 20th century. Men and ships ventured from safety into unknown and often dangerous conditions, casting mariners from home for years and enduring dangers and privations with few or no opportunities for rescue. The ships were self-contained communities, which required crews to keep their vessels seaworthy and work in cooperative and efficient ways to avoid peril. Desertion forced captains to scramble for replacements in foreign ports, often diminishing the crew's effectiveness. And all captains had to maintain discipline aboard ship, which often necessitated punishments that were administered by lash and usually on deck for all to view. Maintaining the crew's health was imperative, and captains paid close attention to keeping water supplies pure and taking advantage, whenever possible, to acquire fresh foods. Most vessels carried live animals to provide milk, eggs, and meat, putting additional burdens on the crew to tend and feed them. James Cook recorded at the start of his third voyage, for example, that the crew "took on board a Bull, 2 cows with their Calves & some sheep . . . with a quantity of Hay and Corn for their subsistence" (Beaglehole 1967, 1:4).

The most fearsome health worry, though, was scurvy, a progressive and debilitating disease caused by inadequate vitamin C ingestion. Scurvy could ravage a ship, leaving afflicted men weak and in danger of death. It plagued all long-distance voyages and encouraged captains to devise an inventive range of precautionary measures. Some of those measures appeared to work, but no one understood where scurvy came from, and its cause was not known until the second and third decades of the 20th century. Long-distance voyages were dangerous. Vitus Bering perished from scurvy in 1741 on an island near Kamchatka, and British mariner James Cook was killed by Hawaiians in 1779. Nine years later, French commander Jean-François La Pérouse lost his life, along with his entire crew, in a shipwreck in the South Pacific.

The inherent peril in these voyages and their enormous expense made them heroic, risky, and always promising for their sponsors, whether they were governments or trading companies. None would have occurred without significant historic changes during the modern era in science and technology and the influence of European nations' political interests. The documents and analyses in this book focus on subjects that help us understand the history of Northwest Coast maritime expeditions and how broad changes in human endeavors made this era of exploration so momentous.

How National Interests Shaped Explorations

Most of the expeditions included in this book were launched by European nations for political and economic purposes. They were part of a larger contest among nations over the world's natural resources, control of territories and ports, and prospects for colonization. In addition, a desire to expand the boundaries of scientific knowledge motivated the sponsors and mariners who participated in what some historians call the Second Great Age of Discovery. Each nation's participation varied in scope, genesis, and aims, voyage by voyage. In some cases there was direct competition on the high seas, but in others there was genuine collaboration, especially when human safety was at stake or when the pursuit of scientific information would profit everyone. By the end of the 15th century, Portugal and Spain dominated long-distance ocean sailing. Portugal had opened trade to the Spice Islands in present-day Indonesia, and both nations had crossed the Atlantic to the Americas, as had Britain. In 1494, the two countries arrogantly divided the world's newly found lands for their own enhancement in the Treaty of Tordesillas. Spain received hegemony over lands west of a line set 1,200 miles west of the Cape Verde Islands off Africa, while Portugal had dominion over territories east of the line. By the mid-16th century, after Magellan's circumnavigation, determining where the Tordesillas line should fall prompted argument, but Spain held firm that it had legitimate entitlement to the Pacific from Mexico to the Philippine Islands. Spain struggled to perfect or even project the claim, but the issue became so contentious with Britain in the 1790s that war nearly resulted. Nonetheless, Spanish mariners referred to the treaty in possession-taking ceremonies of lands they claimed throughout the late 18th century in the North Pacific.

The idea of nations and companionate ideas of national boundaries and interests developed during the 17th century and became codified in the Treaty of Westphalia in 1648, which ended the ghastly Thirty Years' War. That agreement created a modern international system of mutual respect, but it surely did not diminish competition among states, nor did it eliminate illegal activities and a resort to violence and war. In the early 16th century, Spain had established a port on the Mexican coast at Acapulco, where it launched northern voyages to explore and expand to present-day California and to protect its claims in the North Pacific. By the 18th century, Spain had also been profiting from two centuries of trade—the Spanish galleon voyages—that connected Manila in the Philippines to Mexico and Spain and carried gold and silver from Spanish America and Asian goods for European sale. In short, Spain had much to protect in the Pacific, and this explains why its ships played such a central role in exploration of the North Pacific. This also explains why Spain attempted to keep its maps and charts practically under lock and key. In large measure, it would be other nations' explorations that advanced the world's cartographic knowledge of the region.

Other nations pursued their interests in the Pacific, but unlike Spain their purposes were offensive, not defensive. By the time they launched their voyages, the revolutionary scientific, political, and social ideas of the Enlightenment had become a powerful motivating force in European affairs. Russia brought Spain its first direct challenge, after the Enlightenment-informed Tsar Peter the Great had expanded his nation's influence to the Pacific. From a Russian base on the Kamchatka Peninsula, a land appended to Siberia in the far western Pacific north of Japan, Vitus Bering led an expedition across frigid seas in 1741 and attempted to map the strait separating North America from Asia and large parts of the North American coastline in present-day Alaska. In 1752, Joseph Delisle published a map depicting Bering's track as well as an apocryphal inland sea and a water passage across North America. This was a tantalizing suggestion that a long-sought water passage through the continent could offer a dramatic shortcut between Europe and Asia. This pathway, often called the Northwest Passage, had been a holy grail for maritime nations since the 16th century.

Three decades later, Britain brought a second challenge to Spain in the North Pacific when James Cook mapped the region on his third exploratory voyage. Among his primary goals was searching for an inlet opening to a Northwest Passage. His failure to find the passage was a minor disappointment, because his magnificent map became a baseline for further explorations and justified claims of national interest by Britain in the region. In addition, after Cook's death in Hawaii his crew continued on to China, where they discovered that the pelagic furs they had gathered on the Northwest Coast had a high value. News of the profitable trade prompted merchants to sponsor commercial voyages to the region.

The publication of Cook's journals and maps in 1784 also spurred European nations to send scientific expeditions to the North Pacific that included explicit surveys of coastal areas and islands, detailed mapping and charting of harbors, and engagements with Native people. Nations competed for any potential advantage in the resource-rich region, and their governments directed them to continue searching for a Northwest Passage. The French sent Jean-François La Pérouse in 1785, and Spain sponsored a scientific voyage led by Alejandro Malaspina in 1788–1791. Britain continued the work begun by Cook, sending George Vancouver on a voyage in 1791 that carried notable scientists and produced the most complete maps of the region, including the complex waterways of Puget Sound. During those voyages, captains often performed rituals of discovery by burying a bottle or other memento to signify a national claim of possession. These ceremonies announced more than "we were here," because the physical observation and especially mapping of places counted in a competition that ended in Europeans dividing a good portion of the non-European world explored by their maritime expeditions.

An added and important factor in the international competition to explore the North Pacific was Europe's revolutionary political environment during the latter half of the 18th century. The revolutionary wars in America and France were fought on the seas, and their outcomes shook many relationships between key naval powers. The naval wars aided nations by stimulating shipbuilding and training sailors, but they often put ship captains in difficult political positions far from home. In all, the North Pacific exploratory voyages were creatures of national and international politics, while most of their achievements extended far beyond the outcomes of even the most important political events.

A Revolution in Technology and Science Changed Everything

The exploratory voyages to the Pacific were heroic and technological marvels, and their successes were impressive. Some voyages still astonish us for their exceptionally daring and skillful exploits, but behind them were even more profound developments in scientific understanding and inventive technologies. By the 18th century, nearly 200 years of scientific and intellectual discoveries had changed how humans understood their world. Some knowledge in mathematics had been rediscovered from antiquity by Renaissance scholars, but the most powerful discoveries came from controlled experimentation and observation. Isaac Newton's experiments in optics, his creation of calculus for mathematical description, and his explanation of motion, along with study and experimentation by many others in biology, astronomy, mechanics, and electricity, powered a dynamic investigation of how the world worked. The Royal Society for Improving Knowledge, formed in 1660 in London, and the French Academy of Sciences, organized in 1666 in Paris, set the model for international advancement in scientific knowledge. By the 18th century, several European nations had established their own versions of those institutions. The Enlightenment—which challenged religious authority, argued for the primacy of reason and legal process in human affairs, and supported political change—was largely born from these organizations and other centers of learning and by the 18th century had become the dominant intellectual movement in Europe and its colonies.

Two scientific experiments in the 18th century had important implications for explorers: the measurement of the length of a degree of latitude at the equator in Ecuador in 1736 and the observation of the transit of Venus in Tahiti in 1769. Charles de la Condamine, a French scientist, conducted surveys along a longitudinal meridian in an Ecuadorian valley to test Isaac Newton's theory that Earth bulged at the equator and narrowed at the poles, a difference that made the distance between latitude parallels shorter from equator to pole. Condamine confirmed Newton's mathematical theory. The observation of the transit of Venus, which prompted James Cook's first voyage to the Pacific (1768–1771), engaged scientists in different locations to use the transit—seen as a shadow moving across the sun—to measure the distance of the sun from Earth. For mariners, though, experiments that addressed the longitudinal navigation problem made the most difference.

The problem of longitude was the challenge of determining a ship's location east to west on the ocean. Seafarers had long been able to establish latitude, their north-to-south location, by using a sextant to determine the angle of their position in relation to a direct sighting of the sun or a primary celestial object at night. A figurative line drawn from the astronomical object to the ship, like the side of triangle, could locate latitude. Determining longitude—how far they were east or west of a located place—was much more difficult. The problem mattered less in an enclosed sea such as the Mediterranean, but knowing longitude on the ocean could determine success and even survival on a voyage. Cartographers drew meridian lines on a globe connecting the North and South Poles to divide Earth's 360° circumference into 24 equidistant longitudinal sections. Each meridian was 15° apart, which made them one hour different in time, a 24th fraction of Earth's daily rotation. Determining longitude, in other words, meant measuring distance by noting the time difference between a known location and the position of a ship at sea. A clock could provide the answer, of course, but no one had figured out how to build a clock that could function at sea. A standard pendulum clock that used a gravity-driven swinging pendant to move gears in the timepiece clearly would not work on the deck of a pitching ship. One solution was to use the position of the moon and other celestial objects, like hands on a clock, to determine the time on a ship. Astronomers created numerical tables that tabulated daily lunar positions, but this lunar-distance method required several calculations, often taking hours to complete. In addition, mariners could not see the moon clearly for up to 20 percent of the time, and an occluded night sky could make it impossible.

In 1714, following earlier prize rewards for a solution to the longitude problem by Spain, Portugal, and Holland, the British Parliament established a £20,000 payment ($2 million today) for anyone who found a practical way to find longitude within one half of a degree at sea, or accuracy within about 30 nautical miles. The Board of Admiralty in cooperation with the Royal Observatory at Greenwich invited proposals. In one of the more unlikely stories of technological advancement, carpenter John Harrison solved the problem in 1735 by constructing an elaborate time-measuring mechanism—a chronometer—using levers, counterbalances, springs, and special metals that could keep reliable time aboard ship. The fourth version of his chronometer—H4, which he completed in 1759—passed sea voyage tests in 1764 and became the basis for maritime chronometers built by instrument makers. James Cook carried a copy of H4 made by Larcum Kendall on his second voyage (1772–1775) and marveled at its accuracy and utility. All subsequent maritime expeditions to the Pacific carried chronometers, which improved the accuracy of chartmaking and increased the safety of all voyages.

Scientific motivations for maritime exploration were behind many voyages, especially those of Cook, Jean-François La Pérouse, Alejandro Malaspina, and George Vancouver. Scientists accompanied each voyage, their interests covering all facets of natural history, including documentation of flora and fauna species new to science. Carl Linnaeus had published *System Naturae* in 1735, and its logical and formulaic division of all living things into categorical groups, from inclusive types to individual species, gave scientists a way to catalog nature. The journals of the expeditions are peppered with references to Linnaean-based observations, which are mixed in with commonplace descriptions of sea life, terrestrial plants and animals, and measured observations of climate. The journals, in addition to documenting the voyages, offer a time-capsule view of those natural history discoveries. More important to posterity perhaps were the collected specimens that captains and scientists brought home. They carried back flora and fauna from Pacific coastlines and islands, and they also collected Native-made implements, clothing, and art. Some of those collections became part of national museums, such as the British Museum in England, established in 1753, and the Royal Cabinet in Madrid, founded in 1771, thereby extending the influence of these voyages to the present day.

Why Maps Have Power

Maps are among the most powerful documents in the history of human endeavor. Since antiquity they have served as descriptors of places and representations of relationships between specific communities and the larger world. In exploration, though, maps serve the instrumental purpose of projecting knowledge and imagination onto future investigations. Maps show what no person can possibly see: the spatial relationship between one occupied place and a distant range of other places, all of it in a vertical aspect whereby distances are easily comparable. The accuracy of any map, ancient or modern, is a function of acute measurements and clear presentation. The maps that figure in history told in this volume range from global views to nautical charts of coastal bays. The documents included tell the stories of their creation during exploring expeditions. The manuscript maps and charts created during those explorations became incorporated into larger representations of the globe. They were meant to present ideas about the shape of the world as well as the details of specific places. These maps also communicated ideas about imperial dominion and sovereignty. Cartographers borrowed, copied, and creatively manipulated geographical and nongeographical information to produce their versions of the world. The maritime maps in this book offer a visual history of the cartographic process, and that story documents an accretion of knowledge and perception of the world changed over time. Interestingly, cartographers often used much earlier geographical information that had been ignored or cast aside. It was not a strictly linear process.

During the Renaissance, European cartographers rediscovered Claudius Ptolemy's second-century conception of a world map, which changed how they made maps. On his world map, his ideas included the use of vertical meridian and horizontal parallel lines to indicate relative positions of places. Cartographers adopted Ptolemy's system and regularized the representation of space on maps. The printing press in the 15th century, however, truly revolutionized mapmaking by reproducing identical text and images in accurate representations that were broadly disseminated. Printers used woodcuts and copperplates that were either engraved by cutting lines into the metal or etched using metal-incising acid to create the lines and marks to represent continents, oceans, islands, and topographical features. Elaborate legends and decorative cartouches made many of these maps extraordinary works of art. With the use of legends, toponyms, symbols, and pictorial figures, mapmakers often communicated stories of encounters and events or political context and intention. The first widely distributed maps by Sebastian Munster in 1540, Gerard Mercator in 1569, and Abraham Ortelius in 1570 set in motion a process that continually revised views of the world. As observations by explorers expanded knowledge of areas, cartographers made corrections and added new information to each edition. Successively revised printed maps carried forward precise and authoritative representations, but that same precision also retained inaccurate reports of earlier maps and kept imagined geographies alive. By the 18th century, two hoped-for geographical features—a water passageway through North America and a southern continent—remained on some maps and pressed mariners to search them out.

James Cook played an important role on his second and third voyages in diminishing the likelihood of a Northwest Passage in the North Pacific or a southern continent in the South Pacific, and his expertly crafted maps and coastal charts became foundational for subsequent expeditions during a feverish period of exploration from 1785 to 1795. The lack of means to establish longitude at sea had consistently underestimated the size of the Pacific Ocean, and publications such as Munster's 1540 map of the Americas suggested a short sailing time between South America and Asia. Even after the circumnavigations by Ferdinand Magellan (1519–1522) and Francis Drake (1577–1580), published maps misrepresented relative distances between continents and islands. In addition, areas that had not been explored, such as the North Pacific, appeared vaguely represented, with imagined territories and sea passages filling space. By the time Cook made his three great voyages in the Pacific, maritime chronometers made accurate calculations of longitude possible, and a new accuracy revolutionized cartographic representations of the world.

More than increased general accuracy, though, the methods that mariners used to survey coastlines, especially bays and harbors, provided detailed sailing charts. Surveys used depth soundings, astronomical observation, and triangulation with transits to determine distances between locations. Engravers took the careful drawings on manuscript charts that used hachure lines to represent topographical elevations and physical landform relationships. Such improvements made it easier to recognize mapped locations, but more important to mariners, the coastal charts could save them from unanticipated navigational dangers. Each major exploration, from Sebastián Vizcaíno's survey of the California coast in 1602 to George Vancouver's charting of the Northwest Coast in the 1790s, added observed detail that cartographers included on their maps. What appeared on those maps combined the methods and decisions of mariners who created the manuscript charts, the choices that cartographers made to include new information and revise their maps, and the expertise of engravers to create images of diverse and distant places. Those maps successively became baseline references for subsequent surveys. As explorers used the maps generated from earlier explorations, they oriented themselves—for example, when Bruno de Hezeta recognized Point Reyes on the

California coast in 1774, a place that Vizcaíno had named in 1602, or when Cook spotted Mt. St. Elias on the Alaskan coast in 1778, named by Vitus Bering in 1741.

The names they applied always reflected their perspective. The origins of names could express the explorers' purposes and even disappointments in what they found or could not find. The name might express a view of the physical shape of the location or an appropriate commemorative for a historical incident, an individual associated with the date when the explorer first saw the place, or a religious reference. Some names were retained from their predecessors, but many were replaced with new names that honored fellow seamen or politicians. Toponyms (geographic names) are symbolic markers laid down by explorers, a kind of claim on location that orients a cape, bay, strait, or other place to a world of near and far places. The grid of squares, bounded by longitudinal meridians and latitudinal parallels, gives the named place a precise but relative location. More important, though, toponyms create powerful identities that convey cultural ownership. Those spots written down on maps give the named places a memory—where Bering, Cook, or Bodega, for example, encountered a particular place or experienced something notable. And names become utilitarian, because they communicate descriptions or specific ideas—Cook's Cape Foulweather on the Oregon coast and Vancouver's Deception Pass in Puget Sound—that make them important places where readers and users of maps plot their own relative locations and often make their experiences part of the named place's history. Names on maps, in other words, extend well beyond their original notations and affect our view of the world and its places.

How They Sailed the Seas

Shipbuilding technology and materials matured markedly in Britain, France, and Spain during the early modern era, from the 15th to the 18th centuries. The first important developments in shipbuilding that made deepwater sailing voyages feasible was the design of the caravel. Perhaps modeled on an Arab sailing vessel for navigating Atlantic bays, the caravel had two masts, a long sloping bow, and a tall enclosed castle cabin at the stern. It is the type of ship that Vasco da Gama, Christopher Columbus, and Amerigo Vespucci used to complete their voyages. Caravels were fast, maneuverable, and durable. Portuguese carracks, which were large vessels of 1,000 tons and more, dominated shipping in the East Indies in the 15th and 16th centuries. By the mid-16th century, Spanish shipbuilders had designed large galleon ships that were cargo vessels and warships. The ones used in the Manila-Acapulco trade could be 2,000 tons. English shipbuilders also designed exceptional galleons, such as the *Golden Hind* that Francis Drake used in his circumnavigation. The ships that explorers used in exploring the Northwest Coast were mainly frigates, brigantines, and sloops. Full-rigged ships had three masts and larger square sails for power to move the vessels at higher speed and had smaller fore and aft jibs, staysails, and mizzen sails to enhance maneuverability. Some ships had flat single top decks, while others had multiple top decks with castle cabin structures. Gun decks usually were below top decks, with cargo holds in the lowest portion of the ship and crew quarters on midlevel decks. The best of these vessels—Cook's *Endeavor,* La Pérouse's *Astrolabe,* Bodega's *Favorita,* and Vancouver's *Discovery*—were fast, capacious, and very maneuverable.

Maintaining ships on long voyages and repairing them when necessary required that captains and crews be knowledgeable about shipbuilding and carry sufficient materials for fabricating key components. This also required blacksmith and woodworking tools, a forge to shape iron, and tools and gear to manage repairs in unusual conditions. Ship carpenters had to fix leaks, hull damage, and broken rigging, which required tars and oils, oakum for caulking, and metal sheets to repair damage from shipworms. To work on hulls, especially to scrape off barnacles that limited speed, crews careened ships onshore, a procedure that required using an anchor in a shallow bay with a line tied to a mast to winch the ship on its side. Masts and other key wooden ship parts often required replacement from timber found on the Northwest Coast or Pacific islands. Two of the fur trade expeditions covered in this book brought enough materials to build small coasting vessels. John Meares built the *Northwest America* in 1788 at Nootka Sound, and Robert Gray fabricated the *Adventure* at Clayoquot Sound in 1792.

Maintaining the crew proved to be even more difficult than caring for the ships. Decorum aboard ship was essential for the safety of all personnel, and that meant adherence to routine methods of giving and following orders, especially acknowledgment of officer rank and specific duties and proper responses to commands. Captains set the tone for the entire ship, and good morale was usually the result of the captain's balanced application of care for the welfare of the crew and enforcement of discipline. Discipline included nearly every shipboard interaction, from sailor-to-sailor relationships to carrying out commands to management of shipboard guests, especially Native people on the Northwest Coast and on Pacific islands. Discontent from the crew, which could lead to dysfunctional shipboard relations, desertion, or even mutiny, might be the result of everything from poor food to low pay for services. The worst infractions required stern punishments. Captains and officers could "start" a disobedient crew member by striking him once with a short length of rope or a rattan switch. Flogging, though, was a formal affair, with all crew as witness and the captain announcing the rule that the sailor had violated. The offender received up to 12 lashes by a cat-o'-nine-tails (a one-pound device with handle and 9.25-inch cords attached) that could leave the skin looking scorched. Most captains administered floggings in moderation.

In comparison to discipline, however, scurvy was the more potent threat to shipboard safety. The malady had been known in antiquity, but its cause was mysterious. Long sea voyages, where fresh foods were less available and sailors lived on salted meats and coarse breads, seemed to breed conditions for scurvy outbreaks. Per week, sailors consumed about 7 pounds of biscuits, 8 gallons of beer, 4 pounds of salted beef or pork, nearly 20 pounds of oatmeal, and several pints of soup. This amounted to about 5,000 calories a day, but it did not hold scurvy at bay. The human costs were often devastating. Vasco da Gama, for example, lost 100 of 160 crew members on a voyage to the East Indies in 1497, and occasionally up to half died from scurvy on ships in the 17th and 18th centuries. By the mid-18th century, mariners had tried many strategies with limited success. The first important scientific study on the disease came from a surgeon in the British Royal Navy, James Lind, who wrote an influential pamphlet in 1753 on the disease, arguing that eating citrus fruit could prevent the disease. Many captains applied his remedy, while others followed a range of other strategies. The Royal Navy would not make citrus an authorized scurvy preventative on its ships until 1795. James Cook, for example, used sauerkraut, spruce beer, and frequent reprovisioning on his first voyage to the Pacific (1768–1771) to avoid losing a single man to scurvy, but others, including La Pérouse and Malaspina, advocated fresh drinking water or regular airing of bedding and clothing as ways of avoiding the disease.

What Finding New Peoples and New Places Meant

When big sailing ships first cruised the Northwest Coast, as Native oral traditions relate, some indigenous people saw them as floating islands that brought people who fulfilled prophecies of strange individuals coming to their lands. It did not take long, however, for these encounters to open a dynamic exchange between Northwest Coast residents and the occasional visitors who carried with them cargoes of desirable goods, such as copper, woven cloth, iron, and firearms. The mariners exploring the Pacific encountered indigenous people across the ocean, from Hawaiians in the South Pacific to Tlingit in Alaska to Natives in Kamchatka, but the most concentrated and developed relationships took place after the onset of the pelagic fur trade. The trade brought dozens of vessels to the Northwest Coast during the 1780s and 1790s, and this brought Europeans and Americans into the homelands of Native people from the Oregon coast to the Aleutian Islands.

Europeans and Americans came to the Northwest Coast with attitudes about Native people that had matured and become rationalized over more than two centuries. In the 15th and 16th centuries, Spain used physical conquest in the Caribbean, Mexico, and Peru to dominate Native populations, making slaves of people to work in gold and silver mines, while Catholic missionaries forced them to become Christians. English and French colonists in the 17th and 18th centuries had pushed Natives off their lands in North America. And throughout the hemisphere, Europeans had brought diseases to Native populations who had no natural immunities and suffered great loss of life. For justification or as a believed description, Europeans and Americans harbored a general view that they and their culture were inherently superior to any Native person or community. In account after account in expedition documents, captains and mariners express this arrogant perspective, but they also paid attention to what Natives knew and told them about the environment of these newly found lands. The information that mariners collected soon became part of science's description of the world; modern science, in fact, became a hybrid of local knowledge from distant parts of the world that scientists in London, Paris, Madrid, and Amsterdam collated and analyzed.

The relationships between mariners and Natives, though, often centered on economic exchange, because that is where there was common interest. Two ship-accommodating bays on Vancouver Island—Nootka Sound and Clayoquot Sound—became centers of direct engagement between Europeans and Native groups. First seen by Juan Pérez in 1774 and visited by James Cook in 1778, Nootka Sound drew fur traders in the 1780s and became a flashpoint between Spain and Britain in the early 1790s, but it was also a place where European and American mariners had ongoing contact with Native people. Nearly three-fourths of the fur-trading vessels that plied the Northwest Coast between 1785 and 1795 spent time at Nootka. Those encounters were scenes of mutual benefit, especially in Native access to European goods and fur traders' acquisition of valuable pelts, but there were also conflicts and violent exchanges. Mariners represented nationalist claims to territory, without reference to Native rights or desires, and fur traders approached places such as Nootka as unrestricted theaters of operations. And the intrusion of European and American actors in this Native place made more complex and contested existing competitive relations between Native groups. In substantial ways, the encounters between Natives and the visitors began a profound shift in the economy of the region, political relationships, and eventually the colonization of the Northwest Coast. The explorations highlighted in this book did not create the colonized regions in the American Northwest, Canada, and Alaska, but it is not too much to say that the components of those early encounters, including maps bearing an imperial imprint, framed later region-changing developments.

What the mariners learned about Native people we can glean from their logs and journals, but this requires careful reading to penetrate the meaning of their descriptions. While many accounts praise Native skills and art, most included critical and derisive descriptions of Native behavior, appearance, and especially theft of mariners' property. They often saw Natives as fundamentally deceitful. At Nootka Sound, for example, James Cook found that Native people had their own clear understanding of property, such as charging for any harvest of natural resources, but they also had little hesitancy in absconding with items from Cook's ships without payment. He and other ship captains often forbid Natives from coming on board or allowed only chiefs on deck. For Cook and other mariners and traders, economic exchanges had to follow prescribed rules and outcomes, and Native traders seemed to ignore the rules. But Native people approached trade from a much different viewpoint. For them, trade was not an end in itself; it was part of a complex set of political and economic relationships among Native chiefs and groups. The Mowachaht chief Maquinna, for example, managed his meetings with visiting mariners to Nootka, such as Vancouver and Bodega, as opportunities to enhance his power among other local leaders and to prevent a wholesale destabilization of political relations among Indian peoples. European and American observers at Nootka and elsewhere on the Northwest Coast, in other words, often misunderstood the political landscape at places such as Nootka, but there is little question that their ongoing contact with Native groups in the late 18th century was prologue to a later conquest and domination over the indigenous people of the region. Those logs and journal entries describing Native people, though, remain important sources of information about indigenous cultures. When they are combined with oral traditions from tribal people today, we can better understand the deeper history of the region.

Exploring

The exploits and experiences of mariners exploring the North Pacific reveal several important things about that now-distant era. First, and perhaps most important, they lived in times and under conditions much different from today. The world they understood was changing—politically, economically, and socially—while they undertook exceptionally dangerous voyages. The contrast between their views of the world and the perspectives of the Native peoples they met posed significant questions that had no obvious or easy answers. They often reverted to an assumption that their technology and social organization proved them superior to all indigenous people they met. Second, and a contributor to their idea of supremacy, was the European and American expectation that all lands they encountered that had not been claimed or inhabited by a similar maritime nation were open to legitimate claim, conquest, and occupation. Third, and in some ways the most consequential, the maps and charts they created through their use of advanced technology and detailed surveys led directly to an expansion of European and American nationalistic power and economic development in the Pacific. Finally, the expeditions took place during a modern era of change that included unmatched advances in scientific knowledge, revolutionary new political experiments and intercontinental wars, and the expansion of capitalist exchange around the world. The consequences of these voyages extended for decades, and their imprints are still visible today.

Selected Chronology

1494 Spain and Portugal sign the Treaty of Tordesillas, which divides the unclaimed world, with all land in North America assigned to Spain.

1507 German cartographer Martin Waldseemüller creates the first world map to illustrate a conceptualized ocean separating Asian and American continents, effectively fashioning an as yet unexplored western coast of North America. Many maps of the period continue to show North America connected to Asia.

1513 Spanish conquistador Vasco Núñez Balboa is the first European to sight the Pacific Ocean and name it the Mar del Sur (Southern Sea) as he looked south from the Isthmus of Panama.

1519 Spanish explorer Alonso Álvarez de Pineda is the first to explore and map the Gulf of Mexico.

1519–1521 Spanish conquistador Hernán Cortés conquers the Azetc Empire.

1519–1524 Spanish explorers Ferdinand Magellan and Juan Elcano carry out the first circumnavigation and the first European entry into a western ocean, named the Mar Pacifico by Magellan.

1530s The Spanish establish the port of Acapulco on Mexico's Pacific Coast at 16°51′ north latitude.

1535 Cortés explores, briefly occupies, and maps the southern tip of the Baja California peninsula.

1539 Spanish explorer Francisco de Ulloa rounds the southern tip of Baja and sails north to Cedros Island at 28° north.

1542–1543 Spanish explorer Juan Rodríquez Cabrillo sails along the California coast to Santa Barbara at 34°24′ north.

1561, 1565 Italian cartographers Giacomo Gastaldi and Paolo Forlani issue world and continental maps that illustrate a conceptualized narrow water channel between the Asian and American continents and apply the name Strait of Anian, derived from accounts of Marco Polo. Mapmakers subsequently adopt the concept that this strait between Asia and America might connect the Pacific Ocean with another conceptualized strait, the Northwest Passage north of the continent.

1565 Spanish explorers Miguel López de Legaspi and Andrés de Urdaneta, respectively, claim possession of the Philippines and establish a northern return sea route from the Philippines to Acapulco. The first return of the Manila galleon inaugurates nearly 250 years of commerce and immense wealth transfer between Spain and Asia via Acapulco.

1570 Antwerp cartographer Abraham Ortelius publishes the first modern atlas, *Theatrum Orbis Terrarum.*

1577–1580 English explorer Francis Drake completes second circumnavigation and captures the Spanish treasure ship *Nuestra Señora de la Concepcion* off the Pacific Coast of Ecuador.

1586–1588 Thomas Cavendish completes the second English circumnavigation and captures the Spanish Manila galleon *Santa Ana* in November 1587 off the southern coast of Baja California.

1602–1603 Spanish explorer Sebastián Vizcaíno leads an expedition from Acapulco to above Cape Mendocino (~40° north) to locate a suitable harbor to reprovision the returning Manila galleon. Accounts of Vizcaíno and others introduce the idea of California as an island and possible transcontinental water passage (Entrée Martín de Aguilar at 43° north), which appear on European printed maps well into the 18th century.

1603–1769 Europeans abandon voyages of exploration to the Pacific Coast of North America north from Cabo San Lucas at 22°53′ north.

1607 English settlers establish Jamestown, Virginia, as the first permanent English colony in America.

1610 English explorer Henry Hudson explores Hudson Bay in search for a northern route to the Pacific Ocean, the Northwest Passage.

1622–1625 The maps to illustrate California as an island are published, including a widely disseminated map of English mathematician Henry Briggs that he based principally on information from accounts of Vizcaíno's voyage in 1602–1603. For the next 100 years European mapmakers depicted an insular California.

1625 English author Samuel Purchas publishes an account of apocryphal discoveries on the Northwest Coast in 1592 by a Spanish-employed pilot, Juan de Fuca, that suggests the presence of a large inland sea, the "Sea of the West."

1670 The English crown grants a royal charter to the Hudson's Bay Company to trade in North America.

1696 Russian tsar Peter the Great conquers Kamchatka in Siberia, begins modernization of Russian institutions, and lays the foundation for St. Petersburg in 1703, which becomes the capital of Russia in 1712.

1700–1722 French cartographers Claude and Guillaume Delisle and family introduce maps of North America that are among the first to suggest a peninsular California. On his 1722 map, Delisle incorporates a conceptualized "River of the West" running west from the North American interior between 45° north and 50° north latitude. At ~45° north, he introduces the *Entrée découverte par Martín d'Aguilar* derived from the accounts of the 1602–1603 Vizcaíno expedition.

1708 The apocryphal voyage of Spanish explorer Bartholomew de Fonte, who claimed discovery of a Northwest Passage extending northeast beginning at 53° north on the Pacific Coast, is published.

1713 In London, the Board of Longitude is established to promote the development of instruments and techniques to calculate longitude at sea.

1733 Joseph Nicolas Delisle, brother of Guillaume and previously chief astronomer of the Russian Academy of Sciences in St. Petersburg, draws a manuscript map of a conceptualized North Pacific Ocean and adjacent landmasses that significantly influences route of Russian voyage to North America during the second Kamchatka expedition.

1735 English clockmaker John Harrison completes the first reliable maritime chronometer to calculate longitude at sea.

1740–1744 British naval officer and explorer George Anson completes circumnavigation with the loss of nearly 1,000 men from scurvy. Anson captures the Spanish Manila-bound galleon *Cavadonga* off the coast of Chile, including its treasure and maps of the sailing routes of the Manila-Acapulco galleons that were published in London in 1748.

1741 The second Kamchatka expedition, led by Russian naval officers and explorers Vitus Bering and Alexis Chirikov, sails from eastern Siberia to the North American continent at ~60° north latitude. Bering dies from scurvy on December 8, 1741. Following the return of the crew to St. Petersburg, news of potential profit from skins of fur-bearing sea mammals initiates an extensive Russian private fur-trading activity that eventually extends from the Aleutian Islands to the North American continent by the end of the century.

1747 English naval officer and physician James Lind conducts a controlled experiment proving the benefit of citrus sources in treating scurvy, but his 1753 treatise had little impact on how European ship captains provisioned their vessels.

1750–1752 J. N. Delisle returns to France with news of Russian advances in the North Pacific that he presents before the French Academy of Sciences. Delisle and Philippe Buache publish maps that promote major geographic misconceptions about the North Pacific American coast based on information from the apocryphal voyages of Juan de Fuca and Bartholomew de Fonte. These popular representations are not universally accepted but reappear on maps of many European cartographers. Delisle also incorporates Vizcaíno's *Entrée découv. par Martín d' Aguilar en 1603* as a possible entrance to an inland "Sea of the West."

1754 German cartographer Gerhard Müller, employed by the Russian Academy of Sciences, issues a map of the North Pacific refuting some of Delisle's earlier geographical concepts and illustrates cartographical results of the second Kamchatka expedition with more fidelity. Many European mapmakers copy and republish Müller's interpretation.

1766 French cartographer Jacques-Nicolas Bellin issues a map of North America that embraces idea of Rio de Martín de Aguilar connecting with a conceptualized "River of the West" across much of the continent, although Bellin does not fully adopt J. N. Delisle's other concepts. Several subsequent explorers to the Pacific Northwest use Bellin's map as a reference.

1768 Fearing possible Russian expansion into North America from the north, Spain commits to increasing naval and exploratory activity on the Pacific Coast and establishes a naval port at San Blas at 21°32′ north, approximately 5° north of Acapulco.

1772–1775 English naval officer and explorer James Cook demonstrates the accuracy of Harrison's fourth chronometer for calculation of longitude during his second voyage of exploration in the southern Pacific Ocean.

1774 Spanish naval officer and explorer Juan Pérez (first Bucareli expedition) begins the first European exploration of the North Pacific Coast after the Russians, sailing as far as ~55° north.

1775 The British Admiralty offers a prize of £20,000 for the discovery of the Northwest Passage.

Spanish naval officers and explorers Juan Francisco de la Bodega and Bruno de Hezeta (second Bucareli expedition) reach ~58° north latitude. Their notable accomplishments include the first non-Russian European landing on the Pacific Northwest Coast on Olympic Peninsula on July 14, the first fatal encounters with indigenous people, Hezeta's description and map of the mouth of the Columbia River (*Entrada de Ezeta*) on August 17, and several ceremonial acts of possession based on the Treaty of Tordesillas.

1776–1780 Captain James Cook's third voyage is initiated by the British Admiralty to obtain information on geography of the Northwest Coast, discovery of the fabled Northwest Passage, and the collection of observations and artifacts related to flora, fauna, and indigenous people of the Pacific Northwest. Cook is killed on February 14, 1779, at Kealakekua Bay on the island of Hawaii. Results of the voyage published in London in 1784 include much ethnographic information and maps of the coast and several ports that were used during subsequent voyages by other explorers. Reports of great profits from the sale of sea otter skins in Asian markets stimulate English entrepreneurs to develop privately funded fur-trading expeditions to the Pacific Northwest.

1785 English fur trader James Hanna initiates fur-trading voyages of non-Russian Europeans in the North Pacific.

1785–1788 French naval officer and explorer Jean-François de Galaup, Comte de La Pérouse, circumnavigates the globe and explores the Pacific Northwest. The expedition is lost at sea in 1788, but journals and maps are published in Paris in 1797.

1785–1792 One hundred three ships of fur traders from six nations work in Pacific Northwest waters with a primary focus on the regions of Nootka Sound and Clayoquot Sound on the west coast of Vancouver Island.

1789 English fur traders George Dixon, Nathaniel Portlock, and John Meares publish accounts and maps of their fur-trading expeditions. Geographical discoveries of several traders are included in a widely disseminated world map by British commercial mapmaker Aaron Arrowsmith, published in 1790. English statesmen use this map to promote English imperial and commercial hegemony in the Pacific Northwest.

Spanish commander Esteban José Martínez seizes English vessels at Nootka Sound and precipitates the Nootka Controversy between England and Spain. War is averted with the first Nootka Convention agreement in October 1790.

1789–1794 Spanish naval officer and explorer Alejandro Malaspina's expedition explores the Pacific Northwest for purposes of collecting scientific information and discovery of the Northwest Passage. Results of the expedition were not immediately published for political reasons.

1790–1793 American fur trader Robert Gray enters, names, and maps the Columbia River estuary in May 1792 during his second voyage to the Northwest Coast and shares this information with Bodega and English explorer George Vancouver.

1791–1795 English naval officer and explorer Captain George Vancouver's expedition extensively charts the Pacific Coast and Vancouver Island. Vancouver's account and maps are published in London in 1798 and subsequently are widely distributed among European commercial mapmakers, which establishes English-applied place-names and an English identity to much of Vancouver Island and the Pacific Northwest.

1792 English naval officer Lieutenant William Broughton of the Vancouver expedition charts the Columbia River for about 100 miles upstream during October–November 1792.

Spanish naval officers Alcalá Galiano and Cayetano Valdés circumnavigate and map Vancouver Island. Their detailed maps contain mostly Spanish place-names, but long-standing Spanish policy of not publishing results of voyages delays recognition of Spanish maritime results in the region.

1792–1793 English explorer Alexander Mackenzie is the first to cross the North American continent to the Pacific Ocean at 52°38′ north latitude. Journals and maps are published in 1801.

1802 *Relación del Viage hecho por las goletas Sutil y Mexicana en el año de 1792 . . .* is published in Madrid. This publication and accompanying atlas describe and map Spanish voyages to the Pacific Northwest from the early 16th century through the expedition of Galiano and Valdés. However, the limited publication run and the previously published accounts of Cook, Vancouver, La Pérouse, and others remained the principal sources of information of geography, place-names, and imperial claims of sovereignty in the Pacific Northwest.

1803 In January, U.S. president Thomas Jefferson asks Congress to fund the Lewis and Clark expedition of 1804–1806.

EXPEDITIONS OF SEBASTIÁN VIZCAÍNO

CHAPTER 1

Sebastián Vizcaíno, Expedition to California, 1602–1603

Sebastián Vizcaíno is best recognized as a Spanish mariner and especially for his voyage of exploration along the California coast in 1602–1603, when he charted San Diego and Monterey Bays—an expedition that produced far-reaching and enduring cartographic legacies. During the four decades he lived in Mexico, he also served the king of Spain and the viceroy of New Spain in other roles. As a merchant, Vizcaíno helped define the commercial potential of a California pearl industry, and he was the first Spanish ambassador to Japan. Late in life, as a soldier he led a successful defense against a Dutch attack along the southern Mexican coast.

While Vizcaíno was not a nobleman by birth, he achieved a favored status in Madrid and Mexico City because of his keen and intrepid seamanship, his knowledge of Pacific Coast and oceanic routes, his desire to serve, and his zeal in carrying out his assigned missions. His career in Mexico and California touched two periods in the history of Spanish expansion on the North American Pacific Coast: (1) forays to the tip of Baja California by Hernán Cortés and his followers in the 1530s; and (2) the end of active exploration north along the coast for more than 150 years after his own 1602–1603 voyage.

Vizcaíno was born in about 1548, less than 30 years after Cortés's arrival in Mexico and the start of Ferdinand Magellan's circumnavigation of the world and entry into the Pacific Ocean through the straits that now bear his name. Although Vizcaíno's surname suggests that he was from Vizcaya, a Basque region in northeastern Spain, historian Michael Mathes (1972) believes that he was a native of Extremadura in western Spain. In about 1567 when Vizcaíno was quite young, he served in a Spanish military campaign against Portugal, and in 1583 he worked as a merchant in New Spain, but little else is known about his early years. By 1586, less than 20 years after the Spanish conquest of the Philippines and the initiation of the Acapulco-Manila galleon trade, Vizcaíno

DID YOU KNOW?

The Manila galleon ships brought a wide variety of Asian-produced goods to energized markets in Spain and Mexico, but raw and Chinese manufactured silk were the most valuable cargo. One ship carried 50,000 pairs of silk stockings.

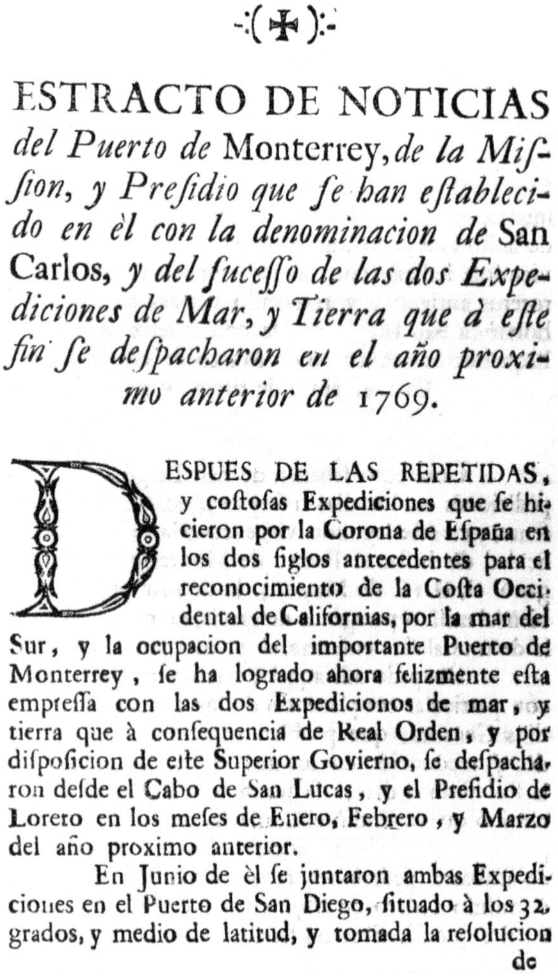
-:(✠):-

ESTRACTO DE NOTICIAS *del Puerto de* Monterrey, *de la Miſſion, y Preſidio que ſe han eſtablecido en èl con la denominacion de* San Carlos, *y del ſuceſſo de las dos Expediciones de Mar, y Tierra que à eſte fin ſe deſpacharon en el año proximo anterior de* 1769.

DESPUES DE LAS REPETIDAS, y coſtoſas Expediciones que ſe hicieron por la Corona de Eſpaña en los dos ſiglos antecedentes para el reconocimiento de la Coſta Occidental de Californias, por la mar del Sur, y la ocupacion del importante Puerto de Monterrey, ſe ha logrado ahora felizmente eſta empreſſa con las dos Expedicionos de mar, y tierra que à conſequencia de Real Orden, y por diſpoſicion de eſte Superior Govierno, ſe deſpacharon deſde el Cabo de San Lucas, y el Preſidio de Loreto en los meſes de Enero, Febrero, y Marzo del año proximo anterior.

En Junio de èl ſe juntaron ambas Expediciones en el Puerto de San Diego, ſituado à los 32. grados, y medio de latitud, y tomada la reſolucion de

Several years after Sebastián Vizcaíno returned to Acapulco, King Philip III and the Council of the Indies attempted to award Vizcaíno for his efforts. In this *cedula* (order), the king recommended that Vizcaíno establish a settlement at Monterey. Succeeding events resulted in suspension of this cedula, and a Presidio was not built at Monterey until 1770. (MPI/Getty Images)

sailed to Manila, probably on a commercial venture. He may have been aboard the treasure-laden Acapulco-bound galleon *Santa Ana* when it was seized by English privateer Thomas Cavendish in November 1587. Vizcaíno returned to Mexico in about 1589, and by 1596 he was married and had three children.

In 1592, Vizcaíno and 10 others petitioned the viceroy of New Spain, Luís de Velasco (1590–1595, 1607–1611), for exclusive merchant company rights to fish and mine for pearls, salt, gold, and silver over an area extending across the Gulf of California from Navidad, Mexico, to California. After many delays caused by internal company dissension and legal actions by competing merchants, Vizcaíno sailed to Baja from Mazatlán in August 1596 at the head of a force of three ships, 230 armed men, 14 horses, and supplies for eight months. During forays from settlements on the southeast coast of Baja, he explored inland and north along the Gulf of California and took possession of the region for Spain. After hostilities with Natives resulted in the drowning of 19 Spaniards as well as other misfortunes, Vizcaíno abandoned his venture and returned to New Spain before the end of the year.

Despite this commercial failure, Vizcaíno petitioned King Philip II (1556–1598) for permission to make a return voyage to the Gulf of California. Vizcaíno asked for five ships, 150 men, and a year's worth of rations to work the pearl beds and to fish, mine for salt, and convert the local residents to Christianity. In August 1598, Viceroy Gaspar de Zúñiga Conde de Monterrey (1595–1603) and the Council of the Indies issued reports favoring Vizcaíno's petition. Two events, however, had changed the direction of Spanish ventures in California. In 1595, Sebastián Rodríquez Cermeño had sailed from north to south along the California coast in search of a safe harbor for returning Manila galleons. He lost his ship, *San Agustín,* near Point Reyes and was unable to complete his mission. A year later, a story circulated that in 1592 Juan de Fuca, a Greek pilot allegedly employed by the Spanish, had sailed into the fabled Strait of Anian and penetrated far into the interior of the continent, somewhere between the latitudes of 47° and 48°. This information helped rekindle interest in the existence of a transcontinental water passage. Combined with the desire to locate a North American coastal port, Philip III (1598–1621) issued an order in late 1599 to sponsor a voyage to chart the California coast.

A great deal of planning and money went into preparing Vizcaíno's voyage. By mid-1601, Viceroy Monterrey had appointed Vizcaíno general of the expedition, a nonmilitary title that conveyed the authority of the principal commander. Aware of the importance of the voyage to Philip, the viceroy ordered the construction of the *Tres Reyes,* a small launch to carry out close coastal reconnaissance, to accompany the two larger ships of the expedition, the *San Diego* and the *Santo Tomás*. In addition to several officers and pilots, the complement of men included "six old experienced soldiers, worthy of confidence" (Wagner 1929, 174), who were expected to inspire a less seasoned crew during times of hardship. Also on board were an official cosmographer who had been trained at the Casa de Contratación in Seville and three Carmelite priests, including Father Antonio de la Ascensión, who would be a principal diarist of the expedition.

Vizcaíno received detailed instructions from the viceroy on March 18, and the three ships left Acapulco on May 5. The expedition spent the next 11 months exploring the coast to nearly 41° north latitude before Vizcaíno, in the *San Diego,* turned back to Mazatlán, where he arrived on February 18, 1603. After returning to Acapulco on March 21, he learned of the earlier arrival of the *Santo Tomás* and the *Tres Reyes.* The crews of all three ships had endured great hardships and loss of life. Nonetheless, both Viceroy Monterrey and King Philip received the logs, charts, and reports of Vizcaíno and others with great enthusiasm.

On May 23, Vizcaíno wrote directly to the king in praise of Monterey as a harbor for the Manila galleon. Vizcaíno also requested compensation for his accomplishments, especially a series of charts of the California coast and a detailed set of sailing instructions, or *derroteros,* that pilots of the returning Manila galleons would use for years. A royal order appointed Vizcaíno to the prestigious and well-paid position of general of the Philippine ships. Also, in June 1607 the crown granted Vizcaíno the tribute and labor of two communities of *encomiendos* (Native laborers), including an annual tribute of 2,000 pesos for two generations of workers and a payment of 10,000 pesos per year until the system was established. The most enduring consequences of the 1602–1603 expedition, however, were cartographic. Many Vizcaíno-applied names along the coast have survived, and some of the accounts of the expedition resulted in the propagation of two long-standing cartographic myths: that California was an island and that there was a water passageway between the Pacific and Atlantic Oceans.

DID YOU KNOW?

In February 1603, Vizcaíno's crew accidentally discovered that eating the fruit of a cactus rapidly relieved symptoms of scurvy. This naturally abundant vitamin C–rich plant was a staple for Natives of the area and is commonly known today as the prickly pear cactus.

Vizcaíno and Monterrey strongly supported establishing a settlement and port at Monterey, but Viceroy Juan de Mendoza (1603–1607) and others disagreed. Mendoza championed an alternative port at one of two fabled islands—Rica de Oro or Rica de Plata. Reputedly, a Portuguese ship had discovered the islands off the coast of Japan, nearly in the same latitude as Monterey but much closer to the Philippines. The possibility of establishing a port close to Japan appealed to the king and the Council of the Indies for more than geographical reasons. Beginning in the 1590s, Spain had attempted to break the Portuguese and Dutch trade monopolies with Japan, and multiple exchanges between Spain and Japan in the first decade of the 17th century had resulted in promises of potentially lucrative commercial treaties. By early 1611, Viceroy Velasco had decided to coordinate an expedition to chart the two reputed islands and to establish an embassy in Japan. He appointed Vizcaíno ambassador to Japan, and on March 22, 1611, Vizcaíno's ship, the *San Francisco,* left Acapulco with a full crew, a group of Franciscan friars, and a returning delegation of 23 Japanese. Before sailing, in a manner similar to his previous commissions, Vizcaíno petitioned the king for license to settle Rica de Oro and Rica de Plata and to be made general of the Manila ships for six years.

In Japan, Vizcaíno's mission began with pomp. On June 22 in Yedo (present-day Tokyo), two Japanese nobles and 4,000 guards escorted Vizcaíno and his retinue of Franciscans and attendants through five gates into a palace courtyard, where Tokugawa Hidetada, the son of the powerful shogun Tokugawa Ieyasu (1603–1616), was attended by 1,000 courtiers. Despite two years of diplomacy, Vizcaíno failed to secure a preference for Spain in Japan over Dutch, Portuguese, and British competition, and the Franciscans met with limited success in their efforts to spread Christian practices in the country. Although Vizcaíno was able to chart a portion of the Japanese coast, his attempts to locate the two legendary islands in September and October 1612 were, of course, fruitless. Beset by mounting financial pressures and increasing anti-Christian and anti-Spanish sentiment, Vizcaíno left Japan on October 16, 1613, and arrived in Acapulco the following January.

Vizcaíno withdrew in retirement to Jalisco, where he was comfortably supported by his *encomienda* in Avalos. His retirement was interrupted by the threat of a marauding Dutch fleet under the command of Joris van Spilbergen, who reached the Brazilian coast in December 1614 and began a series of raids on Spanish ships and towns along the South American and North American coasts. In October 1615, Viceroy Fernández de Córdova (1612–1621) ordered Vizcaíno to command a contingent of troops to defend part of the Mexican coast. On November 12, the 67-year-old Vizcaíno and 200 well-armed men confronted a comparable Dutch landing party at Salagua, near present-day Manzanillo. After a six-hour battle, Vizcaíno's party prevailed. For his efforts, he was made *alcalde mayor* of Acapulco, a position he resigned in 1619 when he retired again, this time to Mexico City.

Until his death in about 1628 at the age of 80, Vizcaíno may have been aware of renewed activity promoting a pearl-fishing industry on the Gulf of California and also of Ascensión's 1620 memorial promoting the colonization of California and his opinion about the geography of California. In 1625 Father Juan de Torquemada published *Monarquía Indiana,* a narrative of Vizcaíno's 1602–1603 expedition, and in 1625 and 1626 maps were published showing California as an island. The Royal Order of 1628 reopened the debate on the development of California and of Monterey as a port. In the 21st century, Vizcaíno's greatest legacies are place-names along the California coast and several locations in Baja California named in his honor, including Sebastián Vizcaíno Bay and El Vizcaíno Desert.

1. A Port for the Manila Galleon, 1585

INTRODUCTION

The protection of economic potential was one of the principal motivators for the Spanish crown in Madrid and the viceroy of New Spain in Mexico City to organize the expedition of Sebastián Vizcaíno. Spanish entry into the lucrative trade in spices and other commodities from the Moluccas (Spice Islands) of the East Indies began after the expedition led by Ferdinand Magellan sailed from east to west across the Pacific in 1520–1522. After Miguel López de Legaspi's conquest and possession of the Philippines in 1565, Spain found a practical route between Mexico (from Acapulco) and the Philippines (from Cebu and later from Manila).

The three- to four-masted galleons were usually constructed in the Philippines, and many were of enormous size compared with the ships of exploration that frequented the Pacific Ocean from the 16th through the 18th centuries. In the late 18th century, for example, the Resolution, *captained by James Cook, weighed 462 tons and was 110 feet long; the* Discovery, *captained by George Vancouver, weighed 330 tons and was 90 feet long. As early as 1589, the Spanish were building galleons of 700 tons and with lengths of 130 feet. Later some galleons reached 1,000–2,000 tons, with deck lengths of 170–180 feet.*

From 1565 to 1815, usually one or two large and heavily laden galleons sailed from Acapulco (16°51' north latitude) to the Philippines in late February to early March. This western route took galleons due west to Manila (15°), somewhere between 10° and 14° north latitudes. This crossing, favored by predictable east-to-west winds, usually took from two and a half to three months. By July of the following year, the galleons, now laden with spices, silks, and other exotics and large numbers of passengers, returned to Mexico by sailing first northeast to ~35–40° north latitude and then east before reaching the North American continent, usually in the vicinity of Cape Mendocino, before turning south for Acapulco. The perilous journey usually took from four to seven months. By the time they sighted the North American coastline, passengers and crew were often nearly overwhelmed with sickness and in dire need of supplies.

Spanish authorities had incentives to protect this valuable trade arrangement. Considerable correspondence between Madrid and Mexico City about the trade route centered on finding one or more ports on the Pacific Coast, where returning galleons might find succor and reprovision before sailing for Acapulco. These documents illustrate this concern.

Correspondence and Reports on the Trade Route

Viceroy Pedro Moya y Contreras to the King, January 22, 1585

When the ships come from China, they sight the coast of New Spain 700 leagues before reaching Acapulco, and from there sail almost in sight of land. Although it may be necessary to land to repair the ships, or to take on a supply of water or other things, they cannot do so as the ports on all that coast are not known, nor have they any place in which to take shelter from the contrary winds which ordinarily blow in that quarter. The same thing happens to ships which before reaching the coast encounter storms or have other necessities; they have to return to the islands from which they sailed, as there is no port on the coast. In order that all this may come to an end, and Your Majesty may have knowledge of all that coast which some say runs on to join the mainland of China. . . . I have ordered two *fragatas* built to search for and discover all the ports, islands, rivers, mountains and settlements which there are or may be on all that coast.

Father Antonio de la Ascensión's Account of the Voyage of Sebastián Vizcaíno

Besides this, the president and the *oidores* of the Royal Council of the Indies advised His Majesty that the ships sailing from the Philippines to New Spain encounter great storms before they come in sight of Cabo Mendocino and in that latitude, and that inasmuch as that country was not explored and the ports on that coast are not known, the ships ran a great risk and were obliged to return to Japan. . . . Great riches belonging to His Majesty and to private persons were lost in this. . . . In order to avoid such losses it was thought advisable to explore all the seacoast from there to the Puerto de Acapulco so that once the ports situated on it were found, the ships navigating along it might find relief and shelter.

Source: Henry R. Wagner, *Spanish Voyages to the Northwest Coast of America in the Sixteenth Century* (San Francisco: California Historical Society, 1929), 132, 181.

ANALYSIS

During the 250 years of what became known as the Manila-Acapulco galleon trade (1565–1815), Spain lost four ships to privateers and nearly 30 more to shipwreck at sea. "In all the seas," historian William Schurz concluded, "there was no line of navigation so difficult, so attended with perils and hardships, as that of the Manila Galleons" (1959, 253). Thousands of passengers and crew died, and millions of pesos in personal property and treasure were lost. Yet a staggering amount of Mexican and Peruvian silver and pesos reached the Spanish colony in the Philippines and the merchants of Ming China. Likewise, enormous cargoes of valuable merchandise from the Moluccas and China (and eventually India) reached Mexico, Peru, and eventually Spain. The trade nurtured the economies of Mexico, Peru, the Philippines, China, and Spain, and the Spanish crown and the Council of the Indies sought to protect this huge economic engine by exploring and mapping the California coast in search of a way station between Manila and Acapulco.

The return leg of the round trip was the most hazardous. Prevailing winds required pilots to sail northeast from Manila (15° north latitude), skirting the coast of Japan near 40° and then steering a course east across nearly 4,500 statute miles of ocean to the North American coast. Frequent storms beginning in July or earlier delayed or prevented ships from leaving the Philippines and nearby archipelagoes. Aborted crossings were nearly as devastating to the Spanish colony as shipwrecks. In 1603, the year Sebastián Vizcaíno completed his expedition, Spain lost a galleon at sea with all passengers and crew, and storms forced another back to port with the loss of masts and a great deal of cargo. Improper loading, inadequate repairs while in Manila, and incompetence in navigation contributed to the loss of ships, but the greatest threat was sickness and starvation.

If the voyage lasted more than four months without reprovisioning, passengers and crew likely developed scurvy. Accounts of voyages of longer than six months recounted hundreds of deaths due to scurvy and occasional infectious disease; most of the dead were thrown overboard. Perhaps the worst such case was the galleon *San José,* which was picked up in 1657 drifting south past Acapulco (Schurz 1959). Everyone on board the heavily laden ship was dead, and records indicate that the ship had been at sea for over a year.

Frequent correspondence between authorities in Mexico and Madrid, including this 1585 letter from Viceroy Pedro Moya de Contreras (1584–1585) to the king, documents Spanish responses to these nutritional, natural, and manmade threats. In 1587 when privateer Thomas Cavendish captured the rich 600-ton Manila galleon *Santa Ana,* their anxieties greatly increased. Later, Ascención's account of Vizcaíno's voyage provided details of Cavendish's seizure of the ship near Cabo San Lucas.

2. Viceroy Conde de Monterrey's Sailing Orders, 1602

INTRODUCTION

Philip III became actively involved with affairs of the Americas when his father died in September 1598. After consulting with his council on September 27, 1599, he sent a royal order to Viceroy Gaspar de Zúñiga Conde de Monterrey in Mexico City to organize and finance a voyage north from Acapulco. The primary purpose was to explore and map the outer coast of California. Twenty-seven months later after preparations were complete and the three ships under Vizcaíno's command were ready to sail, Monterrey issued formal instructions to Vizcaíno. Apparently to prevent Vizcaíno from repurposing the voyage to seek out pearls for his own profit, the viceroy warned him on March 2 that he could not explore the Gulf of California until he had completed the survey of the outer coast of California—on penalty of death. On March 18, Vizcaíno received a full set of 25 orders. These instructions reveal not only what the king and the viceroy expected but also the state of Spanish knowledge about the geography of the region.

Instructions to Vizcaíno, March 18, 1602

No. 22. The precise command which you carry is to discover, inspect and map all the ports, bars, bays and *ensenadas* [coves or inlets], capes and points from the Cabo de San Lucas to Cabo Mendocino, without failing to bring back a fit clear statement of anything important regarding these. In some meetings which I have held with some expert and experienced persons it has seemed advisable that, once you have inspected and discovered Cabo Mendocino, and you have a fair wind and one neither very contrary nor one which by its force might endanger your breaking your masts or losing your rigging, you should employ reasonable diligence in exploring as far as Cabo Blanco, which is in 44°. As the coast is neither known nor has been seen and in consequence there is nothing certain about it on the maps, if you should find that from Cabo Mendocino to Cabo Blanco it runs towards the west you shall reconnoiter it for a hundred leagues [approximately 300 nautical miles] and no more. Having accomplished this, even though the winds be favorable, do not go farther, but take the return route to Cabo San Lucas.

No. 23. Once having reached the Cabo de San Lucas, after accomplishing the discovery which you have under your charge, if you have wind and it is not from the north and you have sufficient food to accomplish an entry into the Californias you will do so, following the coast from cape to cape, examining the ports and *ensenadas* from the Cabo de San Lucas, entering by the mouth and coming back as far as the Puerto de la Navidad, unless the southwest winds do not give out. In such case, you will take advantage of them, and go on as far as the wind permits, inspecting the quality of the country and the climate, and seeing if the inhabitants, of whom I have heard a great deal, wear clothes or make any show of numbers or of quality, until you reach 37° or at most 38°, unless previously the sea had shown signs for ten or twelve leagues of turning to the east or east-southeast or west or west-southwest. In such a case, when you discover this, no matter where you may be, you will turn about and return to the coast of Navidad or Salagua, inspecting everything as far as the well-known ports of the New Spain.

Source: Henry R. Wagner, *Spanish Voyages to the Northwest Coast of America in the Sixteenth Century* (San Francisco: California Historical Society, 1929), 376–377n27.

ANALYSIS

Monterrey's instructions indicate that the place-name Cabo Blanco was in common usage by 1602, but his sources of information for both the name and its supposed location (latitude 44° north) are not known. Monterrey's Cabo Blanco is not the same as Cape Blanco on today's Oregon coast at 42°50′. No extant account of Spanish voyages to the coast prior to 1602 recorded a Cabo Blanco, but the name probably first appeared on Dutch maps of

Abraham Ortelius (1589) and Peter Plancius (1590), which located the cape in latitudes ranging from 33° north to 58° north. Monterrey was certainly correct when he wrote in his instruction No. 22 that "the coast [in this area] is neither known nor has been seen, and in consequence there is nothing certain about it on the maps."

Similarly, Monterrey's instructions on exploring the Gulf of California grossly exaggerate its potential northern latitude to 37–38° north, for the actual coordinate is slightly less than 32°. In late 1539, Francisco de Ulloa was the earliest Spanish explorer to reach the northern limits of the gulf and the mouth of the Colorado River, which he estimated was located at less than 35°. Despite this assertion, extant maps derived from Ulloa's report (no map drawn by Ulloa has survived) placed the head of the gulf at approximately 35–36°. Monterrey's instructions also suggest that the coast might run to the southwest or the southeast, perhaps indicating that he recognized the theory that California might be an island. In both cases, the viceroy based his knowledge about the higher latitudes of the California gulf coast on textual and cartographic information that was misinterpreted, too imprecise, or simply little regarded.

Some maps depicted the Pacific Coast above Mendocino running to the west or northwest. Because Spanish authorities were principally interested in locating Pacific Coast locations that might offer safe harbors and reprovisioning stops for the returning Manila galleons, Monterrey instructed Vizcaíno to explore "for a hundred leagues" (approximately 300 nautical miles) if he discovered that the coast above Cape Mendocino ran to the west.

The rest of Vizcaíno's instructions were similar to those issued to other Spanish maritime explorers of the 17th and 18th centuries. They required Vizcaíno to hold regular council meetings with all officers regarding decisions on navigation, communications, and anchorages; to abide by the majority opinion of the council; and to keep careful minutes of council meetings. He was directed to take frequent readings of coordinates, soundings, weather conditions, and possible anchorage sites and to describe the composition of the sea floor when possible and the landscape along the coast. He was ordered to detail the appearance and manners of any Native people he encountered and to assiduously avoid conflict with them. After charting the outer coast of California, he could explore for pearl beds.

The viceroy also instructed Vizcaíno to append names, particularly those of saints, to all potential ports, but not to change names that had been previously applied. Vizcaíno apparently had no knowledge of previous Spanish explorations along the California coast, such as Juan Rodríguez Cabrillo's journey in 1542–1543, so the surviving copies of maps of the ports and bays that Vizcaíno entered are filled with toponyms, whether or not the site had been previously named. The surviving accounts of the Vizcaíno expedition do not record preexisting Native place-names, and no indigenous name for ports along the coast explored by Vizcaíno survive.

3. At the Port of Acapulco, 1602

INTRODUCTION

In May 1602, Vizcaíno sailed north from Acapulco (latitude 16°51' north), which had become Spain's epicenter of lucrative trade relationships on the American Pacific Coast with merchants from the East. Following his conquest of the Aztecs in 1521, Hernán Cortés quickly turned to the Pacific Coast and within a few years had established port towns in Tehuantepec (16°19'), Zacatula (17°57'), and Acapulco, where a small shipyard was operating by 1528. Later, Spain built ports at Navidad (19°12'), Mazatlán (near Tehuantepec), Huatulco (15°51'), and Zihuatanejo (17°38'). Several early Spanish voyages of exploration to the Pacific sailed from Acapulco, including the first recorded European navigation north along Baja California by Francisco de Ulloa in July 1539. Although Spanish authorities attempted to prevent the dissemination of acquired cartographic information, the port and location of Acapulco appeared on manuscript and printed maps by Italian mapmakers Battista Agnese and Giacomo Gastaldi by the early 1540s.

Acapulco's principal notoriety came in 1561, when chief pilot Andrés de Urdaneta recommended moving the main port activities on the Pacific Coast from Navidad to Acapulco because of a healthier climate, deeper water anchorages, and nearby pine forests for shipbuilding. Beginning in 1565, Urdaneta and Alonso de Arellano established a practical route for return from Manila to New Spain for the Manila galleon after an arduous four- to eight-month voyage. By 1572, Viceroy Martín Enríquez de Almansa (1568–1580) wrote to Philip II that "Acapulco is coming to be the first port for the trade with the Philippines, because of its nearness to the City of Mexico" (qtd. in Schurz 1918, 18). Acapulco soon became the only fortified Spanish Pacific Coast seaport. A regimented administrative order, an efficient transportation and communication system connecting to Mexico City, and a vibrant annual fair were all tied to the annual return of the Manila galleon. This excerpt from the account of Father Antonio de la Ascensión touches on some of these elements of Acapulco life.

Father Antonio de la Ascensión's Account of the Voyage of Sebastián Vizcaíno

When the *Capitana* [Vizcaíno's ship] entered the Puerto de Acapulco all the inhabitants were surprised to see how well and sound all on board her were, as, by what those who had come back alive in the *Almiranta* had told them, they had persuaded themselves that they would never see her again. . . . Those on the *Almiranta* who had come back alive said that on the return they had suffered many hardships and misfortunes, as the sound men aboard her fell sick with the same disease as the others and almost all died before entering Acapulco, only three of those on her being well. . . . Advice had been sent to the Viceroy of the arrival of the *Almiranta* and of what had happened to her on her voyage. His Excellency ordered the royal officials of the port to provide the sick with every comfort and to cure them with care, especially Father Tomás de Aquino, to whom he commanded should be given all the animals and every possible comfort to enable him to reach Mexico. The royal officials and Lorenço Pacheco, the *alcalde mayor*, complied with what the Viceroy had commanded with due care and diligence.

As soon as the *General* [Vizcaíno] landed he gave an order to send a post to the Viceroy [in Mexico City], giving him an account of his arrival and of the condition of his men, so that His Excellency might command what was pleasing to him. . . .

The Viceroy having received the letter the *General* had sent from Acapulco by the post, at once dispatched another in which he ordered the royal officials to pay the soldiers very well what was due them and to take good care of the friars . . . and to give them everything necessary to enable them to reach the City of Mexico. All was carried out with the greatest fidelity and diligence. They allotted what pay seemed due to the soldiers who were alive and to those who had died.

The soldiers having been paid and animals given the friars for their journey, the *General* [and others] . . . departed from Acapulco April 7 for the City of Mexico in company with the friars, and reached there sound and well on Saturday morning, the 19th.

Source: Henry R. Wagner, *Spanish Voyages to the Northwest Coast of America in the Sixteenth Century* (San Francisco: California Historical Society, 1929), 262–264.

ANALYSIS

The most superior political authority in Acapulco was the *alcalde mayor,* who had mostly civil judicial and administrative duties and later served as warden of the fortified Castle of San Diego. Occasionally, the viceroy dispatched a *visitador* (inspector) to assume a temporary but powerful function: to oversee the regulations imposed on commerce and the exchange of money in the city. *Oficiales reales* (royal treasury officials) also had jurisdiction over direct fiscal activity pertaining to the extraction of customs duties on imported merchandise for the royal treasury. The church was represented by a *cura* (parish priest) who was fiscally minded. As a late 17th-century observer noted, "The Curate, tho' the King's allowance to him be but 180 pieces of Eight, makes 14,000 a Year, exacting a great rate for burying of strangers, not only that die at *Acapulco,* but at Sea aboard the Ships from China and *Peru;* as for instance he will expect 1000 pieces of Eight for a rich Merchant" (qtd. in Schurz 1918, 27). The potential for profit encouraged abuse, including frequent smuggling operations such as unloading merchandise intended to be sold for personal gain before royal agents had an opportunity to board the ship.

Vizcaíno's ship was probably first sighted far out at sea by an established sentinel posted somewhere north of Acapulco. From November to March, when sentinels first spotted sails that they determined were not signs of an enemy ship but a returning galleon, messengers carried news of the impending arrival (sometimes with the ship's papers) to Acapulco and the viceroy in Mexico City, nearly 300 miles to the northeast. From the time Acapulco was settled as a remote seaport until the late 18th century, this route over mountains, across rivers, and through forests and dry plains was traversed on foot and with pack animals, usually mules. All who traveled the China Road, as it was known—including single messengers bearing letters and reports and caravans of officials, soldiers, and merchants—suffered harsh conditions and primitive accommodations. In the late 18th century, Simón de Anda y Salazar reported that "all the treasures of this world could not compensate for the necessity of living there [Acapulco] or of traveling the road between Acapulco and Mexico" (qtd. in Schurz 1918, 23).

After confirmation in Mexico City that the Manila galleon had arrived safely in Acapulco, the viceroy proclaimed the opening of the annual Acapulco *feria,* which the great German scientist Alexander von Humboldt later called "the most renowned fair of the world" (qtd. in Schurz 1918, 31). The fair would swell the village of Acapulco, with its 250 households of mostly Indians and Asians (according to the 1598 census), to thousands of people, including traders from Mexico and Peru, royal officials, soldiers, and their entourages. Neither Vizcaíno nor Ascensión described this event in March 1603, but Ascensión reported that Vizcaíno remained in Acapulco for only 17 days before leaving for Mexico City, which they reached in 12 days traveling by mule.

4. Scurvy Interrupts the Voyage, 1603

INTRODUCTION

Spanish authorities had mounted only a few voyages of exploration for geographical discovery along the Pacific Coast of California before Vizcaíno's expedition in 1602–1603. Hernán Cortés took possession of what he called Puerto de Santa Cruz on the tip of the California peninsula on

May 3, 1535, almost 67 years to the day before Vizcaíno left Acapulco on his historic voyage. In July 1539, Francisco de Ulloa left Acapulco in command of the first recorded European voyage of exploration north along the coast of California. In June 1542, Juan Rodríguez Cabrillo left Navidad Harbor and sailed north along the coast to a port later named Monterey Bay; he may have reached as far as latitude 41° north. Other 16th-century expeditions that ran north to south along the coast started from Macao (Francisco Gali in 1584 and Pedro de Unamuno in 1587) and the Philippines (Alonso de Arellano and Andrés de Urdaneta in 1565 and Sebastián Cermeño in 1595).

Some of these voyages were associated with personal tragedies: Ulloa may have been lost at sea, Cabrillo died of an infected leg wound, and Cermeño suffered shipwreck with the loss of many crew members. The lack of surviving detailed narratives and maps of many of these voyages makes it difficult to appreciate the extent of human costs suffered during the early days of coastal exploration. In the case of the Vizcaíno expedition, a large number of extant reports and accounts provide rich documentation of some of these hardships.

Diary of Sebastián Vizcaíno, June 2, 1602–February 18, 1603

[June 2, 1602] Going forward on the voyage [departing the mainland for the California peninsula] . . . the general [Vizcaíno] gave orders to cross the entrance of the Californias to the Cape of San Lucas. This was done, although with much labor, there being westerly headwinds; and going forward by tacking back and forth, after five days we found ourselves to be twelve leagues from the said Cape of San Lucas. Coasting along the land, on Saturday, the 8th day of the month of June, we arrived at the said Cape of San Lucas [Bay of San Barnebé]. . . .

It was agreed to-day that on Wednesday, the 19th of the month, the moon being in conjunction, we should set sail in continuation of our voyage. This we did, and at four o'clock in the afternoon the captain's ship set sail, the others following. After we had rounded the cape, when six leagues from it a northwest wind came up which forced us to take shelter in the same bay, where we remained another two days, until it grew calm. We set sail a second time, but having arrived at the place before-mentioned the same wind struck us again and forced us to put into port. We were there three more days until the eve of the feast of the glorious San Juan Baptista, when, being desirous of going forward, we that night set sail the third time; but after sailing five days, during which we struggled as best we could, we were driven back with much force to the same bay and cape, where we remained until July 5. . . .

We set out from the said cape and bay of San Bernabé on July 5, as has been said, and after going two leagues out to sea, sailing with a strong wind, we met with a moderate headwind, and, tacking against it, sailed with great difficulty. . . .

[December 16 in Monterey Bay] In view of the fact that we had so many sick, that the pilot of the admiral's ship and his assistant were very ill, that there was a shortage of sailors for going forward, and that the supplies were becoming exhausted because of the length of time we had spent in coming, it seemed to the general impossible to complete the exploration this time without a new supply of men and provisions; and he therefore at once called a conference of the admiral and the members of the council to consider what should be done for the best service of his Majesty. . . .

[January 21 during return south] We did not enter it [the port of Monterey] because the state of our health was so bad and the sick were clamoring, although there was neither assistance nor medicines nor food to give them except rotten jerked beef, gruel, biscuits, and beans and chick-peas spoiled by weevils. The mouths of all were sore, and their gums were swollen larger than their teeth, so that they could hardly drink water, and the ship seemed more like a hospital than a ship of an armada. Affairs were in such a condition that anyone who had ever in his life been at the helm steered, climbed to the maintopsail, and did the other tasks, and all who could walk assisted at the hearth, making gruel and porridge for the sick. . . .

[February 18, at anchor off the California mainland] God, like a father of mercy, caused to be provided in these islands a small fruit like agaves, called *jucoystlis.* These, because of their strength, when eaten by the men who had sore mouths caused the ulcers to slough off and bleed profusely; but such was the efficacy of the fruit that within six days there was not a single person whose mouth was not healed.

Source: Herbert E. Bolton, ed., *Spanish Exploration in the Southwest, 1542–1706* (New York: Scribner, 1916), 55–58, 92, 97, 99–100.

ANALYSIS

Vizcaíno's orders required him to sail as close to land as possible to examine all potential bays and harbors north from Cabo San Lucas, but this ensured that he would encounter prevailing adverse headwinds and currents along the California coast. Strong winds from the northwest, more regular during summer months, funnel along the Baja peninsula and create some of the most difficult sailing conditions in North America. Occasionally, seaward-blowing summer storms in the southern Baja area add to the hazards of sailing. Although the accounts of Ulloa and Cabrillo frequently mention adverse wind conditions, especially along the southern coast of the peninsula, it is unclear if Vizcaíno had access to those reports. His chief pilot, Francisco de Bolaños, had sailed with Cermeño in 1595 and may have alerted Vizcaíno.

After leaving Acapulco on May 5, the expedition reached Cabo San Lucas and took shelter in a harbor that Vizcaíno called the Bay of San Bernabé (latitude 22°51′ north) in early June. Contrary winds prevented the fleet from advancing farther north three times before Vizcaíno finally cleared the cape on July 5. During the preceding two months, the ships had covered only about 6° of latitude, or roughly 360 nautical miles. Further progress was burdened by increasing cold weather, more storms, the frequent need to replenish water, and dwindling food stores. On November 10 the expedition reached San Diego at 32°40′, a port that, Vizcaíno recorded in his diary, "must be the best to be found in all the South Sea" (Bolton 1916, 80). After nearly five more months, they had advanced another 570 miles. By this time, many crew members were becoming ill. When the expedition arrived at Monterey on December 16, five weeks later, their situation had become desperate because of the devastating effects of scurvy.

Among early deepwater sailors, scurvy usually occurred after four months among those who had no regular intake of vitamin C. Vizcaíno wrote in his diary that during frequent landings along the coast he was able to procure water and often fish and game from the inhabitants, but he likely did not harvest grasses or fruits, good sources of vitamin C. He may not have had previous experience with scurvy, although at Monterey he had noted a "shortage of sailors for going forward" (Bolton 1916, 92). Father Antonio de la Ascensión attributed the sailors' illness to the "sharp, subtle, and cold wind" that "must carry with it much pestilence. . . . [I]t produces with its subtlety and coldness some corruption of bad humours, especially in persons worn out and fatigued with the hardships of the navigation." He detailed the inevitable progression of symptoms of scurvy, from "pain in the whole body" to the development of "purple spots" (hemorrhages) that expanded to extensive skin and soft tissue bleeding, leaving the "whole body . . . stiff, and sore as a boil." Further symptoms, "more insufferable and loathsome," included swollen gums, loss of teeth, and eventually death (Wagner 1929, 245–246).

In February 1603 on an island off the mainland, Vizcaíno and Ascensión documented the amelioration of even the most frightful symptoms of scurvy by eating a "little fruit" called *jocoistles.* A suffering crew member had discovered the salutary effects of this fruit, which "on eating of it twice, it put the mouth and the teeth in such condition that one could eat any other kind of food without difficulty or pain" (Wagner 1929, 260–261). Of the more than 200 officers and crew on Vizcaíno's ships, 42 reportedly died, the majority likely from scurvy.

5. Navigating with the Bolaños Derrotero, 1603

INTRODUCTION

A principal product of European maritime explorations was the collection of written sailing directions, usually compiled by the pilot of a voyage. Reliable geographical information required both accurate maps and careful observations on how to maneuver around landforms. From the beginning of Spanish exploration along the Pacific Coast of North America, authorities gathered these derroteros, *along with charts, logs, and officers' narratives when ships returned to Mexican ports. The information was usually sent on to the viceroy in Mexico City, who usually forwarded originals or copies to Madrid. Eventually the documents were archived in the Casa de Contratación in Seville.*

This document is from the derrotero *of the chief pilot of the Vizcaíno expedition, Francisco de Bolaños, which he compiled from the time Vizcaíno's ship left Monterey in December 1602 to when the officers eventually arrived in Mexico in early 1603. At Bolaños's request, Ascensión corrected and added to the* derrotero *at the completion of the voyage. Thirty-eight-year-old Bolaños was experienced with the routes of the galleons between Acapulco and Manila, and he and Vizcaíno may have been aboard the* Santa Ana *when it was sacked by Cavendish in 1587. Bolaños also served in 1595 as a boatswain on the expeditions of Sebastián Cermeño, which resulted in shipwreck and loss of the* San Agustín *near present-day Point Reyes, California. His status as chief pilot meant that he was well paid and that the viceroy had confidence in him.*

The Bolaños-Ascensión Derrotero

Whoever comes from six leagues off Cabo Mendocino bound for this port, on a course southeast a quarter south [compass point], will come to the Punta de los Reyes and will see the farallons [headlands or rocky outcrop], a good landmark to recognize it. Here it was that the ship *San Agustin* was lost in 1595, coming on a voyage of exploration. The loss was caused more by the man commanding her than by the force of the wind.

From the Punta de los Reyes about fourteen leagues southeast a quarter south there is a point. Before reaching it the country consists in places of sierra, bare to the sea and of medium height with some cliffs. From here the coast runs southeast, low at the sea with cliffs, but soon the country inside becomes massive and wooded until you reach a point of low land in 37½° named the "Punta de Año Nuevo." From this point the coast runs to the east, forming a large *ensenada* which ends at a point of low land heavily covered with pine groves down to the sea itself. This was named "Punta de Pinos" and is in 37°. From Punta de Año Nuevo it is about twelve leagues southwest to this point.

On nearing this point, a massive range is seen to the southeast, extending from the northeast to the southwest. Steering straight towards the point of this from the northeast, the Punta de Pinos will soon be seen. It is a low hill, about two leagues long, running north-northeast and south-southwest, and well covered with groves of pine trees. Near the point on the south side there is a patch of very white cliffs, a very good landmark by which to recognize it. This Punta de Pinos forms on the northeast side a fine port, and continuing in a straight line, you will enter it and can come close to the land in a depth of six fathoms; all the point and the entrance are of rocks. Steering southeast and east, after you have passed the rocks there is a fine beach. Before reaching this there is a very good clean anchorage, sheltered from all winds except that from the north-northwest.

In this port named "Monte Rey," there are many fine pine trees suitable for masts, spars and yards, all very near the beach. Where the rocks end at the beginning of the beach there is an *estero* of salt water in which the tide enters when high. A musket-shot away from the beach southeast of this *estero* and close to it is a very moist flat place where, in digging down a very short distance, very good sweet water in abundance comes out. The port is in the latitude of 37° and is a very good one for relief of the ships from China [Manila galleons], as this is the first land sighted on their way to New Spain.

Following the coast from the Punta de Pinos, south-southwest of it is another fine port which extends from the north to the south, and has shelter from all winds except the two from these directions. In this empties a river of very good water and little depth, on the banks of which are many very high black ash trees, straight and smooth, and many other trees of Castile [Spain]. It comes down from very high white sierras and is named the "Rio del Carmelo."

Source: Henry R. Wagner, *Spanish Voyages to the Northwest Coast of America in the Sixteenth Century* (San Francisco: California Historical Society, 1929), 435–436.

ANALYSIS

The surviving documents related to the Vizcaíno expedition include two *derroteros,* one written by the chief cartographer, Gerónimo Martín Palacios, and the other by Bolaños and Ascensión. Palacios also drew 32 detailed charts of harbors and bays and 1 coastal profile, all with accompanying explanation. This specific information about geographical formations over relatively short distances supplied Spanish authorities with valuable knowledge of coastal harbors and bays. Within a few paragraphs, Bolaños described the principal features of a coastline that extended less than 2° of latitude, or fewer than 110 nautical miles. His observations included coastal landmarks, latitudes, soundings, distances between points, the nature of and potential resources of the land, sailing hazards, the type of sea bottom, the character of the prevailing winds, and how to safely enter the port (Monterey).

It is unclear how the written directions that Bolaños and Ascensión provided were subsequently used, although historian Henry Wagner (1929, 385) noted that pilot Joseph Cabrera Bueno copied their *derrotero* almost word for word in a guide for navigators published in Manila as late as 1734. Bueno made only a few corrections to the document, and according to Wagner he retained most of the latitudes from the original. This evidence,

Wagner wrote, "raises a doubt whether any new observations had been made on that part of the coast since Vizcaino's time" (1929, 385).

The written practical description of a physical space is one of the oldest forms of navigational convention used by mariners. Over 2,500 years ago, wayfinding at sea using explicit sailing directions preceded the development of sailing charts. Twenty-seven hundred years ago a *periplus,* perhaps the first known surviving Greek sailing instruction, described navigation in the Mediterranean and Black Seas. The earliest known 13th-century sea charts, known as *portolans* in Italian, were accompanied by written sailing guides, or *portolani.* Over time, all maritime nations that were engaged in commerce or exploration depended on sailing instructions as well as charts and maps. The instructions were called *roteiros* by the Portuguese and *rutters* by the English and Dutch (a translation of the French *routier*). In the northern Low Countries in the 16th century, popular forms of these guides became known as a *leeskaarten* or a *Waghenaer.* Pilot guides became so useful and familiar that they comprised a significant portion of the active commercial cartography market in Europe from the mid-16th century, and they remain valuable navigational tools today.

6. Rio de Martín Aguilar, 1603

INTRODUCTION

The process of mapping a little-known region, such as the Pacific Coast of North America in the 17th and 18th centuries, involved incorporating new information from actual observations into a framework of known and conceptualized knowledge. One delayed but important consequence of Sebastián Vizcaíno's expedition was the mapping of a new river on the mid–Pacific Coast at latitude 42–44° north. During much of the 18th century, mapmakers depicted this river, then known as the Rio de Martín Aguilar, as the entrance to an inland sea or as a river that extended far into the interior and connected to an unknown transcontinental water passage. More important, the Rio de Martín Aguilar prominently figured in the objectives of nearly every 18th-century voyage of exploration.

One of Vizcaíno's ships, the Tres Reyes, *was a small frigate or launch whose officers included the commander Ensign Sebastián Meléndez, Pilot Antonio Flores, and Ensign Martín de Aguilar Galeote. During its long voyage, the ship frequently became separated from the other ships during storms. In such a situation on January 19, 1603, Flores sighted the coast at an estimated latitude of 43°. The officers observed and named a river and then set sail for Mexico. Both Flores and Aguilar died en route. In Mexico a surviving crew member, Esteban López, related an account of their discovery, which was assimilated into the reports of Vizcaíno and Ascención. This secondhand information was later recapitulated in* Monarchia, *a book published in Madrid in 1615 by Father Juan de Torquemada. It was through subsequent derivatives of Torquemada's account that news about Vizcaíno's voyage reached Europeans.*

Reports on Mapping Rio de Martín Aguilar

Diary of Sebastián Vizcaíno, January 1603

In latitude 39¼° he [Esteban López] discovered a copious river, and an island at the entrance of a very good and secure port, and another large bay in latitude 40½°, into which another large river emptied. A great number of Indians came out to them in canoes made of pine and cedar planks, but because there were so many people they did not dare to go up the river with the launch, although the Indians invited them to do so, giving them many fish, game, hazel nuts, chestnuts, acorns, and other thing to induce them to go with them up the river.

Father Antonio de la Ascensión's Account, January 1603

After the wind and sea calmed down, she [the *Tres Reyes*] went on close to land, and on January 19 Antonio Flores, the pilot, found himself in 43°. Here the land makes a cape or point, which was named "Cabo Blanco," and here the coast begins to trend to the northeast. Close to it a very copious and deep river was discovered on whose banks there were very large ash trees, willows, brambles, and other trees [similar to those] of Castile. On attempting to enter it, the force of the current did not permit it. Ensign Martín de Aguilar, who commanded the *Fragata* [*Tres Reyes*], and the pilot Antonio Flores decided to return to New Spain, as they had already reached a higher latitude than the instructions of the Viceroy laid down. The *Capitana* [Vizcaíno's ship] did not appear, and they themselves as well as almost all the men on board were sick. Having charted the country, and named the river "Santa Inez," they set out. . . .

No news nor signs of the City of Quivira could be had during the whole course of the exploration, not even in this latitude, and although some have thought that this *Rio de Santa Inez* which the *Fragata* discovered is the one that reaches that city . . . in search of which this fleet set out, I consider it very certain . . . that such was not the case.

Fray Juan de Torquemada's Account, January 1603

The tender [*Tres Reyes*] . . . had left the *Capitana,* and thinking her to be before, followed in search of her: and in the latitude of 41 degrees, the south east wind brought her to the *Capitana,* but the ship not being able to live in the sea, ran before the wind near the shore, and came to an anchor under a huge rock near cape Mendozino: and when the wind abated, continued her voyage along shore. On the 19th of January, Florez the pilot who was in the tender found himself in the latitude of 43 degrees, and the land forming a cape called Cabo Blanco where the coast begins to stretch towards the N.W. and near it a very large river having its banks covered with ash trees, willows, and other Spanish trees. This pleasing appearance rendered them desirous of putting into it, but the currents hindered them. . . .

This river is supposed to be that which runs up to a large city discovered by a Dutch ship, and that it is the Streight of Anian, through which the ship passed from the north sea into the south. The city de Quivira is doubtless in these parts; and to this country the narrative delivered to his majesty, and which first induced him to order an improvement of those discoveries; and a report to be made to him of the several proceedings, relates.

Sources: Herbert E. Bolton, ed., *Spanish Exploration in the Southwest, 1542–1706* (New York: Scribner, 1916), 102; Henry R. Wagner, *Spanish Voyages to the Northwest Coast of America in the Sixteenth Century* (San Francisco: California Historical Society, 1929), 255; Miguel Venegas, *A Natural and Civil History of California,* Vol. 2 (London: Rivington and Fletcher, 1759).

ANALYSIS

Slightly below Cape Mendocino at 40–41° north, the crew of the *Tres Reyes* sighted a large bay and a river entering it from the north. Farther on at 43°, Flores recorded a "very copious and deep river" near a point called Cabo Blanco. This was the river they named Santa Inez because that saint's day was January 21. The expedition's principal cartographer, Gerónimo Martín Palacios, drew a bay with a river entering from the north on his large-scale map of the area, but he did not depict a River Santa Inez on this chart, nor was the name Santa Inez mentioned by Vizcaíno in his account. Historians have labored to identify this river's present-day equivalent in latitudes ranging from 39° to 43°, but there has been no consensus. Part of the problem was the variable reliability of the latitudinal calculations of a crew who, weakened by scurvy and buffeted by storms, had reached the end of their endurance.

In Madrid in 1615, Torquemada published an abridged version of Ascensión's account. Torquemada appeared to extend the significance of the River Santa Inez by suggesting that it somehow connected with "the Streight of Anian"—a connection with a Northwest Passage to the Atlantic—although he attached a footnote in which he expressed some doubt about that possibility. His 1615 published account, which is exceptionally rare today, may not have been available in large numbers at the time. It was reprinted in Madrid in 1723, and Andrés Marcos Burriel included it in Miguel Venegas's *Noticia de la California* in 1757, which was translated into French, Dutch, German, and English (1759) and was widely disseminated. It is likely that most Europeans received their first knowledge of the Vizcaíno expedition and a conjectural river extending far into the interior through these translated and reedited texts.

Earlier work of mapmakers in France also propagated the idea that a Spanish expedition had discovered the entrance to a major intercontinental thoroughfare (Wagner 1968). In 1700, an anonymous French map of the North Pacific may have been the first to depict an Entrée d'Aguilar extending inland from the California coast. It is difficult to trace with certainty the delineation of this toponym. The name of the river observed at 43°, Santa Inez, did not appear on either Palacios's manuscript maps or subsequently published derivative maps of the voyage. At some point after Martín Aguilar died on the return to Mexico, a decision was made to rename the river Santa Inez in his memory. Inexplicably, that name did not appear on a map until nearly a century later. It is equally puzzling what source the anonymous French mapmaker used to derive this information in 1700.

In 1722, the prominent Paris map publisher Guillaume Delisle issued his seminal map, "Carte d'Amerique," on which he engraved a conspicuous legend, "Entrée Découverte par Martín d'Aguilar," that extended through a broken coastline at approximately latitude 45° north above "Cap Blanc de St. Sebastien" (another Vizcaíno-given name). In the interior to the east of this legend, Delisle located "Quivira" (a mythical city to the north of New Spain rumored to be rich in gold) and a river extending to the west with the legend "Grande Reviere coulant a l'Ouest" (Great River Running to the West). This was perhaps the first cartographic illustration of the fabled River of the West that many geographers believed originated from somewhere in the interior of the continent.

In 1722, the extended Delisle family was recognized as among the most authoritative and influential mapmakers in Europe. Joseph Nicolas Delisle's "Carte d'Amerique" was one of the most popular maps of the 18th century, and it was reissued more than 20 times well into the 19th century. Delisle most likely received his information on Vizcaíno's expedition from Torquemada. In 1722, although he did not draw any hypothetical coastline above 45°, he clearly elaborated on Torquemada's (and Ascensión's) suggestion that the entrance of

Martín Aguilar's river might connect with an as yet undiscovered interior thoroughfare. Within the next few years and until nearly the end of the century, European mapmakers seized on this concept. Soon the "Entrée découv. par Martín d' Aguilar en 1603" appeared on maps as "Rio de Martín Aguilar" at 43°. On some maps, this river stretched far across the continent as a "R. de L'Ouest," with headwaters near "L Winipique" (Lake Winnipeg). On others, it extended through a short arm into a large inland Sea of the West.

By the 1750s, the extended Delisle family had incorporated the Rio de Martín Aguilar into maps that also illustrated the conceptualized waterways of Bartholomew de Fonte and Juan de Fuca, which were part of the cartographic debate at the time. The influence of these ideas derived from the authority of the map itself as a representation of reality, the authority of a mapmaking family such as the Delisles who enjoyed royal patronage, and the authority of the apparent Spanish source of the initial discovery. Perhaps just as important was that Joseph Nicolas Delisle's apparent interpretation of the Entrée de Martín Aguilar as part of a more elaborate transcontinental waterway typified the manner in which mapmakers blended observational information with preconceived conceptual information.

Not surprisingly, then, these kinds of maps influenced those who later planned European maritime explorations to the Pacific Northwest. During the second Bucareli expedition of 1775, the Spanish commander Bruno de Hezeta recorded in his diary that during a council at sea held to determine a proper heading to sail, "the judgment of Don Francisco de Bodega y Quadra and pilot Francisco Maurelle was to continue up to the Rio Aguilar located on the map in 43°" (Beals 1985, 61). On May 20, "morning dawned with clear horizons. . . . Considering that in the latitude of 43° is where the maps place Aguilar's river, I made every effort to approach as near as possible to the coast with the purpose of discovering it. I examined the vicinity of this latitude closely, but all I found were some small islands" (Beals 1985, 89–91).

During this voyage, Hezeta and Bodega frequently referred to the maps from the *Noticia* and a map published in Paris in 1766 by Jacques-Nicolas Bellin, "Carte Reduite de L'Ocean Septrionale" (Small-scale Map of the Northern Area). Bellin's map suggested a communication between the "Entrée découverte par Martín d' Aguilar en 1603" and the central part of the continent. A decade later, Louis XVI instructed Jean Gallaup La Pérouse that "it will be noted that Captain Cook . . . was unable to examine this Coast between the 45th and the 50th parallels . . . and between the 50th and the 55th; and that it is in this gap that one can assume would be found the Entrance discovered in 1603 by Juan de Fuca, and the one of Martín d'Aguilar, which might possibly communicate with Hudson Bay through lakes and rivers" (Dunmore 1994, cxvii).

On September 5, 1791, while surveying the coast south of Nootka Sound, Alejandro Malaspina wrote in his diary that "the *Atrevida* sounded fifty fathoms, rock, and, following the course required by the trend of the coast, we began the survey that was to lead us to Cape Blanco and Río de Martín de Aguilar. . . . By this time we were in no doubt that a cape lying to our east, fringed by needle-shaped rocks and not far south of the mouth of a large river, was the Cape Blanco of Martín Aguilar. We therefore saw that Comandante Bodega y Quadra's suspicions that such a river did exist in the short stretch of coast near the parallel of 42°, which he had not been able to examine, were well founded" (David et al. 2003, 195–196). The following year, George Vancouver searched in vain in the same region. In 1798, he noted in his published account that "some of us were of opinion that this was the cape Blanco of Martin D'Aguilar. . . . [W]e ranged along the coast at the distance of about a league, in hopes of determining the existence or non-existence of the extensive river or straits, asserted to have been discovered by Martin D'Aguilar" (Vancouver 1798, 1:187–188).

Finally, in February 1792 in his instructions to frigate captains Dionisio Alcalá Galiano and Cayetano Valdés y Flores, Viceroy Juan Vicente de Revillagigedo (1789–1794) wrote of the importance of carrying out "a detailed exploration from Fuca to San Francisco, sailing into the Boca de Ezeta [present-day mouth of the Columbia River] . . ., determining the true position of the Rio de Martin de Aguilar, and entering ultimately the port of San Francisco" (Kendrick 1991, 51).

7. Port of Monterey, 1603

INTRODUCTION

When Sebastián Vizcaíno and Father Antonio de la Ascensión saw Monterey, they believed they had fully satisfied their mission to identify a harbor for refuge, rest, and reprovisioning of the galleons returning to Mexico from Manila. The present-day Bay of Monterey, which lies between latitudes 36°36′ and 37°06′, was first seen by Juan Rodríguez Cabrillo in November 1542 and by Sebastián Cermeño in 1595; both mariners applied different names to it, although neither explored the bay for a safe harbor. On December 16, 1602, Vizcaíno's three ships became the first European vessels to enter the harbor, which Vizcaíno named Puerto de Monterey in honor of the viceroy of New Spain. He calculated his latitude at 37° (actually 36°36′) and remained in the harbor until January 3, 1603.

On December 28, Vizcaíno dispatched a letter to the king in Acapulco extolling the virtues of Puerto de Monterey. On his return to Mexico on May 23, 1603, Vizcaíno wrote another letter to the king, excerpts of which are included here. Through these letters, Vizcaíno's official journal, and the writings of Ascensión, Spanish authorities learned about Monterey's great potential to become a major port and settlement on the Pacific Coast of California.

Letter to His Majesty from Sebastián Vizcaíno, Dated at Mexico on May 23, 1603

In the past year of six hundred and two, by order of your Viceroy, the Conde de Monterrey, I set out on the discovery of the coast of the South Sea. . . . [I]n conformity with the order and instructions I had, I explored very diligently the whole coast, not leaving harbor, bay, island or gulf without sounding and delineating it in accordance with the rules of good cosmography and the art of demarcation. . . . Among the ports of greatest consideration which I discovered was one in thirty-seven degrees of latitude, which I called Monterey. . . . [I]t is all that can be desired for commodiousness and as a station for ships making the voyage to the Philippines, sailing whence they make a landfall on this coast. This port is sheltered from all winds, while on the immediate coast there are pines from which masts of any desired size can be obtained, as well as live oaks and white oaks, rosemary, rock roses, the rose of Alexandria, a great variety of game, such as rabbits, hares, partridges, and other sorts and species found in Spain. . . . This land has a benign climate, its waters are good, and it is very fertile—judging from the varied and luxuriant growth of trees and plants; for I saw some of the fruits, particularly chestnuts and acorns, which are larger than those of Spain. And it is thickly settled with people whom I found to be of gentle disposition, peaceable and docile, and who can be brought readily within the fold of the holy gospel and into subjection to the crown of Your Majesty. . . . I was informed by them . . . that inland there are great communities, which they invited me to visit with them. They manifested great friendship for us and a desire for trade; were fond of the image of Our Lady which I showed to them and very attentive to the sacrifice of the mass. They worship different idols . . . and they are well acquainted with silver and gold, and said that these were found in the interior.

And, as some port or place on this coast is to be occupied, none is so proper for the purpose as this harbor of Monterey. For the reasons given, this port can be made by ships on the return voyage from the Philippines; and if, after putting to sea, a storm be encountered, they need not, as formerly, run for Japan, where so many have been cast away and so much property lost; and, had this port been known previously, Your Majesty would not have been so badly served.

Source: Sebastián Vizcaíno and George Butler Griffin, ed. and trans., *Publications of the Historical Society of Southern California,* Vol. 2, No. 1, Documents from the Sutro Collection (1891), pp. 68–73. Reprinted in Donald C. Cutter, ed., *The California Coast: A Bilingual Edition of Documents from the Sutro Collection* (Norman: University of Oklahoma Press, 1969), 113, 115. *The California Coast,* translated and edited by Donald C. Cutter.

ANALYSIS

The strictest accounting of Vizcaíno's voyage might conclude that its only unique geographical discovery was Puerto de Monterey, whose suitability as a port and settlement occupied much of the correspondence between Madrid and Mexico City after the expedition returned to Acapulco in 1603. Spanish authorities were understandably impressed by Vizcaíno's description of the harbor's attractive location, its "benign climate," the fertile territory, and the friendly residents who were eager to be converted to Christianity and knew where gold and silver could be found. Ascensión embellished these descriptions, and Vizcaíno's accounts were much exaggerated. Monterey was not as well protected a port as San Diego, for example, and Vizcaíno and his men experienced extreme cold weather during their 18 days in the area and discovered only a single abandoned village in the interior. Nevertheless, Vizcaíno's and Ascensión's overblown and enthusiastic claims were enough to convince Viceroy Monterrey that the harbor should be revisited to develop a port.

In July 1603, the viceroy wrote to the king supporting Vizcaíno's recommendation for a harbor and endorsing a reward for his efforts. After taking up the matter, the king and the Council of the Indies in Madrid issued a *cedula* (official document) in July 1604 to the new viceroy, Juan de Mendoza, 3rd Marquis of Montesclaros (1603–1607), that Vizcaíno should be rewarded for his efforts. On August 19, 1606, the council issued additional *cedulas* with the following orders: "It has seemed advisable to me [King Philip III] that all the ships coming from the Philippines should enter this port [Monterey], as they sight that coast. . . . [I]t has seemed to me very important to endeavor to settle it and that this be done with Spaniards, committing the project to Sebastián Vizcaíno as he by reason of the confidence built in him . . . seems to be more than any other person the most appropriate one to go back and settle it" (Wagner 1929, 276).

Montesclaros, however, was not a supporter of Vizcaíno and was even less enamored of developing a port and settlement at Monterey. A lively debate ensued, and on September 27, 1608, a royal order delayed further action on the Manila-Monterey-Acapulco proposal in lieu of exploring for alternative sites rumored to lie closer to Japan. The order recommended that Vizcaíno lead this exploration by sailing from Manila in two small ships. The issue of developing a port at Monterey was more or less put on hold until October 1620, when Ascensión reinstituted his advocacy for the settlement of California. Renewed interest in the possible benefits of Puerto de Monterey prompted a royal order of August 2, 1628, that resulted in extensive testimony from 1628 to 1632 on the virtues and the drawbacks of Monterey.

8. California as an Island, 1625

INTRODUCTION

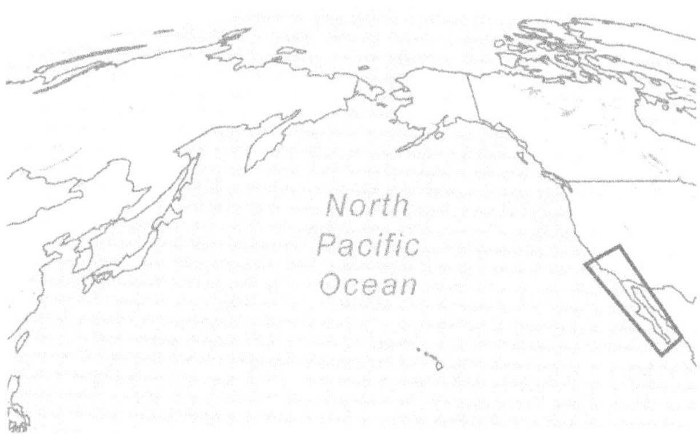

In the history of mapping, iterations of maps of a region based on successive voyages of discovery have not always resulted in more accurate cartography. A most remarkable example is 17th-century mapmakers' embrace of California as an island separate from the North American mainland. The earliest map depicting any part of California based on firsthand observations was the so-called Cortés manuscript map of 1535, which included four place-names on the end of the peninsula where Hernán Cortés briefly resided that year. For the next 90 years, maps of North America depicted California as a broad, lengthy band of land separated from the mainland along its southern extent by a gulf variously known as the Mediterranean Sea of California, the Mer de Cortés, and the Mar Vermejo (Red Sea).

Spanish explorer Francisco de Ulloa first reached the mouth of the Colorado River at the northern extent of the gulf in late 1539. But the geography of the area was not clear, and Spanish authorities did not publish results of voyages that they believed could be of value to foreign explorers. Subsequent Spanish voyages along the Pacific Coast originating from ports in Mexico or the Philippines produced reports of a continuous coastline from approximately latitude 41–42° north to Cabo San Lucas at 22°. The name Mendocino first appeared on a map published in 1587 by Antwerp cartographer Abraham Ortelius. The source of Ortelius's information is unknown, although the name was probably bestowed in recognition of Viceroy Lorenzo Suárez de Mendoza (1580–1583). Thereafter, the name Mendocino was applied to the coast variously between 41° and 43° (most references here to place-name origins are based on Wagner 1968). No firsthand observations above that point were recorded before Vizcaíno embarked from Acapulco in May 1602. Conjectures about the nature of the territory of California above Mendocino included an extension of the continent to the northwest, a coast continuing due north, and a possible connection with the long-rumored and sought-after water passage through or across the North American continent.

One cartographic legacy of the Vizcaíno expedition was this map, one of the first to depict California as a great island separated from the mainland by a broad gulf that could connect with a thoroughfare shortcut to the Northwest Passage. The English polymath Henry Briggs drew this map in London in 1625 to accompany his "A Treatise of the North West Passage to the South Sea through the Continent of Virginia and by Fretum Hudson [Hudson Bay]."

"The North part of America," 1625. (Manitoba Historical Maps)

ANALYSIS

In 1622, Henry Briggs alleged that he had seen a map from Holland that showed California as an island. On the lower third of this image of his copperplate engraved map, Briggs noted that "California sometimes supposed to be a part of ye western continent, but scince by a Spanish Charte taken by ye Hollanders, it is found to be a goodly island: the length of the west shoare being about 500 leagues from Cape Mendocino to the South Cape there of called Cape St. Lucas: as appeareth both by that Spanish Chart and by the relation of Francis Gaule whereas in the ordinarie Charts it is felt downe to be 1700 Leagues." Neither Briggs nor any contemporary ever identified the source of the map from Holland, but most historians believe that it was a copy of a map made by Antonio de la Ascensión. Ascensión's journal of the voyage and his subsequent writing about the nature of the geography of California, the people, and the benefits of the port of Monterey were valuable sources of information for Spanish authorities and, apparently, for Briggs. In an extensive official account written shortly after returning to Mexico in 1603, Ascensión stated that "I hold it to be very certain and proven that the whole Kingdom of California discovered on this voyage, is the largest island known or which has been discovered to the present day" (Wagner 1929, 265).

It is uncertain whether Ascensión submitted a map with this account, but he did write an abbreviated account of the voyage in October 1620, where he more clearly stated his geographical theories. The land of California, he wrote, "has the exact form and shape of a casket, being broad at the top and narrow at the point. It is this latter which we commonly call Punta de la California. From there it widens out to Cape Mendocino, which we will describe as being the top and breadth of it. The breadth of this land from here to the other sea, where the Mediterranean Sea of California comes and connects with the sea that surrounds and encircles Cape Mendocino, must be about 100 leagues. In this part this realm has north of it the Kingdom of Anian, and to the east the land which is continuous with the realm of Quivira. Between these two realms extends the strait of Anian, which runs to the North Sea, having joined the Oceanic Sea which surrounds Cape Mendocino and the Mediterranean Sea of California, both of which are united at the entrance of this strait which I call Anian. . . . [B]ut in order to make myself better understood I made for those who do not understand it a geographic map of it all, which I have with me; and I am sending a copy of it to his Majesty and to his Royal Council of the Indies" (Bolton 1916, 109–110).

Neither Ascensión's original map nor copies of it have been found, and it is unknown how the Dutch could have obtained a copy of his map. Nonetheless, in 1622 Briggs wrote that he had seen such a map. Three years later, the most notable features on his map of North America were the island of California and numerous Vizcaíno-applied place-names, including Punta de los Reyes, P. de monte Rey, and St. Catalina. During his voyage north of Mendocino, Vizcaíno recorded that the California coast tended to the northeast instead of to the west, but he was prevented from continuing his reconnaissance because his crews were decimated by illness. Ascensión used Vizcaíno's observation to conceptualize how the northern border of the island might appear. Beyond this point of observation, Briggs (and presumably Ascensión) left open the possibility of a passage from Hudson Bay to the Pacific Ocean and, hence, to China or Japan.

There is no record of how Ascensión or Vizcaíno responded to the dramatic cartographic renditions of Briggs and his predecessors, but the response of most European cartographers was obvious. They adopted this understanding of the geography of California based on

Briggs's influence and the presumed authority of his source, a Spanish map populated with place-names recorded after firsthand observations of topographical coastal features. In addition, the idea of an insular California somehow supported the theory held by many of a transcontinental water passage. For nearly the next 150 years, European cartographers almost always depicted California as an island on regional, hemispheric, and world maps. Not until the late 17th century would overland explorations prove that it was a peninsula. In 1747, Ferdinand VI of Spain (1746–1759) famously decreed that "California is not an Island."

An intriguing footnote to this story is the role that scurvy may have played in bringing about the cartographic consequences of Vizcaíno's voyage. In January 1603 when Vizcaíno's ship was in sight of Cape Mendocino and Cape Blanco de San Sebastian, where the pilots calculated a latitude of 42° and recorded that the coast trended to the northeast, most of the crew was incapacitated with scurvy and unable to work. Vizcaíno terminated the voyage, leaving to speculation the nature of the coast that lay beyond.

9. Coastal Chart of the Vizcaíno Voyage, 1802

INTRODUCTION

The principal objective of the second voyage of Sebastián Vizcaíno in 1602–1603 was to chart the Pacific Coast of California to at least latitude 40° north and possibly farther if the coast extended to the northwest. Despite extreme hardship, the pilots and cosmographers aboard the three ships accomplished that task. Spanish authorities sequestered the charts and maps in the Casa de Contratación in Seville, the government office that oversaw maritime affairs in the Americas. While Spain had a policy of suppressing information that might benefit foreign powers, such policies rarely succeeded. Within a few years, knowledge derived from the accounts and maps of Vizcaíno, his mapmakers, and Antonio de la Ascensión began to appear in publications, which sometimes introduced mistakes, made changes, or expressed interpretive biases.

This map, which appeared in an atlas published in 1802, 200 years after Vizcaíno left Acapulco, was drawn using the extant copies of the original maps from the expedition. Published in Madrid with the full backing of the king—that is, with a royal order by the Imprenta Real (Royal Publishing House)—it was part of an atlas whose purpose was to provide documentation of the extent of Spanish maritime activity along the California coast in the 17th and 18th centuries. The title on the map can be translated as "Map from the Surveys Made in 1602 by Captain Sebastián Vizcaíno Created from the Plans Made during This Voyage." It was the fourth map in the Atlas para el viage de las goletas Sutil y Mexicana al reconocimiento del Estrecho de Juan de Fuca en 1792 *(Atlas of the Surveys of the Schooners* Sutil *and* Mexicana *to the Strait of Juan de Fuca in 1792), although it had no direct link to that later voyage. The atlas accompanied a textual account of the 1792 voyage,* Relación del viage Hecho por las goletas Sutil y Mexicana, *introduced by Martín Fernández Navarrete with an extensive summary of Spanish voyages from about 1532 to 1792, including an abbreviated summary of Vizcaíno's voyage of 1602–1603.*

ANALYSIS

It is neither possible nor necessary to read the almost 150 names on this small-scale map of the California coast to appreciate that Vizcaíno's foremost cartographic legacy is the retention of the place-names he applied on the landscape between Cabo San Lucas and Cape Mendocino. Some of the more familiar names include Puerto de los Reyes (Point Reyes), named for the Day of the Kings or Three King Day in June 1603; Puerto de Montery (Monterey Bay), named for the viceroy of New Spain; the Canal de Santa Barbara, named on the day of that saint in December 1603; Isla de Santa Catalina, named in honor of the day of that saint in November; Puerto de San Diego, named on November 12 in honor of San Diego de Alcalá; and Esenada de Todos Santos (Port of All Saints) (most references to place-name origins are from Wagner 1968).

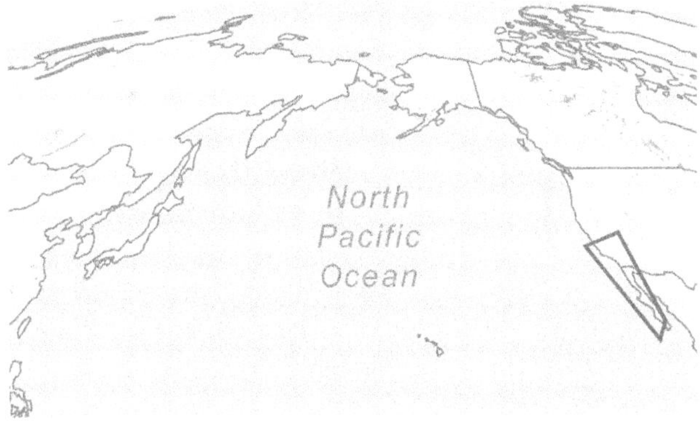

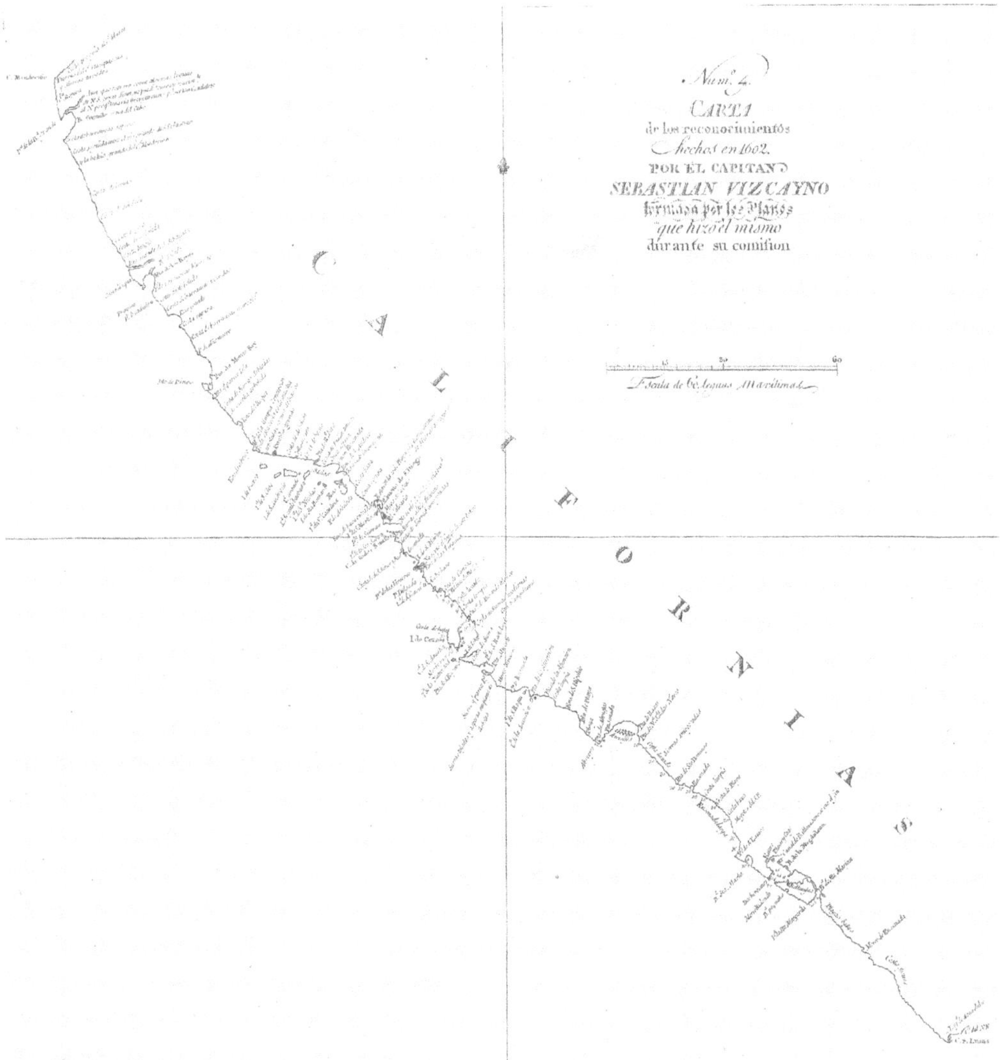

"Map from the Surveys Made in 1602 by Captain Sebastián Vicaíno Created from the Plans Made during This Voyage." (Private collection of James Walker)

Gerónimo Martín Palacios, the primary cartographer of the expedition, produced 1 profile chart of headlands and 32 large-scale charts of the coast, harbors, and inlets from Port Navidad on the mainland to Cape Mendocino; 29 of them document places north of Cabo San Lucas. Palacios also may have drawn a general map. His original charts have disappeared, but the resident official cartographer and printer in Mexico City, Enrique Martínez, made copies of Palacios's 33 charts in November 1603, which are in the General Archive of the Indies in Madrid.

In his introduction to the *Relacíon,* Navarrete recounted that he had reduced Palacios's 32 charts in order to create a single map of Vizcaíno's voyage along the Pacific California coast while "retaining all geographical positions and the name[s] made known" (Milich 1966, lxxxvii). In fact, Martínez's copies of those charts include names that Navarrete was not able to incorporate into his composite map. Navarrete omitted important latitude and longitude gradations, so it is not possible to determine coordinates of specifically named features. The copperplate from which this map was printed, however, was engraved with consummate skill, and it seems likely that any reader in 1802 would have been able to understand names that were authoritatively said to have been located with firsthand observation 200 years earlier.

A curious feature of Vizcaíno's cartographic legacy is that many of the places on the coast that he is credited with naming were visited and named by Spanish explorers before him, especially during the voyage of Juan Rodríguez Cabrillo in 1542–1543. Cabrillo sighted, named, and likely charted most of the same coastline, and Vizcaíno apparently did not consult an account of his voyage or any of his charts and sailing directions. It is possible that authorities intentionally withheld Cabrillo's information from Vizcaíno, or they may have considered that 60-year-old knowledge was no longer useful. It is also possible that Vizcaíno took advantage of his situation to achieve notoriety by claiming primary naming authority. In any event, almost none of Cabrillo's coastal place-names survive.

Even though Spanish authorities in Madrid intentionally withheld or delayed publication of accounts, charts, maps, and sailing directions that were the result of voyages of exploration, place-names from the Vizcaíno expedition began to appear on non-Spanish maps in Amsterdam in 1624 and in London in 1625. Manuscript maps by Portuguese cartographer João Teixeira in 1630 and Englishman John Daniell in 1637 and a published map by Robert Dudley in Italy in 1647 contained some of Vizcaíno's names (or mistranscriptions of them). Subsequently, Spanish cartographers such as Miguel Costansó in 1771 incorporated Vizcaíno place-names in their maps. The widely disseminated *Noticia de la California* (History of California), published in Madrid in 1759, included a map of part of the California coast populated with Vizcaíno names.

In the absence of any official published sources from Vizcaíno's voyage, geographers and cartographers likely obtained their information from captured Spanish material, perhaps by British privateers. They might have derived cartographic information from verbal or published textual accounts of Vizcaíno's voyage, or they may have "borrowed" material from each other. None of these maps credited Vizcaíno as the source of geographical information or coastal place-names. It was not until 1802 that Spanish authorities allowed the map to be issued, a map that focused on the results of the Vizcaíno expedition and acknowledged the lasting contribution of this great explorer to the cartography of the Pacific Coast of North America.

Further Reading

Beals, Herbert, trans. *For Honor and Country: The Diary of Bruno de Hezeta.* Portland: Oregon Historical Society Press, 1985.

Bolton, Herbert E. *Spanish Exploration in the Southwest, 1542–1706.* New York: Scribner, 1916.

David, Andrew, et al., eds. *The Malaspina Expedition, 1789–1794,* Vol. 2, *Panama to the Philippines.* London: Hakluyt Society, 2003.

Dunmore, John, ed. and trans. *The Journal of Jean-François de Galaup de la Pérouse,* Vol. 1. London: Hakluyt Society, 1994.

Kendrick, John, trans. *The Voyage of* Sutil *and* Mexicana, *1792.* Spokane, WA: Arthur H. Clark, 1991.

Mathes, W. Michael. "Sebastián Vizcaíno and San Diego Bay." *San Diego Historical Society Quarterly* 1(2) (Spring 1972), http://www.sandiegohistory.org/journal/72spring/vizcaino.htm.

Milich, Alicia Ronstadt, trans. *Relaciones by Zarate Salmeron.* Albuquerque: Horn and Wallace, 1966.

Schurz, William Lytle. "Acapulco and the Manila Galleon." *Southwestern Historical Quarterly* 22(1) (1918): 18–37.

Schurz, William Lytle. *The Manila Galleon.* 1939; reprint, New York: E. P. Dutton, 1959.

Vancouver, George. *A Voyage of Discovery to the North Pacific Ocean and Round the World, 1791–1795,* Vol. 1. London: G. G. and J. Robinson and J. Edwards, 1798.

Venegas, Miguel. *A Natural and Civil History of California,* Vol. 2. London: Rivington and Fletcher, 1759.

Wagner, Henry R. *The Cartography of the Northwest Coast of America to the Year 1800.* Amsterdam: N. Israel, 1968.

Wagner, Henry R. *Spanish Voyages to the Northwest Coast of America in the Sixteenth Century.* San Francisco: California Historical Society, 1929.

Second Kamchatka Expedition, 1733–1742

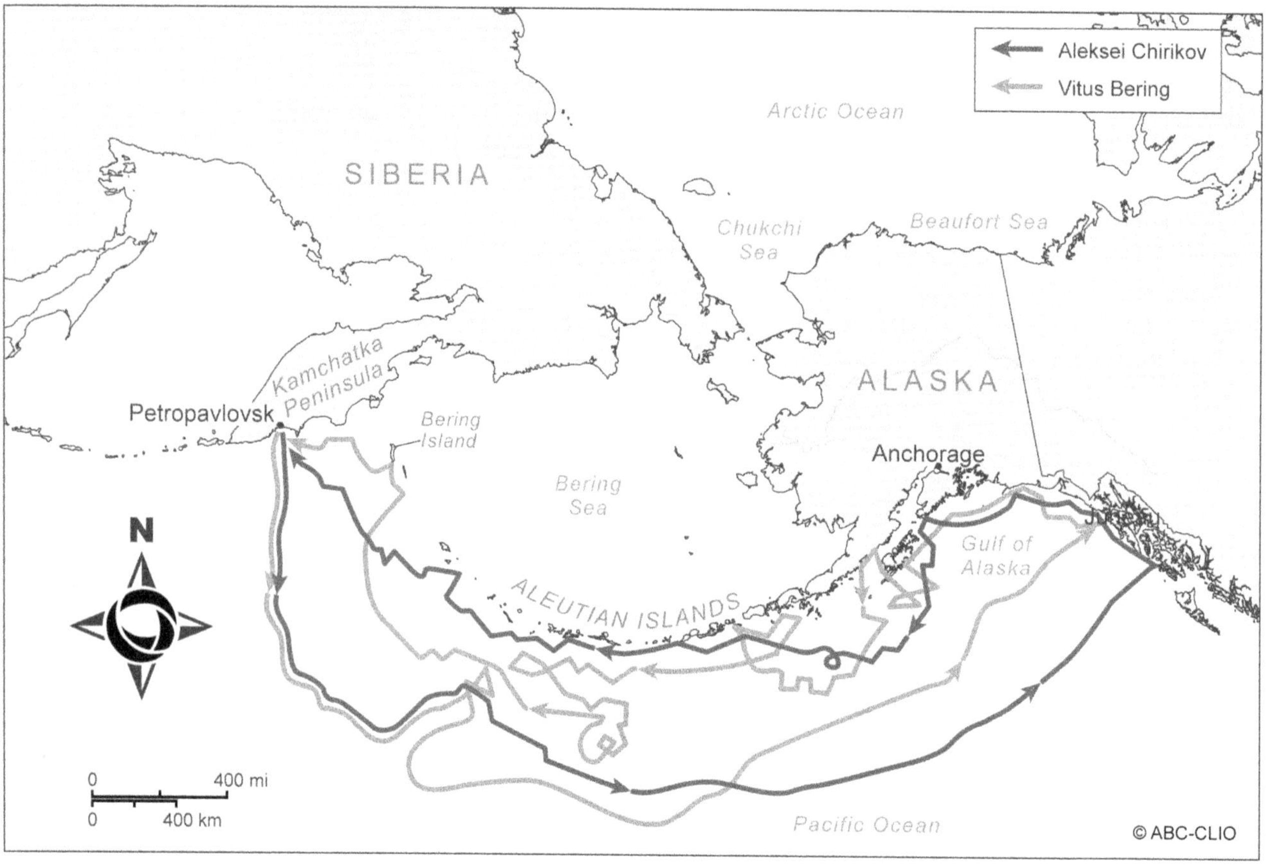

CHAPTER 2

Vitus Bering and Aleksei Chirikov, Second Kamchatka Expedition, 1733–1743

From 1724 to 1741, Vitus Ivan Ivanovich Bering (1681–1741) commanded two of history's largest, longest, and most costly expeditions across a vast expanse of Siberia and eventually to the shores of North America. These expeditions—the first Kamchatka (1725–1727) and the second Kamchatka (1733–1743)—resulted in the first maps of the North Pacific Ocean based on actual observation, the expansion of Russian commercial interests to the North Pacific Rim, and the increase in the scientific knowledge of the region's flora and fauna. They also stimulated competition among European maritime powers to explore, map, trade, and claim sovereignty in the region.

This is an undated purported image of Vitus Bering, although no verifiable portrait of Bering is known to exist. Bering's skeletal remains allowed reconstruction of his facial features and stature. A statue of Bering stands on the island where he died of scurvy and starvation at age 60. (Bettmann/Corbis)

Bering was born in 1681 in Horsens, Denmark, where his mother's family included the town's mayor, administrators of the university, members of the clergy, and attorneys. He was named for a maternal uncle who had been the royal historiographer for King Frederick III (1648–1670). Bering apparently had little or no formal education when he went to sea as a ship's boy at age 15 in 1696. During the next seven years, he reportedly participated in at least one whaling voyage, sailed to India and the Dutch East Indies, and studied navigation, geography, and chartmaking in an Amsterdam navigation school. While in Amsterdam in 1703, the 22-year-old's career trajectory changed abruptly when he was invited to join the Imperial Russian Navy as a sublieutenant. For Bering and for other Western-trained seamen, an invitation to foreign service promised good pay and an opportunity to rise in rank.

> **DID YOU KNOW?**
>
> Peter the Great and his successor Catherine I created the first Russian Academy of Sciences in the early 18th century, but its scientist and academic members were composed almost entirely of non-Russians recruited from France, Germany, Austria, and Switzerland.

Bering's recruitment was part of an ambition of Peter I, also known as Peter the Great, the tsar and first emperor of Russia (1682–1725), to make his country the naval and scientific equal of other European powers. During his first trip abroad in 1697–1698, the tsar had studied shipbuilding in England and Amsterdam and had consulted Dutch geographers about initiating explorations from the northeast coast of Asia to America. He began the construction of the planned city of St. Petersburg in 1703, and the next year he established the Admiralty, which became the principal shipbuilding yard for the Baltic Fleet. Peter subsequently moved the nation's capital from Moscow to St. Petersburg and created the Senate, a nine-member primary council of state.

During his second visit to Western Europe in 1717, Peter became familiar with some of the latest biological and astronomical observations in the Netherlands and Paris and with institutions such as the French Academy of Sciences that were designed to study them. He developed the Admiralty Board in 1718 and created plans for a Russian Academy of Sciences in 1724. These accomplishments were particularly remarkable because Russia was engaged in the Great Northern War between 1700 and 1721, a conflict largely between Sweden and an alliance of Russia and Denmark. The end of the war allowed the tsar to redirect resources to develop the institutions of the Russian Empire, as he named it, and he imported from Western Europe the naval and scientific personnel needed to expand his nation's fledgling navy, shipbuilding industry, and scientific community.

Bering was one of several officers who were attracted by the potential opportunities offered by a well-funded and expanding navy. He served in the Baltic Sea against Sweden and in the Sea of Azov against Turkey in 1710 and achieved the rank of captain-lieutenant in that year. Bering commanded several ships in 1715 and 1717, and on October 12, 1718, he married 17-year-old Anna Christina Pülse. He continued to advance rapidly through the ranks, with promotions to captain–third class and then captain–second class in 1720. But there his career stalled.

With the end of the Great Northern War in 1721, Bering was passed over for promotion, and non-Russian-born officers lost pay and opportunities to command ships. He concluded that he no longer had much of a future in the Imperial Russian Navy. In February 1724, he wrote to the navy explaining his reasons for considering retirement: "I have been denied the raise in pay given to several of my fellow officers who are much junior to me in service as captains, several of which I have actually commanded. I much regret all of this . . . but I very much doubt that I can expect any further advancement" (Urness 1992, 15). Some scholars have speculated that Anna may have been humiliated by her husband's failure to be promoted and that the family's honor and status might be preserved by retirement (Frost 2003, 27–28). After he left the Russian Imperial Navy, Bering and his family left St. Petersburg for his estate in Viborg, Finland, about 75 miles to the northwest. Curiously, at the time of his retirement the Admiralty promoted him to captain–first class, a rank he had not achieved when on active duty. Bering's retirement lasted less than six months. In August, perhaps for financial reasons, he applied to the Admiralty College for reenlistment. Bering's friend, Vice Admiral Peter Sievers, and his own brother-in-law, Rear Admiral Thomas Saunders (a British initiate in the Russian service), supported the application, and Bering returned to military service in the Baltic in October.

Bering's actions were timely. In 1724–1725 the tsar drafted plans for an ambitious overland expedition to the Kamchatka Peninsula in eastern Siberia (the first Kamchatka expedition), followed by a sea voyage to the north in ships that were to be built in Kamchatka. In December, Peter asked the Admiralty College for advice on choosing officers to lead the expedition. With recommendations from Vice Admiral Sievers and Rear Admiral Naum Seniavin, Bering was chosen as commander, apparently because of his experience in the East Indies. Peter also chose another Dane, Martin Spanberg, and a Russian, Aleksei Chirikov, to serve as Bering's lieutenants. They would later also serve as Bering's associate officers for the second Kamchatka expedition.

Bering worked with Peter the Great until the tsar died on January 28, 1725, and then with Empress Catherine I (1725–1727) to finalize preparations for the expedition, which left St. Petersburg on February 6, 1725. It would take nearly three years of arduous travel, crossing almost 6,000 miles of Siberian wilderness, to reach the lower Kamchatka Peninsula. On July 14, 1728, Bering sailed northeast from Kamchatka in the newly built *Archangel Gabriel,* and on August 16 he reached latitude 67°24′ north. He returned to St. Petersburg in February 1730, having been absent from his family for five years.

Not everyone was satisfied with the outcome of the expedition, including perhaps Bering himself. He had not reached the North American coast, and the question of the relative locations of eastern Siberia and America remained unresolved. Although Bering expressed no regrets in his March 1730 account of the voyage to Empress Anna Ivanovna (1730–1740), Gerhard Friedrich Müller, a firsthand observer at court, later wrote that "Captain Bering, sensible of the deficiency of his first discoveries, had offered himself to prosecute them, and so did his Lieutenants" (Müller 1754, 12). Whatever Bering's own judgment, the voyage resulted in a new map of the extent of Russian Siberian landholdings and a reasonably accurate charting of the northeast Asian coast. The map, which was soon available to West European governments, brought prestige to Imperial Russia, and the Admiralty and the Senate gave Bering a substantial monetary award and a promotion to captain-commander.

Within a month of returning to St. Petersburg, Bering initiated the planning for a second and more ambitious expedition from Kamchatka. He presented two sets of proposals to the Admiralty College and the Senate. The first included suggestions on how to develop the administration and infrastructure of eastern Siberia and how to explore and utilize the resources of the region, particularly the mines. The second advocated for three voyages, including one to search for and explore the North American continent. For three years the documents were reworked in the Admiralty College, the Senate, and the Academy of Sciences until Empress Anna issued the final orders. Favoritism in the court continued to work to Bering's benefit. He was supported by three prominent men: Ivan K. Kirilov, senior secretary of the Senate; Vice Admiral Nikolai F. Golovin, who soon would become president of the Admiralty College; and Andrei Osterman, a member of Anna's three-man Cabinet of Ministers.

During this time, Bering's noble rank and increased pay enabled his wife Anna to improve her social status, and she befriended a representative of Russia's closest European ally, Austrian minister Nikolai Sebastian von Hohenholz. On April 29, 1733, Vitus and Anna Bering and two of their children left St. Petersburg on the first leg of the second Kamchatka expedition, accompanied by many servants and the clothing and household goods that were worthy of their increased respectability. The nearly 6,000-mile trek to eastern Siberia (Okhotsk) would take over four years, at considerable cost in animals, supplies, and people, including those inhabitants of eastern Siberia who were made responsible for supplies and transportation.

On August 19, 1740, Anna Bering and her children left Siberia to return to St. Petersburg. On June 4, 1741, Bering on the *St. Peter* and Chirikov on the *St. Paul* set sail from Kamchatka

DID YOU KNOW?

Vitus Bering's officers desperately but unsuccessfully tried to treat their scurvy-stricken captain by partial burial in earth, a well-known treatment method that was reportedly used by both land and maritime explorers well into the 19th century.

on a course to America. Buried halfway in a sandpit, covered with lice, and suffering from scurvy and perhaps other diseases on an island that would be named for him, Bering died on December 8, 1741. He was 60 years old. Anna Bering did not learn of her husband's death until September 4, 1743. Her status and income diminished, and in March 1745 she wrote to the Admiralty College for monetary assistance. By then Russia had a new empress, Peter's daughter Elizabeth (1740–1762), who was not as supportive of foreigners in influential positions as her father had been. In January 1742 Andrei Osterman, a longtime friend and promoter of Bering's, was threatened with execution and exiled to Siberia. Anna Bering's influential group of friends at court gradually passed from the scene, and her star faded.

Vitus Bering's repute also declined. The Senate officially terminated the second Kamchatka expedition in September 1743, and Elizabeth ordered that all records be transferred to her private archive. Russian policy forbade publication of any information relating to the expedition except for brief official reports, until the Academy of Sciences was forced to respond in 1752 when Joseph Nicolas Delisle made public disclosures in France about the journey.

Because Bering left little written material, historians have relied on the accounts of the men who served with him to learn about the commander's character, navigational skills, and leadership qualities. Fellow officers Georg Steller, Sven Waxell, Safron Khitrov, and Aleksei Chirikov were critical of Bering's command, judging him as being too cautious, plodding, and indecisive, but his legacy as a principal participant in the development of Imperial Russia and the initiation of European exploration of the North Pacific Rim remain unchallenged.

In August 1991, Russian and Danish archaeologists excavated six grave sites on Bering's Island and identified one set of remains as those of Vitus Bering. In September 1992, a military honor guard reburied the captain-commander near a simple monument on the island where he died.

10. Joseph Delisle Map, 1733

INTRODUCTION

In 1731, the Admiralty College instructed Joseph Nicolas Delisle (1689–1768) to prepare a map of the North Pacific Ocean and contiguous lands in planning for the second Kamchatka expedition. In an accompanying memoir, Delisle wrote that "this map was made for the purpose of helping in the discovery of the shortest route between Asia and America." A French astronomer and cartographer, he had been recruited in 1725 by Peter the Great to become chief astronomer of the newly formed Russian Academy of Sciences. In 1733 Delisle prepared an updated version of his map, which is shown here.

Delisle's sources for "A Map Representing the Location and Distance of Eastern Tartary Extending to the Neighborhood of America" included maps drawn by his brother Guillaume in Paris and the cartographic results of Bering's first voyage of 1728. From other Russian, Chinese,

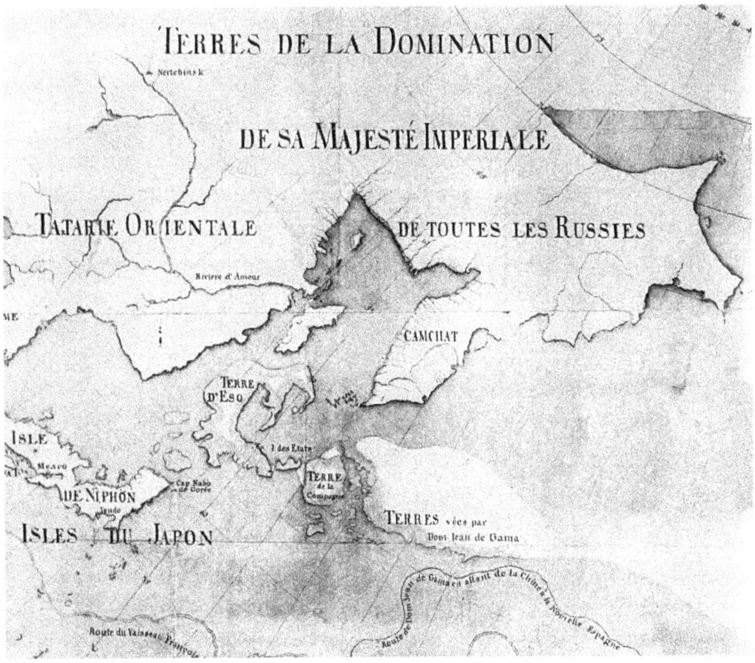

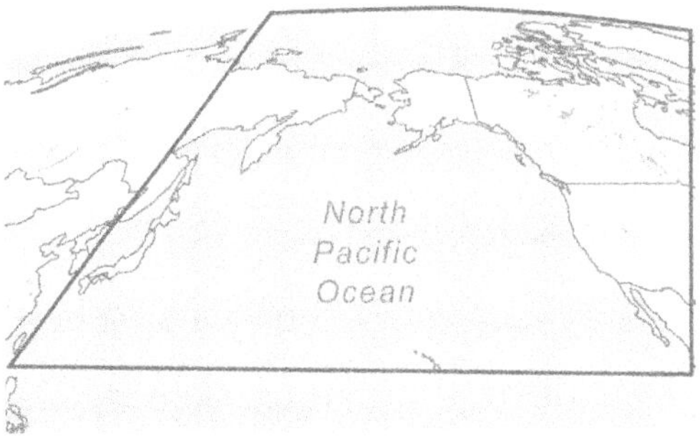

Detail of "A Map Representing the Location and Distance of Eastern Tartary Extending to the Neighborhood of America," 1733. (A. V. Yefimov, ed., *Atlas of Geographical Discoveries in Siberia and North-Western America XVII–XVIII Centuries*, Moscow, USSR: Publishing House Nauka, 1964, map no. 78)

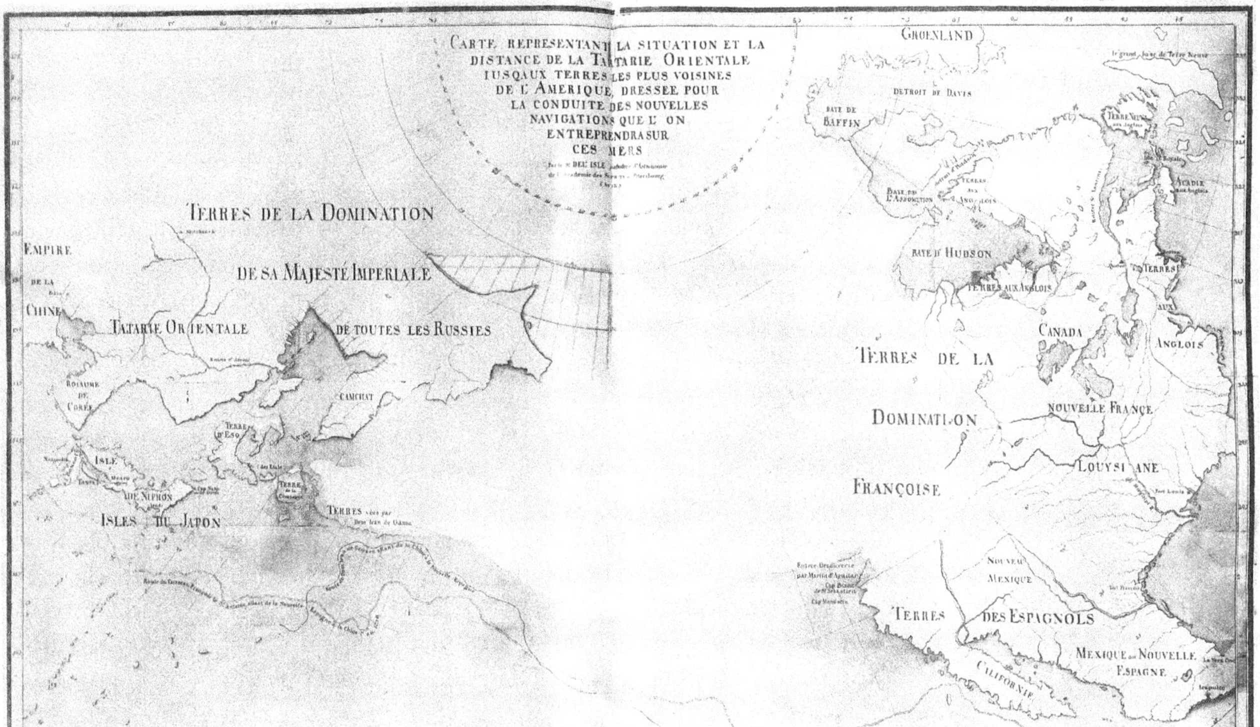

"A Map Representing the Location and Distance of Eastern Tartary Extending to the Neighborhood of America," 1733. (A. V. Yefimov, ed., *Atlas of Geographical Discoveries in Siberia and North-Western America XVII–XVIII Centuries*, Moscow, USSR: Publishing House Nauka, 1964, map no. 78)

and published Dutch maps, Delisle confidently included speculative information on the islands of Esso Land, State Land, Company Land, and, to the southeast of Kamchatka, land seen by Don Juan de Gama. His map tantalizingly suggests that Gama Land might be connected to an unexplored western extension of the American continent. The central part of the map between continents is blank, reflecting the lack of direct knowledge about the geography of the area and suggesting that obtaining that information was one of the goals of the expedition. Delisle also accurately depicted the lack of observational information on the geography of the American coastline above about latitude 45° north. Elsewhere on the map, he used large type to prominently depict Russian, English, French, and Spanish dominions in Asia and America.

ANALYSIS

Delisle's map has to be understood for both the geographical and nongeographical information it contains. Conveying information that was known, unknown, and conjectural, the map included a legend on the North American coastline at approximately 43°, "Entrée Découverte par Martin d'Aguilar," a title Delisle borrowed from a map published in 1722 by his brother Guillaume. This important conceptualized entrance to a transcontinental water passage was derived from the expedition of Sebastián Vizcaíno in 1602–1603. Delisle showed that even after more than 225 years of mapping the North Pacific, cartographers were still unsure of the geography of the North American continent above latitude 45° north and the relationship of Asia to the northwest part of North America. The inclusion of an empty area in the middle of the map implied that there might be an opportunity to discover and claim unexplored territory in the western parts of the American continent.

Delisle incorporated several speculative islands (Esso Land, State Land, Company Land, and Gama Land) from published Dutch maps onto his 1733 map. His authoritative conviction that these lands existed may have been an unfortunate contributor to the delay in fulfilling one of the expedition's goals: to reach and explore the North American coast. The revised map of 1733 was the principal document that Vitus Bering and Aleksei Chirikov consulted when they determined which sailing route to take in order to reach America from Kamchatka. Based on the map, they chose a southeast track to Gama Land, which appeared to be very close to or even part of America. The fruitless search for Gama Land took several days, and with that delay "the expedition was left with less time than anticipated to carry out its assignments" (Fisher 1977, 148).

Delisle's map also conveyed information about the multinational nature of scientific endeavor in the early 17th century and forecasted the international players who would converge on the North Pacific and the Pacific Northwest during the next 60 years. He had been one of many French and German scientists recruited by Peter the Great and his successors to help establish the Russian Academy of Sciences, and it is clear from the archival record that the massive second Kamchatka expedition could not have occurred without their participation. Notably, the title, names, and legends on the map are in French, which had become the preferred language of the Russian court and the scientific community.

As depicted on Delisle's map, Asia includes the dominions of China and Imperial Russia. In America, Delisle located the supposed dominions of Spain, France, and England. It is clear that the Russian planners of the second expedition anticipated encountering other Europeans on the journey, and Bering and Chirikov were instructed to avoid confrontation. Delisle suggested, however, that the islands to the southwest of Kamchatka had not been claimed by any other nation, thus raising the possibility that these lands might be folded into Russia's domain.

The objectives outlined by the Admiralty College and the Academy of Sciences focused on both science and potential profit for the state. Profit might come from either finding territory unexplored by Europeans that held precious metals or taxing indigenous people. Delisle's map suggests that acquisition of land with potential resources was possible both in the "known" lands of Company Land, Gama Land, and elsewhere and in territory contiguous to America. Discovery and acquisition were linked (Fisher 1977).

The Admiralty College made the same implication in a February 1733 report to the Senate regarding instructions to Bering: "According to the scholarly map which Professor Delisle has prepared, and in the judgment of both Bering and Chirikov, it appears that the distance of a direct route between that province [America] and Kamchatka is some 5,000 versts. . . . There is no information about settlements closer than that province in America belonging to a European power" (Dmytryshyn et al. 1988, 129–130). In a report of December 7, 1741, Chirikov acknowledged that he had willingly agreed with others that, according to Delisle's map, "a land called Juan de Gama, which we expect is part of America" was located at a latitude that was easily reached from Kamchatka and that was the route chosen after nearly 10 years of planning. He later acknowledged that the map was in error. As a result of the discoveries of this undertaking, the islands of Gama Land and others quickly disappeared from maps, and the vast blank spaces of Delisle's map were filled in by the exploratory efforts of multiple seafaring nations.

11. Orders from the Empress Anna, 1732

INTRODUCTION

The second Kamchatka expedition included three voyages of exploration: one to the Arctic, one to Japan, and one to North America. These instructions, from Empress Anna Ivanovna to Vitus Bering, relate only to the plans and orders for the voyage to North America. They derived from recommendations of officials of the Admiralty College and the Senate, two bodies established by Peter the Great to oversee governmental affairs.

Bering had returned to St. Petersburg after the first expedition in March 1730 and presented his report to the empress. Within two months, he outlined a proposal for a second expedition that included expanding administrative control over eastern Siberia and mounting a more ambitious voyage of exploration to North America from Kamchatka. In late December 1730, the Senate accepted Bering's proposals for a second expedition. Administrators, naval personnel, and scientists participated in planning for a multiphase, multiyear expedition. Large contingents of men, materials, and supplies had to be moved over thousands of miles of Siberia, where there was little administrative infrastructure or reliable transport and provisioning opportunity.

Three separate voyages of exploration were planned in ships to be built and launched from Kamchatka, and scientific observations were planned of the indigenous people, flora, fauna, and geological resources of Siberia and North America. The purposes of the expedition were to expand geographical knowledge and to expand the Russian Empire. Over the next two years, members of the Senate, the Admiralty College, and the Academy of Sciences exchanged correspondence with Bering, the empress, and each other until Ivanovna issued the final instructions in December 1732. Unlike the sparse set of orders that Bering had received from Peter the Great for his first Kamchatka expedition, these instructions contain many details about how the voyage was to be implemented.

Instructions from Empress Anna Ivanovna to Vitus Bering, December 28, 1732

The island opposite the mouth of the Kolyma River in the Arctic Ocean is to be located on a map. This has been known as Velikaia Zemlia [Great Land]. Some Siberiaks have been there and have seen people there. They have reported about this in detail to Bering and his men in Iakutsk [Jakutsk]. If this proves to be true, then the sloops [Bering's ships] are to put ashore on this island or on other islands or land which they have been sent to explore. On shore they are to investigate as much as possible. If they encounter people, they are to treat them well and not antagonize them in any way. They are to obtain information as to how large these islands and lands are, how far they extend and what they have in abundance.

They are to find an opportunity to entice the natives by giving them small presents similar to the ones given to Siberian princelings and other natives at the time of their first encounter. If these people express willingness to become [Russian] subjects, they are to be accepted, especially those who have been enticed. . . . Encourage them to pay iasak [tribute] but do not interfere much with them. . . .

[They] are to search most diligently and zealously for these American shores or islands. . . .

[They are to] proceed toward America and explore islands and lands, because with the exception of Captain Pavlutskii's report, there is no other information as to whether there are lands and islands between the known parts of the Kamchatka and American coast, or whether there is simply one sea, which, according to the map of the learned Professor Delisle, extends from Chukotsk Cape about 45 degrees to the Spanish province of Mexico. . . .

They are to keep these instructions secret and in safekeeping, especially during the voyages. For public disclosure, Bering and Chirikov will sail toward America and Spanberg toward Japan. Special instructions will be issued from the College for these ships and for the sloop which will sail east from the Lena [River] toward Kamchatka. These instructions will indicate that as a result of the request and desire of the Academies [of Sciences] in St. Petersburg and in Paris and other places, the Emperor Peter the Great, of blessed and eternal memory, dispatched an expedition out of curiosity to discover from His own shores whether the shores of America are joined to those of Asia. . . .

In order to maintain this secrecy the professors and everyone with them on the expedition, as well as professors in the Academy of Sciences and everyone else involved in any way with the expedition, are to be informed in writing that if they reveal any information contrary to this order they will be punished.

Source: Basil Dmytryshyn, E. A. P. Crownhart-Vaughan, and Thomas Vaughan, ed. and trans., *Russian Penetration of the North Pacific Ocean, 1700–1797,* Vol. 2, *A Documentary Record* (Portland: Oregon Historical Society Press, 1988), 113–115, 120, 125.

ANALYSIS

In 1730, Russian and European information about landmasses and bodies of water between Asia and America from latitudes above ~42° north was at best conjectural, as can be appreciated from Joseph Delisle's map and Anna Ivanovna's instructions to Bering. Native inhabitants of Siberia (Siberiaks), for example, referred to a Velikaia Zemlia (Great Land) that lay to the northeast of the Kolyma River in northern Siberia. This suggested the possibility that to the north a land bridge might connect Asia with a western extension of America, a possibility that Bering's first voyage had not excluded. In addition, early 18th-century Dutch maps depicted several speculative islands near Kamchatka, including Esso Land, State Land, Company Land, and Gama Land to the southeast. Based on verbal reports and these published maps, Russian authorities hoped that Bering might find that America was located to the east or southeast, not far from Kamchatka.

Although these instructions encouraged Bering to settle the question of the geography of the area between the two continents, more focus was on the goal to "search most diligently" for America. Certainly there were scientific objectives in this instruction, but Bering was also encouraged to make contact with people and to encourage them to "become subjects." Furthermore, they might be made to pay *iasak,* or tribute. Since the late 16th century, Russian authorities in Moscow had administered the expanding Siberian frontier through a complex bureaucratic system whose economic basis included exacting tribute from local inhabitants. The policy of taking tribute was harshly maintained and often enforced by holding family members hostage until payment was received. The possibility of encountering a new territory of potential tribute payers must have been appealing to the authorities in St. Petersburg and Moscow.

The official instructions to Bering also emphasized the importance of secrecy, even to the point of threatening punishment if anyone, including members of the Russian Academy of Sciences, violated the order. Other nations that had launched voyages of discovery to the Americas had adopted similar policies. Bering's instructions also suggested that he might encounter another European power and be forced to disclose the nature of his endeavor. In a case of "public disclosure," he was directed to describe his objective as an "expedition out of curiosity to discover" and to invoke the authority of Peter the Great and the Academies of Science in St. Petersburg and Paris. Other goals of the expedition, such as the expansion of Russian sovereignty in North America, were not to be disclosed.

12. Imperial Ambitions, 1732, 1734

INTRODUCTION

Nikolai Golovin and Ivan K. Kirilov wrote these two private memoranda about the intended accomplishments of the second Kamchatka expedition. They expand on the information in the official correspondence among Bering, Empress Anna Ivanovna, the Admiralty College, and the Senate. The empress did not exercise the same personal initiative in planning the second expedition as Peter had for the first, and she relied on her close advisers to carry out most of the strategies, plans, and implementation. Each of the principal planners of the expedition, including Golovin and Kirilov, had achieved positions of authority under Peter the Great, and each was a strong supporter of Bering.

These private communications reveal that the goals of the expedition in the minds of senior administrators were substantially more ambitious than the acquisition of knowledge through scientific exploration. The second Kamchatka expedition was also intended to expand the Russian Empire, which helps account for why Russian authorities were willing to expend so many financial and human resources to fund it.

Memoranda of Nikolai F. Golovin and Ivan K. Kirilov

Memorandum, Count Nikolai F. Golovin, Vice Admiral and Member of the Admiralty College, October 1, 1732

In the exploration of America there may be the following great gains for the state: very rich mines, both gold and silver, are there which are still unknown, [and it is known] what profit the Spanish, English, and Portuguese kingdoms receive and how important the commerce and navigation to these regions are to these kingdoms now.

Memorandum, Ivan K. Kirilov, Senior Secretary of the Senate, to Empress Anna Ivanovna before May 1734

In the present prosperous reign of our sovereign empress . . . two undertakings are developing, not only of great and immortal glory, but for the expansion of the empire and for inexhaustible wealth. . . .

This sending of an important expedition, the like of which there has never been before, consists of various investigations: (1) to find out for certain whether it is possible to pass from the Arctic Ocean to the Kamchatka or Southern Ocean sea . . . ; (2) to reach from Kamchatka the very shores of America at some unknown place about 45 degrees of longitude [*sic*]; (3) to go from Kamchatka to Japan, between which the distance is only ten degrees of north latitude; (4) on that voyage and everywhere to search for new lands and islands not yet conquered and to bring them under subjection; (5) to search for metals and minerals; (6) to make various astronomical observations both on land and sea and to find accurate longitude and latitude. . . .

The benefit to be expected is that from the eastern side Russia will extend its possessions as far as California and Mexico, although it will not immediately receive the rich metals which the Spanish have there. However, without preparing for war we can in time acquire [them] through kindness though I know that the Spanish will not be pleased. Besides, the local people are greatly embittered against the Spanish and for that reason have to escape to unknown places farther away (and it appears that there are no other places but closer to us). . . . To the south all the islands as far as Japan do not really belong to anyone, and already four islands have been brought under subjection by Pavlutskii's command. The same has to be reckoned about Esso and the land called "Company," [i.e.,] that they can not escape Russian possession if only assistance for the present beginning does not slacken in the future. Although God will nowhere disclose in any way what wealth the Japanese have, nonetheless the Japanese will not reject trade here for it is better for them to buy needed goods from us directly than to purchase them through the Chinese at higher cost. . . .

Russia is especially endowed by God with natural resources for the development of this new eastern trade. . . .

Here we have a great advantage over the Europeans who sail to the East and West Indies for it is not necessary [for us] to go to the equator and suffer from intense heat.

Source: Quoted in Raymond H. Fisher, *Bering's Voyages: Whither and Why* (Seattle: University of Washington Press, 1977), 131, 184, 186, 187. Used with permission, The University of Washington Press.

ANALYSIS

The second Kamchatka expedition included three separate voyages, including one that sailed south beyond the Kuril Islands to Japan, where it was hoped that contact with a highly structured society might generate trade and economic gain to Russia. These memoranda relate only to Bering's and Chirikov's voyages to North America.

On December 28, 1732, Empress Anna Ivanovna instructed Bering to seek out land and islands that lay beyond the farthest eastern reaches of Siberia's Chukotsk and Kamchatka Peninsulas. These lands or islands, Ivanovna hoped, would be proximal to or part of the continent of North America, a goal repeated in section two of Kirilov's memorandum. But Kirilov went further, suggesting that the goal was to "everywhere . . . search for new lands and islands not yet conquered and . . . bring them under subjection." He added that "the benefit to be expected is that from the eastern side Russian will extend its possessions as far as California and Mexico." His interest in geography seemed to be only partly scientific; he was also interested in extending Russian presence and sovereignty into the American continent, even to the point of antagonizing the Spanish.

The empress instructed Bering to bring under Russian subjection any people he might encounter, with the expectation that they would be required to pay a tribute. Kirilov went further when he anticipated that even those people living in areas under Spanish control might be brought under Russian dominion. The memoranda also suggest that Golovin and Kirilov had in mind even greater benefits from gaining a foothold in America. A previous communication from the Senate had noted that one of the possible nonhuman resources of newly encountered land or islands might be animals such as sables, foxes, beavers, and lynxes, all valuable for their fur. Both Golovin and Kirilov raised the possibility of finding rich sources of minerals and mines, including gold and silver, bringing profit to the nation similar to that already enjoyed by Spain, England, and Portugal. As Golovin noted, the exploration of America may generate "great gains for the state."

All of this suggests that ministers at the highest level of the Russian state understood the connections between exploration, knowledge of geography, and economic gain. They also understood that other European powers would be competitors in the Pacific Northwest for geographical knowledge and potential economic reward. Kirilov noted that the Russians enjoyed an advantage over other exploring nations because of the proximity of Siberia to America. For him, the potential reward of "the expansion of the empire and . . . inexhaustible wealth" may have easily justified the enormous expense and years of preparation for the second Kamchatka expedition.

13. Reports from Kamchatka, 1742

INTRODUCTION

Several documents written by crew members of the St. Peter *and the* St. Paul *chronicle the sequence of observations along the American coast from July to November 1741. One of these is the narrative of Lieutenant Sven Waxel, the officer in command of the* St. Peter *after Bering's death on December 8. Waxel reached Kamchatka on August 26, 1742; on November 15, he sent his succinct report by special messenger to the Admiralty in St. Petersburg. These translated excerpts from his account help us understand the course of the* St. Peter *and the identity of the places he saw. Logbooks might provide more data, such as time of day and the speed and heading of the ship, but Waxel's narrative describes how he and others interpreted what they saw and how they reacted to the conditions they encountered. His account is also a reminder of how important it was to early explorers to apply place-names with a familiar identity to the landscape they encountered.*

To the Imperial Admiralty College, a Most Humble Report, November 1742

It was agreed [in a meeting in May 1741] to steer, after going out of Avacha Bay [Kamchatka Harbor], SE by E, true compass, to the 46th degree in order to see whether there is any land on that parallel. . . . If, however, no land is seen on that parallel, it was concluded to sail from there on an E by N course until land is found, and when found to follow it in a northerly direction for the purpose of observation as long as time will permit so as to return during the last days of September. . . .

Between N and E we sailed until July 16, on which day we saw land [Mt. St. Elias] in the northwest. We steered for it and on July 20 so anchored alongside one of the islands, which we named St. Elias [Kayak]. . . .

When we had provided ourselves with water from the island of St. Elias, the Captain Commander, mainly because it was dangerous to lie out in the open sea, departed July 21, and we steered between S and W, following the lay of the land, for the purpose of observation . . . and came to latitude 56° 54′ by July 26. Here we sighted that same coast [Sitkalidak Island] . . . about seven miles distant. . . . It was quite unsafe to keep near the land because of the sand banks, continuous heavy fogs, and violent winds. . . .

During that night [August 2], as the fog lifted, we saw quite close to us an island [Chirikov Island]. . . . We named this island Tumannoi [Foggy]. . . .

August 4 we sighted five islands [Semidi], and from this time on we kept farther and farther out to sea. . . .

[I]t was agreed on this 27th day of August to go in search of water on the land which according to our calculation was about 60 miles away. On the 29th we saw many islands [Shumagin], towards which we steered closer and closer. . . . Near one of the islands [Nagai] we anchored. . . . The mainland was seen behind these islands about 12 miles away. . . .

[On September 3] we weighed anchor and tacked among the islands, going around the first island [Nagai] to the island [Bird Island] which was to the east of it. . . .

All these islands are uninhabited and treeless.

On September 24 we came to latitude 51° N where we sighted several islands [Adak, Atka] and, behind them on the WNW, the American mainland [Great Sitkin]. At this place there blew an indescribable gale from the west which continued until October 13. . . . In order that the Admiralty College may form some idea of the storm it should be pointed out that the old navigators said that they had seldom seen anything like it. . . .

We continued under these conditions until November 4, on which day we sighted land [Copper or Medny Island] which we looked upon as a godsend because it was beyond our strength to keep any longer at sea. By observation this land is in latitude 54°, but by dead reckoning we thought we were in latitude 53°. . . . [W]e hoped that the land in sight might be Kamchatka. . . . Our faulty reckoning was due to the fact that we had for a long time . . . thick weather and no sunshine to set us right. . . .

The members of the crew announced [on November 5] that because of their sickness and feebleness they were no longer able to do their work at sea. Taking this into account the Captain Commander and the higher and lower officers agreed to look for an anchoring place [Bering Island]. . . . [W]e took the sick Captain Commander ashore on November 8.

Source: F. A. Golder, *Bering's Voyages: An Account of the Efforts of the Russians to Determine the Relation of Asia and America,* Vol. 1 (New York: American Geographical Society, 1922), 270–273, 275–277.

ANALYSIS

On a present-day map, the latitude of the harbor on Kamchatka from which the *St. Peter* and the *St. Paul* sailed is approximately 53° north, and the southernmost latitude of the Aleutian Island chain is approximately 51°. When Bering and his officers decided to sail southeast to 46° and then east with the hope of encountering the islands on Joseph Delisle's map, they were well out of sight of the Aleutians. The explorers soon understood that the conjectural islands of Company Land and Gama Land did not exist. It took 6 weeks to reach Kayak Island, near Mt. St. Elias, and then 16 weeks to reach Bering Island. During that time, the expedition sailed southwest and west along the Aleutian Islands chain and did a reconnaissance of the islands and coastal areas while enduring threatening sandbanks, gale-force winds, fog, inadequate provisions, and a crew whose health was declining. Living conditions on board the 80-foot-long vessel must have been appalling.

Those conditions did not allow the crew to carry out a detailed exploration of most of the land they encountered. This explains why Waxel might have concluded that he saw "the American mainland" in the distance on September 24, when in fact he was seeing another island (Great Sitkin) in the Aleutian chain. When these kinds of observations and interpretations are repeated and eventually reported, it is understandable how mapmakers could depict landmasses as extensions of mainland North America instead of as islands.

Waxel's account illustrates both the considerable navigational skills of experienced deep-water sailors and the potential hazards associated with navigational technology. Frequently his reported latitudes are within a few minutes of a degree of present-day determinations. When possible, available instruments for celestial observations must have been used to determine latitude (at the time, there was no reliable technology to determine longitude at sea). Under adverse weather conditions that precluded the use of instruments, latitude and longitude were determined by dead reckoning, as Waxel mentioned on November 4. This method had been used for centuries by pilots who were sailing out of sight of observable and identifiable landmarks. It required estimating the ship's speed and elapsed time along a specific compass course and then plotting the information on a chart from the previously determined location. As Waxel reported, the result was often in error. Copper (or Medny) Island, for example, is actually at approximately 55° north, not at 53° as plotted by Waxel, an error of close to 140 statute miles.

From their first encounters with the landscape in the Americas, Europeans affixed names to geographical locations and topographical features that often reflected an event, an associate or patron, or a known European location. In this way, a national identity became associated with the so-called discovery of a place, serving to increase familiarity with and legitimize imperial sovereignty in the area. In the concluding paragraph of his report, Waxel informed the Admiralty that he "wished to name [the American coast], after the manner of other European discoveries 'New Russia,'" but that he did not feel empowered to do so without orders of the Admiralty College.

14. Georg Stellar on Bering Island, 1743

INTRODUCTION

Georg Wilhelm Stellar (1709–1746) was a German-born botanist, zoologist, and physician who moved to St. Petersburg and became a member of the Academy of Sciences in 1734. He joined the second Kamchatka expedition as a botanist, and in March 1741 Commander Vitus Bering personally invited him to sail with him on the St. Peter *to America. After returning from the journey in August 1742, Stellar spent four years exploring, studying, and collecting specimens in Kamchatka, the Kuril Islands, and Siberia, and he compiled a lengthy journal of his experiences during the voyage of the* St. Peter. *He died in Siberia in 1746, and his journal remained in manuscript form until it was edited and published in St. Petersburg by naturalist P. S. Pallas in 1793.*

G. W. Steller, Topographical and Physical Description of Bering Island

With warm-blooded sea animals the Bering Island region is more copiously provided. When we arrived there the sea beavers (or sea otters, *Lutris*) were present in large numbers. In November and December [1741] we killed them 3 to 4 versts [2–2.5 miles] from our quarters. . . .

[F]rom November 6, 1741, to August 17, 1742, over seven hundred otters were . . . killed by us, eaten, and their skins taken along to Kamchatka as tokens. . . .

The sea otter is usually 5 feet long and 3 feet in circumference at the breast bone, where the body is thickest. The largest weighed, with the entrails, 70 to 80 Russian pounds. In shape it resembles an otter, the hind feet only excepted, which are smooth and agree in structure with the hind flippers of seals. The entrails are likewise conditioned as in the otter. The skin, which lies as loose on the flesh as in dogs and shakes all over when [the animal is] running, so far surpasses in length, beauty, blackness, and gloss of the hair [that of] all river beavers that the latter cannot be compared with it. The best pelts bring in Kamchatka 20 rubles, in Yakutsk 30, in Irkutsk 40 to 50, and at the Chinese frontier, in exchange for their wares, 80 to 100 rubles.

Source: F. A. Golder, *Bering's Voyages: An Account of the Efforts of the Russians to Determine the Relation of Asia and America,* Vol. 2 (New York: American Geographical Society, 1922), 214, 216–217, 219–220.

ANALYSIS

Unlike the first Kamchatka expedition, the second expedition included at least two scientists, botanist Georg Stellar and astronomer Louis Delisle de la Croyère. Stellar, who joined at Bering's request and had no instructions from the Academy of Sciences, compiled his observations with dogged opportunism during the voyage and while shipwrecked on Bering Island from November 1741 to August 1742. On July 20 he was part of the first documented European landing on the American Northwest Coast. For 10 hours, he surveyed and collected flora and fauna and traces of human habitation. On August 30, he repeated his observations and collecting on Nagai Island.

Stellar made his greatest contributions while on Bering Island, and this account of sea otters was part of his extensive descriptions of plants and animals, which became an important scientific archive. While Stellar's entry on sea otters illustrates the disciplined, descriptive manner of a trained scientist, it also notes that the animals provided a frequent meat source for the shipwrecked company, supplying not only protein and caloric nutritional value but also vitamin C. Eating sea otters probably contributed to easing the symptoms of scurvy among the men.

The value of sea otter pelts was also a primary factor in Russian expansion east toward the American continent and the eventual Russian claims of sovereignty along the North Pacific rim. The crew of the *St. Paul* and Stellar and the crew of the *St. Peter* took nearly 2,000 sea otter pelts back to Kamchatka that they had collected among the Aleutians and on Bering Island. They found an unanticipated lucrative market for the smooth, lustrous skin, especially among Chinese buyers on the peninsula and inland. News of this profitable venture prompted private entrepreneurs to move into eastern Siberia to try to exploit the resource. Extended fur-hunting voyages farther into the North Pacific, along the Aleutian Island chain and eventually to the Alaskan Peninsula, required capital, ships, crews, and eventually outposts on distant islands. Some historians estimate that over 40 companies and 100 Russian ships operated from the Aleutian Islands to Cook Inlet and the Gulf of Alaska during the second half of the 18th century.

15. Tragic Cost of Scurvy, 1741

INTRODUCTION

The American voyages of the second Kamchatka expedition came with great human cost. Scurvy killed 31 of the 77 men on the St. Peter *and 21 of 75 men on the* St. Paul. *These translated excerpts are from a series of documents that highlight some of the events relating to scurvy aboard the* St. Peter. *No diary or logbook in Bering's hand survives, so several different documents must be consulted to appreciate the various aspects of the disease—the role of provisioning the ships, the timing of the onset of the symptoms, the relentless course of the disease, potential treatments, and the ultimate toll. Histories of the second Kamchatka expedition have given little attention to the role of scurvy, perhaps because the disease so commonly affected the crews of extended deepwater sailing voyages from all maritime nations for over three centuries.*

Scurvy on the *St. Peter*

Materials, Supplies, and Provisions Taken on Board the *St. Peter* for the Sea Voyage, May 31, 1741

Provisions	Units	Poods
Groats { Sacks	32 }	Total of 217 [7,834 pounds]
{ Bags	52 }	
Barrels { Beef	15	165 [5,956 pounds]
{ Butter	11	80 [2,888 pounds]
{ Pork	8	70 [2,527 pounds]

(*Continued*)

Provisions	Units	Poods
. . .		
Flour	100	250 [9,025 pounds]
. . .		
Hardtack (sacks)		382 [13,792 pounds]

Journal of Fleet Master Safron Khitrov aboard the "St. Peter," August 10, 1741

The assistant surgeon, Betge, has submitted a report in which he says there are five men on the sick list, totally unfit for duty and that, of the others, sixteen are badly affected with scurvy and if we continue at sea until the late autumn these men too will be unfit for service.

G. W. Steller's Journal of His Sea Voyage from the Harbor of Petropavlovskin Kamchatka to the Western Coasts of America, August 30, 1741

Aside from the red whortleberries and the black crowberry, or *shiksha* (*Empetrum*), which are found here in the greatest abundance, the products of the vegetable kingdom of greatest use to us were the glorious antiscorbutic herbs . . . which I gathered solely for my use and that of the Captain Commander. For although I had made representations that our medicine chest, from the very beginning, had been miserably supplied . . . [with] none whatever of the medicines most needed on sea voyages and serviceable against scurvy . . . and although I had therefore requested the detail of several men for the purpose of collecting such quantity of antiscorbutic herbs as would be enough for all, nevertheless even this proposition, so valuable to all . . . was spurned. Later, however, there were regrets enough, and when we had scarcely more than four able-bodied men left on the vessel, I was tearfully begged to help and assist, which then, though with empty hands, I did to the utmost of my strength and means. . . . the Captain Commander, who from scurvy and confinement had entirely lost the use of his limbs, was restored by me to such an extent simply by his partaking of the fresh spoonwort [scurvy grass] that within eight days he could get out of bed and on deck again . . . and also when the use of *Lapathum* [an herb] for only three days, according to my direction, made the teeth of most of the sailors firm again.

Logbook of the "St. Peter," November 5, 1741

We had few men to manage the ship; 12 of our number were dead, 34 were totally disabled from scurvy, only about 10 were able with great difficulty to get about at all and they were not fit for all kinds of sea duty. Among them was Lieutenant Waxel. We have little fresh water. Taking all these factors into consideration it was decided not to go [back to Kamchatka] . . . but to take advantage of the wind and steer for the shore in sight in order to save the ship and the men.

Lieutenant Waxel's Report on the Voyage of the "St. Peter"

On December 8, 1741, by the will of God, Captain Commander Bering died of scurvy from which he had been suffering for four months, and we buried him on the island where we lived. I succeeded him in command, and, with the assistance of Fleet Master Khitrov, every effort was made to find out where we were; but, owing to bad weather and other serious obstacles, it was not before the month of April that we knew anything certain.

Source: F. A. Golder, *Bering's Voyages: An Account of the Efforts of the Russians to Determine the Relation of Asia and America,* 2 vols. (New York: American Geographical Society, 1922), 1:48, 1:120, 1:209, 1:278, 2:84–86.

ANALYSIS

From 1500 to 1800, scurvy caused the deaths of over 1 million seafarers (McCord 1971). A nutritional disease caused by inadequate amounts of vitamin C, scurvy leads to defects in collagen, a protein that is necessary for the formation of connective tissue and is critically important for the support of many bodily structures. Lack of vitamin C results in a predictable series of symptoms that starts insidiously, progresses with devastating consequences, and, if untreated, ends in death.

Maritime scurvy—also known as the explorers' sickness—occurred when shipbuilding technologies and navigational improvements made protracted voyages of more than three or four months at sea possible, but captains could not adequately reprovision fresh food supplies. The journals of Vasco da Gama's 1497–1498 voyage—from Lisbon, around the Cape of Good Hope, and to Calcutta—describe not only the disease but also the successful use of oranges to treat its symptoms. Scientists did not discover the biochemical nature of scurvy until the 1930s, but many voyagers from the 16th century on—including Georg Stellar, the botanist on the second Kamchatka expedition—recognized that certain grasses and fruits cured symptoms of the disease.

The nutritional quality of the provisions on the *St. Peter* was similar to that found on most extended sea voyages of the 16th to 18th centuries; that is, while food supplied as many as 4,000–5,000 calories a day for each sailor, it was usually almost completely deficient in vitamin C. With the continued absence of fresh vegetables or certain plants, vitamin C levels in previously healthy individuals would be expected to be minimally detectable after three weeks and undetectable earlier in those whose levels were low to begin with. Early symptoms of lassitude at 8 to 12 weeks were followed by progressive weakness, swollen gums, skin and joint hemorrhages, loss of teeth, and eventual death, starting as early as 12 to 16 weeks.

When the *St. Peter* left Kamchatka on June 4, Bering calculated a return by the end of September. We do not know if Bering or any of his officers had previous direct experience with scurvy, so a four-month voyage with no assurances of resupply may have been a calculated risk or a decision made with inadequate knowledge. The decision to chart an initial course to the southwest from Kamchatka based on Joseph Delisle's map certainly led to several days of delay at sea and a longer distance to reach American shores.

Fleet master Khitrov's description of advanced scurvy on August 10 after approximately seven weeks at sea strongly suggests that many of Bering's crew had depleted levels of vitamin C before the ship's departure. The first to die from the disease was sailor Nikita Shumagin on August 30, when the *St. Peter* took on water on present-day Nagai Island, south of the Alaskan Peninsula in the Gulf of Alaska at 55°. In his journal Stellar recorded his understanding of the severity of the situation facing the crew, who were still some 1,000 statute miles from their base in Kamchatka. When he asked for help in collecting certain plants as potential treatments for the sick, his request "was spurned."

The documents further record disaffection between Stellar and other officers, which may have contributed to the deteriorating health of the crew. By November 3, although storms and headwinds prevented the *St. Peter* from making rapid progress in its return from America, it was the toll of scurvy that led to the decision to beach the ship on an island that Bering could not have known was only several miles east of his base in Kamchatka.

16. Social Costs of Natives of Kamchatka, 1743

INTRODUCTION

The second Kamchatka expedition, consisting of several detachments of nearly 600 people, left St. Petersburg between February and April 1733 on an arduous, multiyear, overland trek to the eastern reaches of Siberia. In August 1737 Bering's party arrived at the Siberian coastal city of Okhotsk, where the crew and local laborers constructed two double-masted brig-rigged ships, the St. Peter *and the* St. Paul, *both of which were 90 feet in length and 23 feet wide and displaced 211 tons. The ships sailed on September 8, 1740, to the departure port on Avacha Bay, Kamchatka, for the voyage to America. For more than seven and a half years, this large contingent of men, animals (horses and cattle), materials, and goods was supplied and financed by people in the region, whose strained resources were soon exhausted. These translated excerpts from an official report of the expedition offer a contemporary accounting of some of the human costs to the inhabitants of Siberia.*

Ekspeditsiia Beringa, October 5, 1743

Beyond the irretrievable outlay of 360,659 rubles [from one province] expended on that expedition (besides the still unknown [cost of] provisions, materials, and other supplies), as indicated on the list from the Admiralty College, it is already well known that the aforesaid Irkutsk Province—because of the distant hauling of provisions and all kinds of materials and supplies and the taking of many peasants as laborers to such remote places, resulting in harvest failures of grain—has already fallen into extreme ruin, for as the aforesaid Irkutsk Province has reported, it ordered the annual delivery to this expedition of up to 50,000 puds [approximately 1.8 million pounds] of rye flour, three thousand puds of groats [grain], and not a little hemp and other supplies, but in many *ostrogs* [settlements] and *slobodas* [villages of nonserf peasants]—especially in Ilimsk [a Siberian village] and all along the Lena [River], owing to the above grain harvest failures—the peasants are starving greatly and are eating grass and bark and hence are unable to send provisions to the expedition. Eniseisk Province was asked what provisions were purchased and dispatched, and it replied that in that province, owing likewise to a grain harvest failure, the inhabitants and peasants were eating grass and roots, mixed one-third with flour, and there are no provisions to deliver. Thus, there is no hope of Irkutsk Province supplying this expedition with provisions. From the beginning of the [second] Kamchatka Expedition peasants were found in remote and very difficult and stressful jobs and suffered every want, and many of them did not return from these expedition jobs for 10 years or more, so that the houses of those who had been taken for such heavy labor had gone to ruin and their plow land had been abandoned. . . .

Although at present it is impossible to know exactly what sum this expedition cost, nevertheless in truth it can be accepted that it is no less than one and one-half million rubles because: for ten years forty thousand rubles in cash were sent annually from Irkutsk Province alone, as well as sixty thousand puds of provisions annually, which cost two rubles per pud for delivery to Okhotsk [shipbuilding site], from which both the Iakuts of Iakutsk and the plowland peasants of Irkutsk Province had to endure great burdens up to the present. And because for the transport of these and other provisions six hundred plowmen have been maintained constantly, and for the same transport five to seven hundred Iakuts with horses—and sometimes one thousand—per year have been sent with horses carrying packs about three hundred verstas [200 miles] from Iudoma Cross to Okhotsk, and not half of those horses returned, and because of the lateness of the season they did not have time to move the horses, and then every man had to pull five puds on a sled seventeen times in one winter over the seventy verstas [47 miles] from Iudoma Cross to Urak Landing.

Source: James R. Gibson, "Supplying the Kamchatka Expeditions." In *Bering and Chirikov: The American Voyages and Their Impact,* edited by O. W. Frost (Anchorage: Alaska Historical Society, 1992), 112–113.

ANALYSIS

Undertakings as large as the second Kamchatka expedition have important backstories. These documents, for example, describe the consequences to indigenous people and resident Russians, who were required by officials in St. Petersburg to supply and support the expedition. They provide a grim perspective to the staggering logistics of the expedition—the movement of hundreds of people and materials over 6,000 miles, over terrain without established roads, down rivers in scores of newly constructed boats, and at a cost of nearly one-sixth the total Russian state income in 1724.

The movement of the expedition from St. Petersburg to Kamchatka, geographer James Gibson wrote, "not only depleted the treasury, but also decimated the economy of Eastern Siberia, where most of the supplies and workers were requisitioned" (1992, 113). Manpower and a means of transport were needed, and Gibson estimated that in 10 years members of the expedition appropriated 1,786 peasants and over 5,000 horses in eastern Siberia, amounting to 10 percent or more of the existing population of able-bodied men and horses in the region (1992, 113–114).

Bering and many of his officers had taken this journey once before—from February 1725 to August 1727 during the first Kamchatka expedition—and they had suffered obstacles and hardships then as well. Bering and government officials made attempts to prevent similar experiences during the second expedition. Nonetheless, inadequate administrative infrastructure throughout most of eastern Russia, poor management by officials who were sent early to remote areas to prepare for the main body, inhospitable travel conditions, and a dispersed population of mostly subjugated and sometimes hostile people who were facing their own hardships (e.g., crop failures) resulted in logistical problems and repeated delays in the expedition's progress across Siberia.

The economic and administrative relationships between powerful central authorities in St. Petersburg and the inhabitants of the nation's eastern possessions had long been based primarily on the extraction of valuable fur-bearing animals. This industry brought such large numbers of Russians and Cossacks to the region that from 1700 to 1730 the ratio of indigenous people to Russians changed from 2:1 to 1:4. Native people lost land to the Russians and were required to hunt and trap instead of farm. When they resisted enforced labor, such as supporting the transport of the main contingent of the expedition, many fled. Those who were captured were harshly punished.

Finally, these documents provide glimpses into the stratified society of St. Petersburg and the Russian frontier, which may have contributed to the resentment felt and the oppressive servitude endured by many laborers during the overland journey. Vitus and Anna Bering, historian Orcutt

Frost wrote, "brought Western European privilege, culture, and class into Siberia" (2003, 80). Bering and his wife and daughters traveled with a retinue of household servants, a clavichord, a large collection of silver and porcelain service, and fashionable clothing.

17. Joseph Delisle's Map of New Discoveries, 1752

INTRODUCTION

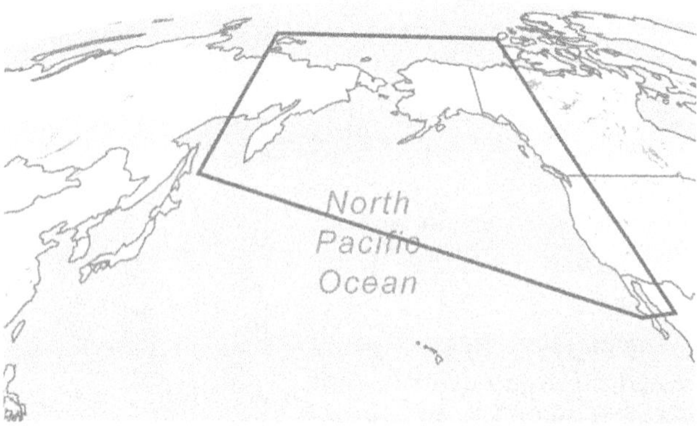

In 1733 in St. Petersburg, Joseph Nicolas Delisle drew the map that Vitus Bering and Aleksei Chirikov consulted when they departed from Kamchatka in June 1741 on their voyage to North America. Delisle did not accompany the explorers, but his brother, astronomer Louis Delisle de la Croyère, sailed with Chirikov. Delisle de la Croyère died of scurvy shortly after the St. Paul *returned to Kamchatka in October 1741. Even though reports of the expedition soon revealed the misinformation on his earlier map, Joseph Delisle maintained his influence at the Russian Academy of Sciences, and he continued to make significant contributions to Russian cartographical practices. He was a principal author of the official* Atlas of Russia, *published in 1745, and his general map of Russia was one of the last finished for that atlas. Delisle frequently quarreled with other officials in the academy, however, and some raised suspicions that he had violated the mandate of secrecy and had sent copies of maps to France. He soon lost control of the Geographical Department and became increasingly disenchanted with his position, refusing to cooperate with several of his assistants.*

In January 1747, after nearly 22 years in the service of Imperial Russia, Delisle returned to Paris, where he maintained his role as a preeminent astronomer and cartographer. He worked in his own small observatory in Cluny and was a member of the French and the Swedish academies of science. In April 1750, Delisle presented to the Royal Academy a paper and manuscript map about the Russian discoveries of 1741. In 1752, he and his brother-in-law, Philippe Buache, engraved this first widely circulated map to illustrate the routes of the Russian Kamchatka voyages.

ANALYSIS

Russian authorities were furious because Delisle had left Russia with unauthorized possession of a map and had published state secrets. Officials of the Russian Academy of Sciences soon arranged to publish an official map that they thought more accurately depicted the discoveries of Bering and Chirikov and the geography of the western coast of North America. Delisle's prestige in the French Royal Academy, however, ensured that his map was extensively distributed. Until publication of the maritime surveys of the Pacific Northwest of America late in the century, many mapmakers copied Delisle's conceptualized geography. The map is noteworthy in the cartography of the Pacific Northwest not because of observed information by the Russians but because Delisle included information derived from apocryphal travel accounts.

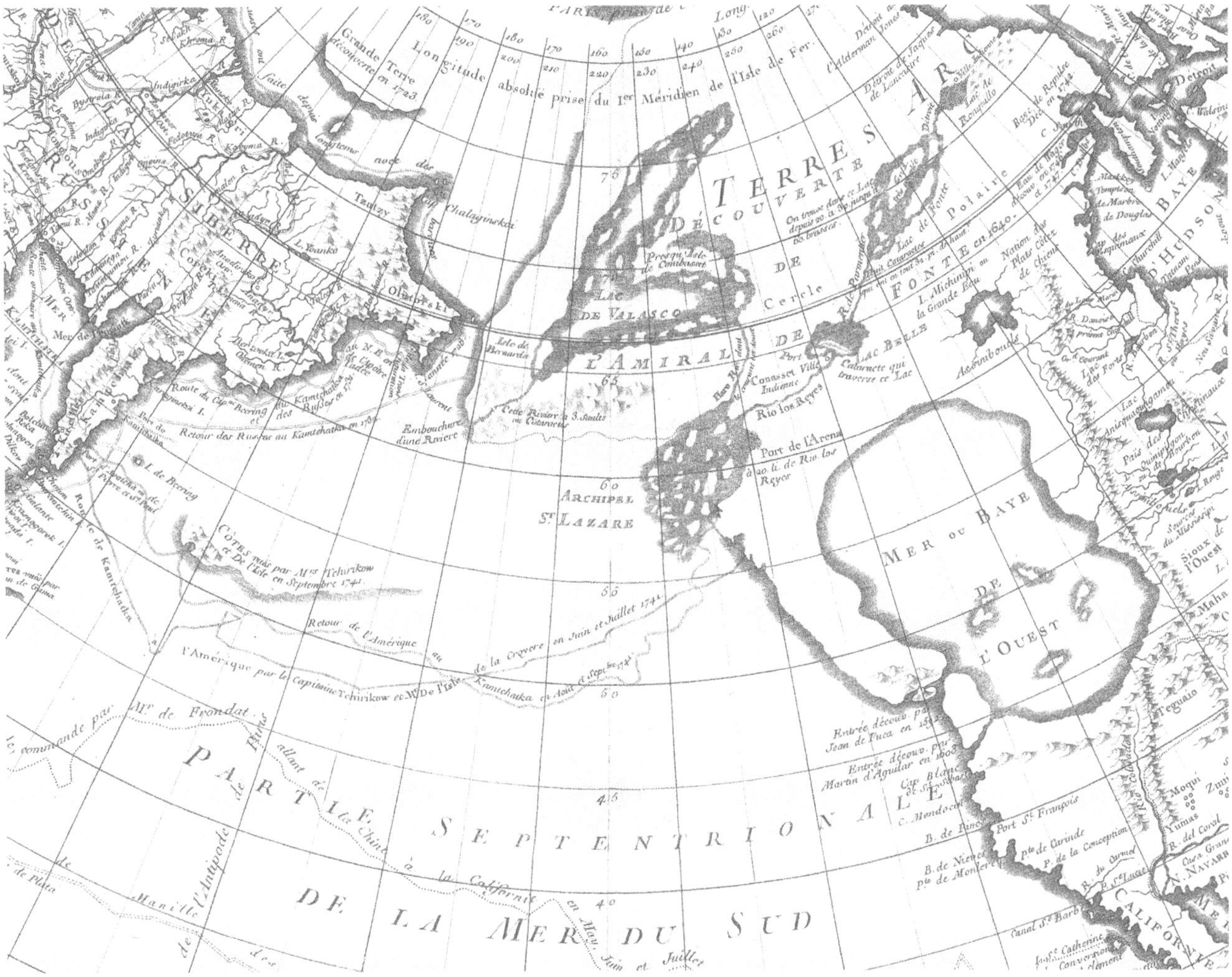

"Map of New Discoveries in the North Part of the Sea That Is East of Kamchatka in Siberia and West of New France." (Private collection of James Walker)

On his map, Delisle drew the outgoing and return tracks of the first Kamchatka expedition, north along the coast of the Kamchatka Peninsula and the more northern Olutorski Peninsula (today's Chukchi Peninsula) to just above latitude 66° north. On the southeast side of the peninsula he noted the Port of Avatchaou de St. Pierre et St. Paul, which was the departure port for Bering and Chirikov in 1741. From there, Delisle tracked the route to America by Chirikov and Delisle de la Croyère to a point of contact on the North American continent at about 56° and then back to Kamchatka. Inexplicably, he did not record Bering's track in the *St. Peter* and prominently included his brother's name next to Chirikov, even though Croyère had not been a cocaptain. Just north of 50° and east of the Kamchatka Peninsula, Delisle noted that Chirikov and Delisle de la Croyère saw the coasts in September 1741, further crediting his deceased brother with a prominent role in the expedition.

The most notable features on the map were on the western coast of North America. An enormous Mer ou Baye de l'Ouest (Sea of the West) occupied much of the inland area from latitude 43–60° north. A large rocky pillar stood at the entrance to the inland sea, the Entrée découv. par Jean de Fuca en 1592 (Entrance Discovered by Juan de Fuca in 1592) at approximately 47°. De Fuca was probably a Greek seafarer named Apostolos Valerianos who may have served as a pilot with the Spanish. In the 16th century many people believed in the existence of a navigable water passage, called the Northwest Passage, above or across the North American continent to the Pacific Ocean. Englishman Michael Lok, for example, was convinced of the accuracy of de Fuca's tale of having passed through a strait at 47–48°, past a pillar, and into an open body of water, where he sailed for 20 days. No record exists of any such Spanish voyage to this region of the Northwest Coast, but Lok's published account of de Fuca's story was picked up by Delisle and Buache, who converted the tale to a cartographic illustration on this 1752 map. A few previously published maps hinted at a "Sea of the West," but Delisle's depiction was by far the most definitively and boldly drawn and the most influential to subsequent mapmakers.

Farther north, from approximately 57° to 78°, Delisle and Buache engraved a complex system of islands (Archipel St. Lazarus), lakes (Lac de Valasco, Lac de Fonte), and waterways (Rio los Reyes, R. de Parmentier) that appeared to extend from the ocean across the interior of the continent nearly to the Baye de Baffin in the far north. The source of this highly conceptualized piece of geography was a story that had appeared in 1708 in an English publication, *Monthly Miscellany, or Memoirs for the Curious.* The story told of a letter from Admiral Bartholomew de Fonte, who alleged that he had sailed from Peru seeking a Northwest Passage between the Pacific and Atlantic Oceans and had discovered several of these lakes and waterways, which he had named. Although both de Fonte and his voyage are now recognized as fabrications, the story was picked up by Arthur Dobbs, a wealthy and influential Englishman who was an ardent supporter of the idea of a Northwest Passage. Dobbs published an abstract of de Fonte's letter in 1744 and actively pushed for repeated English voyages into the western reaches of Hudson Bay to explore for the passage. Such was the enthusiasm to locate a Northwest Passage that in 1745 the English Parliament offered a reward of £20,000 to the discoverer. The de Fonte story had nothing to do with the Kamchatka expeditions, but Delisle nevertheless included the geographical myth in his memoir to the Royal Society in 1750, and he and Buache created its first cartographic rendition on this map, although the location of Rio los Reyes did not correspond to its putative latitudinal location in the de Fonte letter.

Delisle and Buache were quickly assailed not only by Russian detractors but also by members of the French scientific community, but many others were impressed by the reputation and authority of a man who had been a prominent member of the Russian Academy of Sciences and whose mapmaking family was among the most highly respected in France. The Delisle-Buache cartographic myths lived on in controversy for decades, and maps such as this one are evidence of the often nonlinear course of mapping the Pacific Northwest.

18. Gerhard Müller's Attack on Delisle, 1754

INTRODUCTION

Within a few months of the publication of Joseph Delisle's book and maps, which included secret information about the expedition, the president of the Russian Academy of Sciences, Count Kirill Razumovskii, ordered the academy's principal historian, Gerhard Friedrich Müller, to prepare a response "immediately and secretly." Müller wrote a detailed rebuttal in French, which was published in Berlin in 1753 and again in Amsterdam later that year. In 1754, an English translation was published anonymously in London as "A Letter from a Russian Sea-Officer, to a Person of Distinction at the Court of St. Petersburgh." Most historians have attributed the letter to Müller, who wrote in the first person as if he had been with Bering on the St. Peter. *These excerpts from the English edition are an attempt by the academy to use an international forum to discredit Delisle's information and to emulate Bering's achievements in the service of Imperial Russia.*

A Letter from a Russian Sea Officer to a Person of Distinction at the Court of St. Petersburgh

The Sieur de l'Isle further pretends, that the materials, with which he would enrich his work, he had collected from other channels, nay, even from other countries: if so, why has he deprived us of them, as they, of right, properly belonged to us? Why was he not pleased, at least, before his departure, to acquaint the [Russian] academy of his writings, that a copy of them might have been taken: so far from it . . . there is not a single member of the academy, to whom he has vouchsafed to give a sight of the relation of the Admiral de Fonte, which, he says, he received from England in the year 1739. This is a piece of jealousy, utterly unbecoming a man of letters, but is greatly aggravated by his being a member of a learned society, where all discoveries ought to be common. . . .

[Delisle's] map and the observations . . . were both delivered to the senate, by the academy; . . . and then, whether this performance of his did more good, or harm to the expedition, is another question which shall be discussed in the sequel. . . .

These observations I do think are sufficient to shew, that it is a vain and fruitless attempt to make use of the pretended account of Admiral de Fonte, for filling up the vast and hitherto unknown space betwixt California and Kamtschatka. And now from the premises, I am led plainly and impartially to deliver my opinion on Mr. de l'Isle's map, which, as may already be perceived, will not perfectly harmonize with the description he gives us of the coasts of America. . . .

This Mr. de l'Isle knew nothing of, as having no knowledge of Mr. Beerings voyage; yet does Mr. de l'Isle turn it to the praise of the brother, to shew that he has done all he could to immortalize his dear name: A country, these are his words while he is speaking of his brother Mr. de l'Isle de la Croyere, which was seen by Mr. Tschirikov, and by Mr. de l'Isle de la Croyere; again you find the way from Kamtschatka to America by Capt. Tschirikov, and Mr. de l'Isle de la Croyere. This might pass well enough, if Mr. de la Croyere had been any ways instrumental in finding that way; but having been no more than an idle spectator, what title has he to any mention in that affair? He must have been blind not to have seen the coasts, which were visible to every one else, and, not to speak of us officers in both ships, who, as such, had our share in the execution of the affair, I take liberty to affirm, that our common sailors deserve much more than Mr. de la Croyere to have their names perpetuated. . . .

I conclude with a general remark upon the part of Siberia, as it appears in Mr. de l'Isle's map. It is entirely taken from the Russian Atlas. . . . It would therefore be an egregious mistake to presume, that in France better and more exact maps should be made of our country than we ourselves are able to execute.

Source: [Gerhard Friedrich Müller], *A Letter from a Russian Sea-Officer, to a Person of Distinction at the Court of St. Petersburgh* (London: Printed for A. Linde, 1754), 3–4, 12, 29, 31–32.

ANALYSIS

In 1741, surviving members of the Chirikov voyage provided St. Petersburg with maps and geographical information of the North Pacific region based on firsthand observations. The following year, survivors of the Bering voyage brought back valuable sea otter pelts that generated interest in the resources of islands between Kamchatka and North America. Russian officials recognized the potential economic and geopolitical advantages that this information provided. With some exceptions, they continued the policy of secrecy that was part of Bering's original instructions. In 1748 an anonymous brief report of the second Kamchatka expedition was published in Frankfurt, and incomplete reports surfaced in Denmark, England, and Germany.

Delisle's reputation and his forum at the French Academy drew the expedition into a central focus of the international scientific community, and Müller was well qualified to respond. German born and educated, he had been recruited to help initiate the Russian Academy of Sciences in 1725, and he quickly had developed a reputation for historical scholarship through the use of original documents. One of the principal scientific architects of the expedition, he had left St. Petersburg in 1733 as part of its scientific entourage. Although Müller did not reach Kamchatka, for 10 years he traveled throughout Siberia and compiled extensive topographic and ethnographic information. The academy gave him access to the logbooks of the *St. Peter*, and Razumovskii instructed him to correct "all of Mr. de l'Isle's dishonesties." In an untruthful manner of his own, Müller wrote as if he had sailed with Bering so that his account would appear more authoritative and persuasive.

In his *Letter*, Müller recounted details of Bering's voyage and accomplishments that Russian officials had not published previously. He attacked Delisle's integrity and ethics and was scornful of any significant role played by Delisle's brother, Delisle de la Croyère. Müller expressed particular disdain for Delisle's inclusion of suspect information from Admiral de Fonte but also pointed out that Delisle had erroneously depicted the de Fonte geography on his map. In the same year that the *Letter* was published in London, Müller also published a map. With these detailed and scathing rejoinders to Delisle and his work, he hoped to re-establish Russian prestige on the international stage. By the time these documents were published, Russian traders and fur hunters were expanding eastward from Kamchatka, not only generating profit but also increasing knowledge of the land from the Aleutian Islands to the Alaskan Peninsula.

19. The Russian Map, 1754

INTRODUCTION

In 1723, representatives of Peter the Great recruited Gerhard Friedrich Müller and other West European intellectuals to form the Academy of Sciences in St. Petersburg. By the end of 1725, 16 European scholars lived in the Russian capital, including Müller, Joseph Nicolas Delisle, and his brother Louis Delisle de la Croyère, a collegial group who planned and implemented the second Kamchatka expedition. When Joseph Delisle published a book and maps in 1752 that disclosed information about the findings of the expedition, he violated the policy of secrecy he had agreed to. Müller wrote the Russians' official response to Delisle in 1753, and the academy published a new map in Berlin in 1754.

Most scholars have credited Müller as the author of this map, although his name does not appear in the title. The Academy of Sciences issued a revised version in 1758, and over the next 20 years the map was reissued by the academy and foreign cartographers. In 1761 Thomas Jefferys, the geographer to King George III (1760–1820) in London, included the map in his Voyages from Asia to America for Completing the Discoveries of the North West Coast of America, *a translation of an earlier work by Müller.*

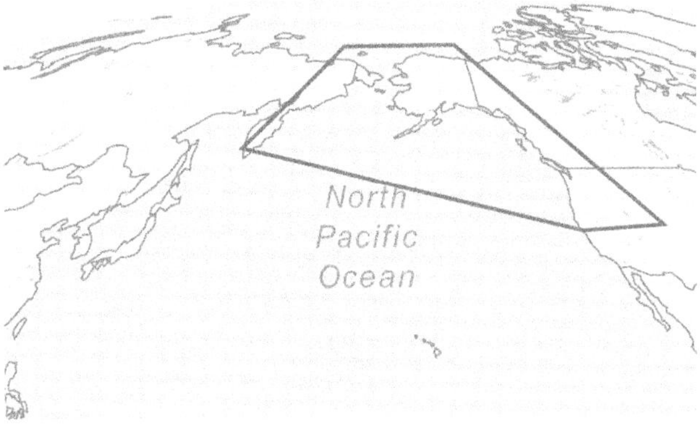

"New Map of the Discoveries Made by Russian Vessels of Unknown Coasts of North America and Adjacent Lands Based on the Actual Memoirs of Those Who Were Present," 1754. (Private collection of James Walker)

ANALYSIS

Although Müller did not sail with the *St. Peter* or the *St. Paul,* the Russian Academy of Sciences gave him access to closeted documents of the Kamchatka expedition, including the ships' logbooks. The title of the map makes it clear that the credibility of the information was based on these documents and on the authority of the Russian Academy of Sciences. In 1754, it was the most up-to-date and revealing description of the results of recent Russian discoveries in the North Pacific.

Unlike Delisle, Müller drew the tracks of Bering's ship from Kamchatka to America and back, in addition to those of Chirikov. Müller located Bering Island relatively accurately, and he eliminated any reference to Gama Land, which Delisle had depicted on his map. Müller included several islands in a long string that extended southwest from the American coast and had been recorded by officers on both ships. Because Bering and Chirikov had not explored farther north, Müller could not know the nature of what lay beyond the islands, and he drew a large hypothetical peninsula, using a dotted line, suggesting that the North American mainland was just beyond them.

Müller used logbook information to establish place-names on the islands: Mont. de St. Elie (Mount St. Elias), which Bering had sighted on July 16 and named for the feast day of St. Elias (although the name was first given to a nearby island Cap de St. Elie, present-day Kayak Island); I. de Schoumagin (Nagai Island in the Shumagin Islands), named for Nikita Shumagin, who was the first to die from scurvy and was buried on the island; I. de St. Abraham, which was noted in the logbook on October 29; and Mont. de St. Iean, probably the volcano on present-day Great Sitkin Island, which was sighted by the crew of the *St. Peter* on the church day of the conception of John the Baptist. Observers on the *St. Peter* thought that this mountain was likely part of the American mainland. Farther north in the Mer d' Anadir (the Bering Strait between Alaska and Siberia), Müller located and named two islands that Bering had observed in August 1728 during the first Kamchatka expedition: I. de St. Deomid (Big Diomede or Ratmanov Island), named for the martyr St. Diomede, and I. de St. Ilarion (St. Lawrence Island), named by Bering when he passed there on St. Lawrence Day.

Just to the east of these islands on the hypothetical peninsular coast, Müller added a legend that the coast had been discovered by the surveyor "Gwosdew." This was a reference to the voyage of a Russian officer, Michael Spiridonovich Gvozdev, who had sailed in this area in 1732 and may have sighted the American mainland to the east. Although Müller erred in his dating of the event—he recorded the year as 1730—he was clearly depicting a Russian claim of discovery.

Müller also engraved more detailed legends relating to Russian discoveries along the Siberian coast, from the Kamchatka Peninsula north into the strait between the continents and then east along the northern coast. One inscription claimed that three Russian ships had navigated the strait in 1648. Many subsequent cartographers would replicate Müller's misconception of a large landmass projecting northeast from the tip of eastern Siberia.

On the American continent side of the map, Müller almost completely rejected the concept of a huge inland Sea of the West, which Delisle had included on his map. Müller's only reference to the story of Juan de Fuca was a break in the coastline and a brief legend noting de Fuca's supposed entrance. Farther north, Müller engraved a legend at latitude 53°, "the pretended Rio de los Reyes according to Admiral de Fonte in 1640, as described by Mr. Delisle." Notably, Müller situated the entrance at the approximate location that de Fonte had claimed (53°) rather than at the higher location (63°) Delisle had used in his

1752 map. Müller avoided charting any of the complex inland water passageways that Delisle had used to illustrate de Fonte's claims.

The maps of Delisle and Müller, which offered very different opinions about conceptual geographies along the North American coast, cast the Russian explorations of Vitus Bering and Aleksei Chirikov onto an international stage. The de Fonte and de Fuca controversies lasted for decades, until meticulous surveys of the British, Spanish, and French charted the true geography of the coast later in the century. Müller's map in particular caught the attention of these European powers and helped stimulate plans to explore and promote their own interests in the Pacific Northwest. The Spanish felt vulnerable to Russian expansion because of the proximity of their outposts in California to Asia, as delineated on maps such as Müller's. One map historian noted that the Spanish minister to St. Petersburg, Vizconde de la Herreria, had sent the 1758 version of Müller's map to Madrid on September 18, 1764 (Cook 1973, 46). Spanish naval officers recorded in their journals that they frequently referred to Müller's map during initial voyages of exploration along the Pacific North American coast in 1774 and 1775.

Further Reading

Cook, Warren L. *Flood Tide of Empire: Spain and the Pacific Northwest, 1543–1819.* New Haven, CT: Yale University Press, 1973.

Dmytryshyn, Basil, E. A. P. Crownhart-Vaughan, and Thomas Vaughan, eds. *Russian Penetration of the North Pacific Ocean, 1700–1797,* Vol. 2, *A Documentary Record.* Portland: Oregon Historical Society Press, 1988.

Fisher, Raymond H. *Bering's Voyages: Whither and Why.* Seattle: University of Washington Press, 1977.

Frost, Orcutt. *Bering: The Russian Discovery of America.* New Haven, CT: Yale University Press, 2003.

Gibson, James R. "Supplying the Kamchatka Expeditions, 1725–1730 and 1732–1742." In *Bering and Chirikov: The American Voyages and Their Impact,* edited by O. W. Frost, 90–116. Anchorage: Alaska Historical Society, 1992.

Golder, F. A. *Bering's Voyages: An Account of the Efforts of the Russians to Determine the Relation of Asia and America,* Vol. 1. New York: American Geographical Society, 1922.

Golder, F. A. *Bering's Voyages: An Account of the Efforts of the Russians to Determine the Relation of Asia and America,* Vol. 2. New York: American Geographical Society, 1922.

McCord, Carey P. "Scurvy as an Occupational Disease." *Journal of Occupational Medicine* 13(9) (September 1971): 441–447.

[Müller, Gerhard Friedrich.] *A Letter from a Russian Sea-Officer to a Person of Distinction at the Court of St. Petersburgh.* Translated. London, 1754.

Urness, Carol. "Captain-Commander Vitus Bering." In *Bering and Chirikov: The American Voyages and Their Impact,* edited by O. W. Frost, 11–36. Anchorage: Alaska Historical Society, 1992.

Juan Francisco de la Bodega y Quadra and Bruno de Hezeta, Second Bucareli Expedition, 1775

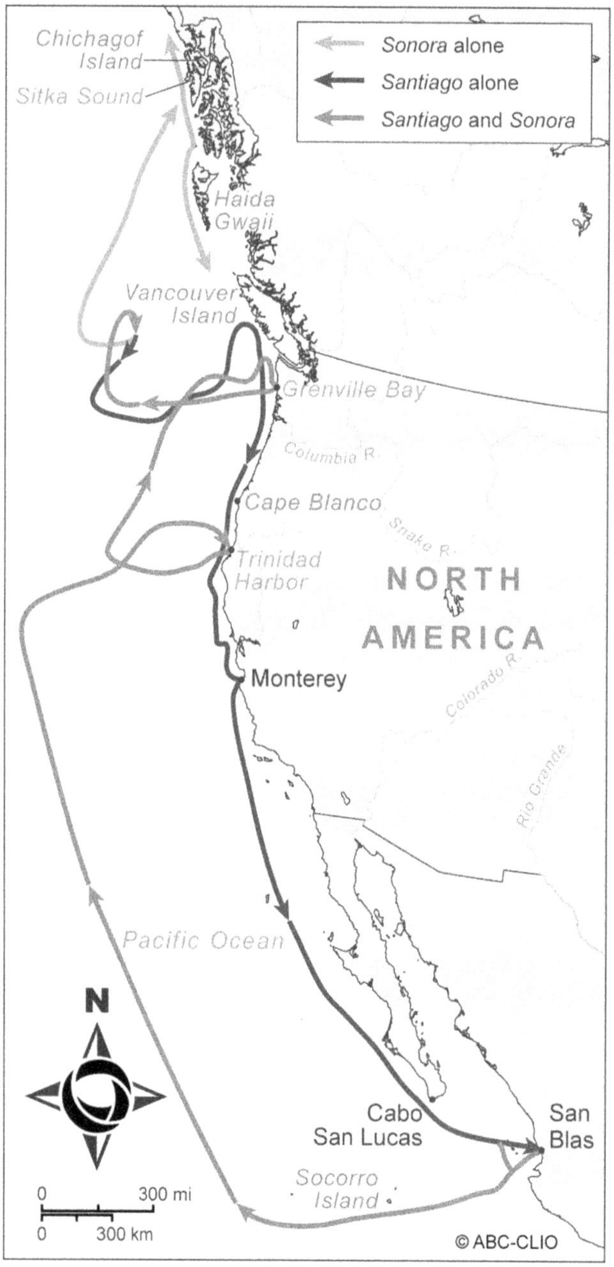

CHAPTER 3

Juan Francisco de la Bodega y Quadra and Bruno de Hezeta, Second Bucareli Expedition, 1775

When Juan Francisco de la Bodega y Quadra arrived at the newly established Pacific naval base at San Blas, Mexico, in 1774, most of the North American coastline above latitude ~40° north had not yet been explored by Europeans. Although cartographers had published maps with conceptualized navigable passages into and across the continent from several points along the Pacific Coast, there were no maps with observational information below ~60°, the farthest reached by Russians Vitus Bering and Aleksei Chirikov in 1741. Other than Russian encounters with Native peoples far to the north, Europeans had had no contact with coastal inhabitants, nor were they aware of the potential commercial value of the region's natural resources. By the time of Bodega's death in 1794, explorers from four nations had thoroughly mapped the coast, debunked myths about the existence of a transcontinental water passage, and had encounters with many Native peoples, often with tragic results. They were also in competition for sovereign and commercial control of the region. Throughout this dynamic and tumultuous period, Bodega represented Spain as explorer, administrator, diplomat, and cartographer. He was a major participant during Spain's ascendency to the farthest reaches of the nation's empire and oversaw its initial dismantling.

Bodega's lineage was from a long line of Bodega and Quadra families of Basque nobility who lived near Bilbao, Spain. In 1720 his father had moved to Lima, Peru, where he was a wealthy merchant and married the daughter of an established aristocratic Peruvian family. Bodega was born on May 22, 1744, in Lima, which had been the capital of the Spanish viceroyalty of Peru for 200 years. He probably attended one of several Jesuit-run educational institutions in his early years, but at age 18 he broke with family tradition and decided to enter a maritime career in the Spanish Navy. In the social hierarchy of New Spain, the families of mixed European and New World ancestry were *criollos* (Creoles) whose careers often centered on commercial and management occupations, not the military.

In 1762 Bodega entered a prestigious naval training academy in Cádiz, Spain, where he studied naval science, mathematics, astronomy, and cartography. After graduation in 1767 he saw service in the Spanish Royal Navy in the Mediterranean and the South Atlantic, sailing out of Lima. In 1768 Spanish administrators, concerned about possible Russian encroachment on

Captain Juan Francisco de la Bodega y Quadra (1744–1794), the foremost Spanish naval officer on the Pacific coast of North America during the last quarter of the 18th century. Within a few years after this portrait was completed, Bodega was appointed commandant of the naval port of San Blas, reoccupied the fort at Nootka Sound, negotiated with British captain George Vancouver to defuse the Nootka Crisis, and facilitated additional Spanish explorations in the Strait of Juan de Fuca. (PjrStudio/Alamy Stock Photo)

the unexplored northern region of the North Pacific, established a naval base at San Blas at latitude ~21° north.

In 1774 Bodega's career in the Pacific Northwest began when he was appointed one of six new officers to San Blas. Just before leaving Cádiz the six received training in mapping and navigation, including the use of a recently available chronometer for telling time at sea. In 1771, Antonio Maria de Bucareli y Ursua had been appointed viceroy of New Spain (1771–1779) and had initiated plans to explore and take possession of as much of the coast above 40° as possible. The first Bucareli expedition of 1774 was quickly followed by the second in 1775, to which Bodega was assigned. Despite Bodega's seniority, Bucareli did not give him command of one of the three expedition ships, which some scholars have speculated was because of Bodega's Creole heritage.

Shortly after the ships departed San Blas on March 16, the captain of the 36-foot schooner *Sonora* became deranged and was relieved of command. Bodega seized the opportunity and accepted command of the ship, which was described as barely seaworthy enough to make short coastal trips, much less the arduous open-water voyage called for in Bucareli's orders. After over six months at sea and having reached a latitude of ~58° north, the *Sonora* entered Monterey Harbor. Bodega had participated in three possession ceremonies on the trip (including one at present-day Kruzof Island west of Sitka, Alaska, at 57°20′, on August 18) and had explored, navigated (without a chronometer), and mapped an extensive section of the Northwest Coast under extreme conditions and circumstances. Near Cape Elizabeth on the coast of present-day Washington, a confrontation with local residents had resulted in the deaths of several of Bodega's crew and many Natives. By the time his ship arrived in Monterey, all of the crew, including Bodega, were ill with scurvy.

After returning to San Blas, Bodega compiled the first of several maps of the Northwest Coast. This map and many others that he created during his career are a testament to his precision and skill as a cartographer. Unfortunately for subsequent Spanish claims of right-of-possession by discovery, the manuscript maps that Bodega and others created were tightly held by officials who feared that their publication would invite explorations by competing imperial powers.

Bucareli and King Carlos III (1759–1788) praised Bodega for his accomplishments, and he was promoted to *teniente de navío* (full lieutenant). In

DID YOU KNOW?

Juan Francisco Bodega and Bruno de Hezeta departed San Blas, Mexico, on an exploratory and possession-taking expedition to the Pacific Northwest at almost the same time the Battle of Lexington in Massachusetts between British and American forces initiated the American Revolutionary War.

1779 the viceroy ordered a third exploration north to explore the coast more thoroughly as far as 70° north and to intercept the expedition of Captain James Cook, which was rumored to be exploring in northern waters. Bodega's ship, the frigate *Favorita,* and the companion ship *Princessa* left San Blas on February 12, 1779, just two days before Cook was murdered in Hawaii. The crews spent nearly two months charting the commodious Bucareli Bay on the west coast of today's Prince of Wales Island, Alaska, at ~55°. Bodega recorded his observations of Native society and culture, including his first attempts at trade with them. One aspect of this activity was the purchase of several children by Bodega and members of the crew. The ships' chaplains condoned this practice and overcame Bodega's initial reluctance because they felt that it was an opportunity to Christianize the children and remove them from what the chaplains considered a heathen environment.

After departing Bucareli Bay and using the maps of Gerhard Friedrich Müller (1758) and Jacques-Nicolas Bellin (1766) for reference, the Spaniards sailed north to the vicinity of Cape St. Elias, which they recognized from Müller's map of Vitus Bering's voyage (1733–1743). They entered Prince William Sound, and somewhere to the northwest, at a latitude of ~61°17′, the ships' officers performed a possession ceremony that later became a justification for Spanish sovereignty over the Northwest Coast. These ceremonies were a part of rituals used to claim possession of newly encountered territory, and most of the time they followed a prescribed set of instructions that were recorded and notarized as reported proof that the proper procedures had been carried out.

DID YOU KNOW?

Bodega navigated the treacherous North Pacific Coast with instruments and estimated the proximity of land based on a number of indicators, such as water color and the presence of whales, jellyfish, sea otters, and different species of birds.

Farther to the west, the expedition sailed partway into present-day Cook's Inlet in the Gulf of Alaska bordering the Kenai Peninsula, but the advancing winter and illness mandated their return for recuperation to the bay of San Francisco, which Juan Manuel de Ayala had first examined during the voyage of 1775. After 11 months at sea the expedition returned to San Blas on November 21, having covered an estimated 9,348 nautical miles. Bodega quickly made adjustments to his many astronomical observations and sent his completed journal to the viceroy within a week of his arrival. He also helped draw an updated map of the Northwest Coast. His ethnological observations and the artifacts of the voyage comprise some of the earliest historical records of Native peoples in the area. The achievements of the expedition were all the more remarkable because the navigation was based almost entirely on dead reckoning.

Bodega was promoted to *capitán de fragata* (commander) and in 1780 was given temporary command of San Blas. Spanish officials in Madrid and Mexico City, however, delayed additional expeditions to the north because naval resources were heavily committed elsewhere. In 1779, Spain had joined forces with France in support of the Americans' war with Britain. Over the next nine years Bodega was assigned to Peru, Havana, and Spain, but his career languished. Through letters and a personal audience with the minister of the navy, he pleaded his case for promotion and improved pay and volunteered to return to San Blas as commandant.

Bodega's efforts on his own behalf, his excellent service, and a developing international crisis at Nootka Sound on Vancouver Island combined to launch the next phase of his career as administrator and diplomat. By mid-1789, Commandant Bodega and the new viceroy in Mexico City, Juan Vicente de Güemes, 2nd Count of Revillagigedo, faced challenges engendered by Russia's and Britain's increased focus on the Pacific Northwest. The documented presence of Russian fur-hunting outposts in the far north and an increase in vessels engaged in the sea otter trade along the coast prompted the viceroy to once again send Spanish ships to the region. In the meantime, from May to November 1789 at Nootka Sound, Spanish officer Esteban José Martínez seized British ships and imprisoned Captain James Colnett and his men on a charge of violating Spanish sovereignty. During the last four years of his life, Bodega vigorously responded to both challenges.

During an extraordinary period of renewed maritime exploration, from 1790 through 1793, Bodega sent out seven expeditions from San Blas, including that headed by Alejandro Malaspina, the only voyage not directly under his command. At the completion of the expedition led by Francisco de Eliza and Martínez y Zayas in 1793, the Spanish had explored and mapped with great accuracy the Northwest Coast and the complex archipelago of islands that stretched from Mexico to the Gulf of Alaska. In late 1792 Bodega and English captain George Vancouver, who was engaged in the dual missions of explorer and diplomat, exchanged charts of accumulated geographic information, and Bodega subsequently completed a manuscript chart that included the findings from all previous Spanish voyages. Vancouver's cartographic information was published in 1798, but Bodega's 1792 map was kept under wraps by orders from Madrid. It was not until 1802 that a map incorporating Spanish and British knowledge became widely available.

Bodega's final efforts in the service of his king, however, were certainly not kept from international scrutiny. At San Blas, he negotiated the release of James Colnett and his ship after over 10 months of captivity; when Colnett left San Blas in 1790, he offered praise for the treatment he had received under Bodega's command. By that time the friction between Spain and Britain over the seizures of ships in 1789 had assumed dangerous proportions, and the two nations avoided armed conflict by negotiating the First Nootka Convention in Madrid in October 1790. While the agreement mandated that Spanish authorities return seized British property, the fundamental issue of which nation held sovereignty in the region was ignored.

To interpret and implement the articles of the convention, Bodega was selected to represent Spain, and George Vancouver represented Britain. It was one of the high points of Bodega's career, and he assiduously prepared for his new role. In April 1792, he arrived at Friendly Cove in Nootka Sound; Vancouver arrived in August. After a month of conversing through an interpreter, neither man was willing to capitulate on the key point—that by abandoning occupancy of Nootka to the British, Spain was not also yielding its declared sovereignty in the region. Bodega's firm position on this point was later rewarded with high praise from his viceroy and his king, Carlos IV (1788–1808).

While at Nootka, Bodega entertained Vancouver and officers of other foreign vessels in port with lavish meals served on his personal silver. Although this action may have been strategic, Bodega's generosity contributed to his reputation as a thoughtful and respected officer. He also earned high marks from local residents, particularly Chief Maquinna, who had managed encounters with foreign traders and naval officers for many years.

In January 1793, Bodega returned to San Blas to an enthusiastic welcome by Viceroy Revillagigedo. Bodega was promoted but not to flag status, a coveted rank in the Spanish Navy that eluded him throughout his career. For unknown reasons his health deteriorated quickly, and he died in Mexico City two months short of his 50th birthday. Bodega did not write a personal diary, and extant official records do not clarify his relationship with María Marín del Valle, who after Bodega's death claimed to have been married to him and to have borne his son. Records do suggest that Bodega had expended a considerable sum on his personal household, which included several servants and may have included members of María's family. He also had amassed large personal debts "on behalf of the Naval Department during the years he was commandant at San Blas" (Tovell 2008, 321). His estate was heavily in debt, and officials in Mexico City made a special inquiry about the extent of his possessions. Among his belongings was extensive correspondence, but his legacy rests in the commentary of those who shared his experience as a naval officer during a formative period of European expansion in the Pacific Northwest. When Vancouver learned of Bodega's death, he wrote that "the unexpected melancholy event of his decease operated on the minds of us all, in a way more easily to be imagined than described . . . [and] produced the deepest regret for the loss of a character so amiable, and so truly ornamental to civil society" (Beals 2007, 84).

20. Defense of Empire, 1757

INTRODUCTION

Noticia de la California *(News of California), published in 1757, contained up-to-date information on the religious and geopolitical status of western North America. Compiled from first-hand observations of Jesuit priests and explorers, the publication helped influence authorities in Madrid and Mexico City to direct attention toward expanding Spanish frontiers in western North America. While* Noticia de la California *was published anonymously, we know that the compiler was Andréas Marcos Burriel, a Jesuit priest and director of the Spanish archives and libraries. He drew principally on the unpublished work of Father Miguel Venegas on the status of the Jesuit mission experience in Baja, California. Burriel added recent information about British attempts on the Atlantic Coast of North America to find a Northwest Passage to the North Pacific and about Russian expansion from the west. He probably finished the book in 1754 and was authorized to publish it after two years of royal and Jesuit review.*

The content of Noticia de la California *was mostly a defense of Jesuit activity and a plea to extend missions to the north. This was considered necessary both to convert Indians and to counter the British and Russians whose potential expansion into northwest America might threaten the Manila galleon routes and Spanish hegemony on the California peninsula.*

Noticia de la California, 1759

It is the advantageous situation of California which renders it so valuable. This makes the conquest of it preferable to that of any other country in America. Though so wretched and poor, its situation alone renders it more important than all these, both with regard to the propagation of religion, and the good of the state. . . .

[I]t cannot be denied but that if any foreign power should find means of building fortifications in California and maintain its superiority there, the empire of Mexico would be in the utmost danger. . . .

Thus California, by its situation, has been, and still is, of the highest importance for the preservation and advancement of the christian religion, and the augmentation of the Spanish sovereignty in America. . . .

The Russians, or Muscovites, whose vast empire extends to the northern extremities of Asia, and even near to the South Sea, are not only endeavoring to civilize the natives of those countries, but are actually erecting forts and planting colonies; and have already made docks and arsenals in several parts . . . taken surveys of their own coasts, sailed as far as the islands of Japan, and, crossing the South Sea, landed in several parts of Spanish America.

In one voyage made in 1741, the Russians landed on this very coast in the latitude of 55 degrees 36 minutes, not above 12 degrees from cape Blanco, the most northern part of California hitherto known. And is it not natural to think that the Russians in future voyages, will come down as low as cape Blanco: and if California be abandoned by the Spaniards even as far as cape San Lucas? and we may well suppose that they who to-day take a view of the coasts and country, may to-morrow determine to plant colonies there. The last Spanish garrison on the Northern American coasts of the South-sea, is that of cape San Lucas at the southern point of California. . . . But along the immense extent of the coast northward, Spain has not one single fortification. How shall we hinder the Russians from making settlements there, unless we be beforehand with them? would it be proper that the Muscovites should become our neighbours and rivals, and establish the Greek church in California? . . .

The publick papers have informed us that the English had a design of crossing the South-sea from the East Indies, erecting fortifications, and making settlements on the coast of America above California towards Hudson's bay; that is, in those very parts which the Russians have visited. Whoever is acquainted with the present disposition of the English nation, and has heard with what zeal and ardour the project for a North-west passage has been espoused by many considerable persons, will be convinced that the scheme is not romantick; and it would not be surprising if the execution of it should one day come under deliberation. If this should ever happen, I would ask, what would be the condition of our possessions?

Source: Miguel Venegas, *A Natural and Civil History of California Containing an Accurate Description of that Country,* Vol. 2, *Appendices to the Account of California* (London: Printed for James Rivington and James Fletcher, 1759), 215–216, 219, 221–223.

ANALYSIS

Widespread accounts of Russian advances in the North Pacific concerned Spanish authorities in both Madrid and Mexico City. In the mid-1750s, Spanish settlements on the western coast of North America included Jesuit missions in Baja, California, only to latitude ~27–28° north, far south of the most northern extent of the Spanish exploration of Sebastián Vizcaíno to ~43° in 1602–1603. At midcentury, advances from the imperial powers of Britain and France threatened Spanish hegemony in eastern North America, the Caribbean, and the Pacific Ocean, but the primary challenge to authorities in the American western portion of New Spain was from conflict with indigenous groups.

Burriel's *Noticia* was one of the earliest publications written in Spanish that helped catalyze the nation's expansion to the Pacific Northwest. It was followed in 1759 by the publication in Rome of *Moscovitinella California* (The Muscovites in Russia) by José Torrubia, a Franciscan priest who gave an informed report of recent Russian discoveries, largely based on the published accounts and maps of Joseph Nicolas Delisle and Philippe Buache. Previously strained diplomatic relations between Madrid and St. Petersburg were reestablished in 1760. Soon after, Spanish ambassadors began forwarding to Madrid reports of

Russian advances in the North Pacific, including copies of Gerhard Müller's new map of the region.

The defense and expansion of New Spain's northwestern frontier was initiated by José de Gálvez, who arrived in Mexico City in 1765 as inspector general. He determined to extend settlements as far north as Monterey (at latitude 36°36′), which Sebastián Vizcaíno had identified as a suitable harbor in 1602. In early 1768, Gálvez received orders from Madrid to attempt to thwart Russian expansion in the north. Shortly thereafter, he and others—including the viceroy of New Spain, Carlos Francisco de Croix (1766–1771)—devised a plan to extend Spanish dominion in the Pacific Northwest by establishing a naval base at San Blas on the Mexican coast, at 21°32′, significantly farther north than the current base of Acapulco. San Blas served as the principal base for shipbuilding and Spanish maritime activity in the Pacific Northwest for the next three decades.

21. Viceroy Antonio Bucareli's Orders, 1775 (1773)

INTRODUCTION

These documents are part of a larger series of instructions that Viceroy Bucareli in Mexico City submitted to Juan Pérez in San Blas prior to the first Bucareli expedition of 1774. Bucareli resubmitted these instructions, with little change, to Bruno de Hezeta, who commanded the second expedition in 1775. In many respects, these directives resembled those that Vitus Bering received from Empress Anna Ivanovna and her advisers prior to his voyage of 1741. Both Bering and Hezeta had been commanded to proceed to a location that had not been explored by Europeans, to treat the Indians "affectionately," to discern any evidence of a "foreign" presence, to look for valuable resources, and to take possession of territory for their monarch. The instructions to Pérez were apparently drawn up within a period of a few months, almost entirely by Bucareli.

Instructions of Viceroy Bucareli to Ensign Juan Pérez, December 1773

I. The kindness of the King, who entrusted this government of New Spain to my [Bucareli's] charge, not only imposes on me the obligation of preserving these vast territories for him but also that of endeavoring to enlarge them, as much as I am able, through new discoveries in unknown areas, so that their numerous Indian inhabitants, attracted to the kind, mellow, and desired vassalage of His Majesty, may receive by means of the spiritual conquest the light of the Gospel which will free them from the darkness of idolatry in which they live, and will show them the road to eternal salvation. These are the true motives that move the pious royal heart of His Majesty in these undertakings. . . .

VII. For his [Pérez's] course he is to ascend to the latitude which he considers suitable, keeping in mind that the landing is to be made at sixty degrees of latitude. . . .

X. In the above-mentioned places, which he [Pérez] may consider suitable for settlement, he can and must take possession in the name of the King according to the terms which he will receive from the draft or formulary. . . .

XI. All places where possession is taken will be marked with a large wooden cross, making its pedestal from rocks and concealing in it a bottle or glass flask in which he will place a copy of the testimony of possession signed by him, the chaplain, and the two pilots. And in order that in the days to come this document may better be preserved and may serve as an authentic testimony, the bottle will be sealed tightly with pitch. . . .

XIII. In the . . . event of discovering a settlement . . . he will endeavor to avoid intercourse with it; but from a distance he will observe whatever he can, noting its location, the landmarks they may have built, the number of inhabitants, and if they have or do not have vessels to depend upon.

XIV. If another vessel is encountered, he must be very careful and must use all the means available to avoid communication. In the event that this should prove inevitable (which must be proved on his return here) he will endeavor to conceal the purpose of his voyage. . . .

XVIII. It is also desirable to investigate whether there is some kind of precious stones or articles which our nation considers valuable. . . .

XXI. For gifts to the Indians, there are four boxes of beads containing 468 bundles, which he may distribute equitably, being careful to give more to the principal Indian who ranks higher among them, than to the lesser ones. . . .

XXX. . . . [H]is principal concern is limited to exploration of the coast as far north as he can sail, in order to learn if there is a foreign settlement. . . .

XXXI. Notwithstanding that the charts published in St. Petersburg in the years 1758 and 1773, concerning the alleged voyages of the Russians, may be of little use to him, copies are included so that he may not be without this information.

Source: Manuel Servin, ed., "The Instructions of Viceroy Bucareli to Ensign Juan Perez," *California Historical Society Quarterly* 40 (1961): 237–248.

ANALYSIS

In April 1773, Bucareli received high-level correspondence from Madrid that the Russians were planning new encroachments into the North Pacific. The Spanish ambassador to St. Petersburg had relayed some of this information to Madrid following his recent appointment. Bucareli reacted quickly and initiated plans to send a naval expedition of exploration to the North Pacific. In July he asked the pilot and ensign Juan Pérez to command the voyage, and by December Bucareli had completed a detailed list of 32 articles of instructions, including an extensive "formulary" for taking possession of territory.

A critical goal of the expedition was to locate evidence of other European settlements or commercial or maritime activity to the north of the Spanish presence in Monterey. Although Bucareli repeatedly used the term "foreign" in his instructions, it seems clear that he was thinking "Russians." The "charts published in St. Petersburg" that Bucareli included in his instructions were revised editions of Gerhard Müller's map of the Russian discoveries of

1741. Pérez's voyage of 1774 did not accomplish all that was intended, so Bucareli quickly planned a more expansive second expedition, to be commanded by Bruno Hezeta.

In his otherwise explicit set of instructions, it is surprising that Bucareli did not specifically mandate Pérez or Hezeta to prepare maps of their discoveries. In fact, no map of Pérez's voyage was known until recently. In contrast, Bodega, Hezeta, and Francisco Antonio Mourelle drew several maps after the second expedition of 1775, but the charts remained in manuscript and were not widely disseminated due to the Spanish policy of secrecy.

Bucareli's orders directed Hezeta to sail to latitude 65° north, apparently because that was above the landmass of the North American continent that had been reached by Bering's vessels in 1741, according to the published maps of Müller and French cartographer Jacques-Nicolas Bellin. Limitations of provisions, the arduousness of such an extended voyage, and illness among the ship's crew, however, induced Pérez to turn back at 55° in 1774 and Bodega to turn back at 58° in 1775. The instructions are particularly significant because they illustrate the paramount importance to the Spanish of taking possession.

22. Shipboard Conditions, 1775

INTRODUCTION

A large volume of primary source material relating to 18th-century Spanish activities in the Pacific Northwest exists in public and private archives. Officers were required to maintain journals and send reports, and those documents and others, including orders and letters, were copied and forwarded to officials in Madrid and Mexico City. A small proportion of the documents has been translated into English, and occasionally a rich and less restrained description of the state of affairs emerges from an examination of several sources.

These four documents are drawn from different primary sources, although all are the views of Juan Francisco de la Bodega and Francisco Antonio Mourelle, officers of one of the three ships of the expedition. Their observations of the daunting conditions on the Sonora *and their resolute determination to carry out their orders despite the circumstances provide a valuable perspective on these two important early maritime explorers in the Pacific Northwest.*

Bodega's and Mourelle's Reports on Conditions on the *Sonora*

Francisco Antonio Mourelle, "Compendio de noticias"

The deck and tiny cabin were all it [the *Sonora*] had to offer for security or living quarters, with no chests or other baggage than a bed and what could be contained underneath in a box; the height and width of the sleeping space necessitated remaining in a sitting position; the small deck did not offer the possibility of a walk, and in this inaction one lived for the space of ten months.

Journal of a Voyage in 1775 by the Second Pilot of the Fleet, Don Francisco Antonio Mourelle

To say the truth, we could not but be sorry to observe the horror that the crew conceived of the bad condition of the schooner, which afforded miserable quarters for the sick, as the seamen could not do the business without being thoroughly wet, except when it was calm. These distresses would have become insufferable, had not the commander [Bodega] behaved with the greatest kindness to the crew, he encouraged them to persist also, by giving them frequently small presents, and reminded them of the glory they would obtain on their return, if they reached the proper latitude. He added also, that the risk was nearly equal to both vessels [*Sonora* and *Santiago*], and that as each ship's company valued their lives, they might be sure that is [sic] would not be attempted to proceed further than was consistent with their mutual safety. This interposition of the commander had at length the proper effect, and we agreed to live and die together.

Report of Mourelle, "Segunda exploración de la costa septentrional," 1791

[Bodega and Mourelle realized that] they would navigate unmapped seas; perhaps find themselves in an archipelago where it would not be easy for a small, lone boat to emerge felicitously; that any illness would put them in the ultimate risk at such an advanced season; that from the moment of separation [from the *Santiago*] they would have to stretch their rations to last the rest of the trip; and finally, that if they returned to port without progress worthy of consideration, at once there would be leveled against them accusations of insubordination. . . . But none of this could overcome the shameful feeling which they could imagine if two young men turned back toward San Blas from that latitude; and hoisting all their sails at ten that evening, by navigating toward the west they found themselves in control of their own destiny by the following day.

Journal of Juan Francisco de la Bodega, 1775

Considering that the sailing qualities of the schooner differ from what they say, having all of its futtock [curved] timbers split, completely useless and incapable of holding a nail, and that this damage, with strong and steady winds, could result in some bad consequences. In all [however], it does not appear to me sufficient cause to delay the expedition. Command having been thrust upon me, I must perform with honour corresponding to one's birth. I am mindful of the advantages that can accrue to one's commission, if I succeed as I hope to in convoying with the frigate to 61 degrees or more. The exposure to danger and the arduousness of the enterprise are no secrets to me, owing to [the schooner's] smallness and poor steering, her little ability to carry a stiff sail or sail swiftly.

Sources: Francisco Antonio Mourelle, "Compendio de noticias," Folio 8iv–82v, quoted in Warren L. Cook, *Flood Tide of Empire: Spain and the Pacific Northwest, 1543–1819* (New Haven, CT: Yale University Press, 1973), 71; Francisco Antonio Mourelle, *Journal of a Voyage in 1775* (London, 1781), 478–479; Francisco Antonio Mourelle, "Segunda exporación," Museum Naval, Madrid (331), fol. 61–63v, 82v–83v, Daines Barrington, trans. Miscellanies, 478–479, quoted in Cook, *Flood Tide,* 79; Herbert K. Beals, ed. and trans., "The 1775 Journal of Juan Francisco de la Bodega y Quadra," in *Four Travel Journals: The Americas, Antarctica and Africa, 1775–1874,* edited by Herbert K. Beals et al. (London: Hakluyt Society, 2007), 93.

ANALYSIS

The two-masted tiller-steered *Sonora* was far shorter in length along the keel (34 feet) and more narrow (12 feet abeam) than either of its consorts, the frigate *Santiago* (77 and 27 feet) or the supply ship *San Carlos* (64 and 23 feet). By comparison, Vitus Bering's ships, the *St. Peter* and the *St. Paul,* were each 80 feet in length and 22 feet abeam; James Cook's *Resolution* measured 110 feet in length and 35 feet abeam; and the *Discovery* was 91 feet in length and 28 feet abeam. Of the *Sonora*'s original complement of 16 men, only 4 of the 14 members of the crew had ever been to sea; the rest were recruited from the population of nearby farmworkers.

The *Sonora,* which was built in 1769 at the newly constructed San Blas shipyard, was intended to sail on short coastal voyages and on mail and cargo delivery runs between mainland Sonora and missions on Baja California, not on extended voyages to the North Pacific. With an additional cargo of one year's provisions and a four-month supply of water, the shallow-drafted and slow-sailing *Sonora* labored to keep up with the *Santiago* and occasionally had to be towed by the larger ship.

The marginal living conditions on the ship were worsened by high seas, storms, adverse winds, scarce provisions, and sickness. With remarkable fortitude, Bodega and Mourelle maneuvered the *Sonora* from San Blas on March 6 and north to the Gulf of Alaska at latitude ~58° by August 23, the last 23 days separated from the *Santiago.* The voyage extended over 2,600 miles of coastline, although the ship observed part of it from a distance far out at sea. On October 8 nearly seven months after its departure from San Blas, the *Sonora,* with a greatly sickened crew, landed at Monterey.

It is likely that the *Sonora* was the smallest sailing vessel in length to explore the North Pacific Ocean during the 18th century. Documents such as these illustrate that however the results of the voyages are measured in geopolitical terms, the accomplishments of individual sailors were sometimes astonishing.

23. Hazards of Navigation, 1775

INTRODUCTION

These documents are from the published diaries of Bruno de Hezeta, captain of the frigate Santiago, *and Juan Francisco de la Bodega y Quadra, captain of the schooner* Sonora. *Diaries and logbooks contained detailed daily accounts of the ships' maneuverings, observations relevant to the ships' locations, and records of weather and ocean conditions. This information can be used to understand the technologies of deepwater and coastal navigation in the mid-18th century.*

In the first document, Bodega had recently departed San Blas and was attempting to situate the position of one of three prominent islands to the west and to reconcile his observation of latitude against one of two maps he carried on the voyage, the "Carte Reduite de l'Ocean Septentrional" by the Royal French hydrographer Jacques-Nicolas Bellin, published in Paris in 1766. Later in the voyage, Bodega had reached his highest latitude at ~58° and noted discrepancies in his observations of North American coastal positions, compared to the alleged courses of the ships of Vitus Bering and Aleksei Chirikov in 1741, as portrayed on Bellin's map. In the second document, Hezeta carefully recorded signs relating to a ship's position relative to land. Later, he made observations of compass headings, latitudes, and positions in the vicinity of San Francisco and Point Reyes.

Navigation and Technology

Journal of Juan Francisco de la Bodega y Quadra, 1775

The 24th [of March] at twelve o'clock [noon] the southernmost of the Islas Marías [Isla Marías Cleofas, off the coast from San Blas] was surveyed at a distance of three leagues, which put the location of the ship in 111 degrees, 20 minutes, longitude west of Paris. Although by solar observation [its latitude] was 21 degrees 14 minutes, north, the French map [by Jacques-Nicolas Bellin] shows 21 degrees 40 minutes, with the result that said islands are situated on this map somewhat farther north of their true positions. . . .

As for knowledge of the coast, I applied myself to do my best with orders to fix its location exactly, the results of which offer new objections to the map M. Bellin published in the year [17]66. Because his map locates the two tracks of the Russians, Bering and Chirikov, which they followed on similar voyages of discovery [17]46 [*sic,* actually 1741], it ought to have been able to locate their landfall. But from [the tracks] themselves, one may infer the error that [the coast] suffered in its placement, because if it were correct I would have had to encounter the westerly lands before those to the east.

Diary of the Voyage Made by First Lieutenant of the Royal Navy, Don Bruno de Hezeta, 1775

[June 18] Landfalls on these coasts are not dangerous to make—although the calculation of longitude suffers from great error—if one is warned before approaching it that about one hundred or more leagues out one encounters some *aguas malas* [jellyfish], as they are called in Europe, and which here have the quality of caravels, being rhomboid-shaped, with a crest that looks like a lateen sail. From seventy leagues out other signs are observed that are named *porras* [kelp], whose shape is that of a white squash. . . . Drawing much nearer the land, the sea keeps losing its color, becoming muddy in the vicinity of the coast, so that it looks like a mixture of earth and grasses, some connected to floating roots and woven together with other things into what is called *balsas* [vegetative debris from coastal streams]. . . .

At noon on the 27th, the latitude was observed to be 37°51′, with the horizon clear, and strong winds out of the NW. I set a course to the NNE and made landfall at eight in the evening very close to land in latitude 38°16′, according to my dead reckoning. With the same clear horizons, I stayed off a little all that night tacking so as not to separate myself from the land. . . .

At the point that forms the S. part of this port [of San Francisco], I observed the latitude of its entrance to be 37°54′, although I am not completely certain of it. The Farallones are at an angle of 82°30′ of the third [SW] quadrant. The Punta de los Reyes is 59° of the 4th [NW quadrant], and the Punta de las Almejas is 6° of the third [SW] quadrant.

Sources: Herbert K. Beals, trans. and ed., "Bodega y Quadra: The Journal of Juan Francisco de la Bodega y Quadra," in *Four Travel Journals: The Americas, Antarctica and Africa,* edited by Herbert K. Beals et al. (London: Hakluyt Society, 2007), 89, 115; Herbert K. Beals, trans. and annotations, *For Honor & Country: The Diary of Bruno de Hezeta* (Portland: Oregon Historical Society Press, 1985), 72, 93, 96.

ANALYSIS

The highly trained and skilled navigators of the three ships of the second Bucareli expedition included three line officer graduates of the Spanish Naval Academy in Madrid, Senior Lieutenant Hezeta and Junior Lieutenants Bodega and Juan Manuel de Ayala, and two San Blas pilots, Francisco Mourelle and Juan Pérez. In 1774 Pérez had sailed in the *Santiago* to ~55° north, but he had made no landings and performed no rituals of possession.

By modern standards, the methods and materials used by Spanish navigators in 1775 were rudimentary. In the 21st century, anyone attempting to negotiate the natural hazards of over 2,600 miles of coastline and deep water from Mexico to the Gulf of Alaska would depend on mechanical power, accurate maps, and advanced instrumentation technology. In 1775, navigators on the Bucareli expedition depended on antiquated or conjectural maps of the North Pacific, a magnetic compass to determine direction, a sextant or octant and noon sightings of the sun to determine latitude, a lead line to determine ocean depths, an azimuth compass to measure variation from true north, and keen lookouts to determine proximity to land. These documents illustrate each of these methods and materials.

Determining longitude or a position at sea east or west of a standard point, such as San Blas, was based on dead reckoning. A pilot charted his daily course by compass bearing, and he judged the ship's speed to estimate the distance traveled each day. The marine chronometer that recently had been invented to accurately determine longitude was not yet widely available. Latitude could be determined at sea by using an octant or sextant to observe the angular position of the ship compared to the sun directly overhead at noon. Bruno Hezeta's calculated latitudes of various locations, for example, were remarkably accurate within 1 to 10 minutes (approximately 1 to 10 miles) of the true latitudes.

As these documents illustrate, compass bearings were given in degrees (e.g., Point Reyes was sighted at 59° in the northwest quadrant of the compass) or by reference to specific wind directions (e.g., NNE, or north by northeast) that represented distinct directions on the compass. In addition, mariners used ocean *senas* (signs) for centuries, and Hezeta's careful notation of recognizable vegetation and sea life would have been welcome to weary and anxious sailors as they neared land. These navigational methods and materials were not greatly different from those used by Christopher Columbus in 1492.

Bodega's journal and Hezeta's diary refer to three maps used on the voyage: one of California drawn by Father Miguel Venegas to accompany his 1757 *Noticia;* a map of the North Pacific Ocean published by Bellin (who never left Europe), which included information on the 1741 Russian voyages, copied from Gerhard Müller's 1754 map; and Pérez's information from the previous year. Bodega's frustration with the inaccuracy of Bellin's map is apparent.

24. Ceremonies of Possession, 1774

INTRODUCTION

This established formulary (a collection of formulas) for how to take possession of land was appended to the set of instructions to Pérez in 1774 and Hezeta in 1775—and presumably to those others who led Spanish explorations in the 18th century. The text, which was twice as long as this excerpt, was written in both present and past tenses, as if the notary had been observing the ceremony. The comprehensive nature of the document left little chance for something to be left out of the process; all one had to do was fill in the blanks of the form, most of which related to dates and the names of those present. Three blank spaces were reserved for the names the captain bestowed on the newly possessed territory. Commonly, such names had religious associations such as saints' days, or territories were named for individual crew members or officials in Mexico or Spain (two of those spaces are included in this document). A sentence elsewhere in the document includes a blank space for "Bay, port, coast, anchoring place, or whatever it may be," followed by the phrase "now newly renamed _____." It was clear that a critical piece of taking possession of a place was identifying it with a Spanish name.

Testimony of Possession from Instructions of Viceroy Bucareli to Ensign Juan Pérez

The formulary which must be used as a model in writing the testimonies of possession-taking in the discoveries of which Ensign Don Juan Pérez is in charge. . . .

In a loud voice he [the captain] said that in the name of His Majesty the King, Don Carlos III, Our Sovereign (whom may God Our Lord keep for many years with an increase of greater states and kingdoms for the service of God and the well-being and prosperity of his vassals and of the very powerful lords and monarchs, his heirs and successors in the future) he, as captain of this frigate and by virtue of the order and instructions which in the royal name were given to him by the aforesaid Most Excellent Viceroy of New Spain, was taking and took possession, was seizing and seized possession, of this land where he has at present disembarked, which he has discovered forever and ever in the said name of the Royal Crown of Castile and León . . . as its own property, which it is and which really belongs to it [the crown] by reason of the Donation and Bull of the Very Holy Father Alexander VI, Supreme Pontiff of Rome (which he issued *motu proprio,* giving one half of the world to the Very High and Catholic Lords Fernando V and Isabel his wife, of glorious memory, and to their successors and heirs), given at Rome on May 4 of the year 1493. By virtue of this, these lands belong to the said Crown of Castile and León, and as such, he takes and took the aforesaid possession of these lands and their adjoining districts, seas, rivers, coves, ports, bays, gulfs, archipelagos, and this said _____ where this frigate is presently anchored; and he was placing and did place them under the power, possession, and dominion of the said Royal Crown of Castile and León . . . as its own property, which it is. As a sign of possession, *vel quasi,* drawing out his sword which hung at his side, he cut trees and branches and plants, and he moved stones, and he walked back and forth on the fields and beaches without encountering opposition, asking those present to be witnesses of it and me _____, being the notary appointed by the captain of this expedition, to draw the deposition of witnesses in public form. Then immediately taking a large cross on their shoulders, the men of the frigate, having been disposed in battle order with muskets and other arms, carried the cross in procession while the religious, _____, chanted a litany with everyone answering the responses. The said procession having ended, the aforesaid captain planted the cross and placed a pile of stones at its foot as a monument and sign of possession of all these discovered lands, seas, and their limits, continuous and contiguous; and he named this _____. . . . After the services had ended, the captain as a more lasting sign and monument of possession, *had a tree stripped which he ordered made into a cross on which was engraved the Most Holy Name of Our Lord Jesus Christ with the four initials I.N.R.I.; and at the foot of the cross he put* "Carolus tertius Rex Hispaniarum" [Carlos III King of Spain].

Source: Manuel Servin, trans., "The Instructions of Viceroy Bucareli to Ensign Juan Perez," *California Historical Society Quarterly* 40 (1961): 243–246.

ANALYSIS

From 1774 through the end of the century, Spanish explorations resulted in acts of possession in nearly 20 locations from San Francisco north to Prince William Sound. As stated in the formulary, Spanish officials interpreted these territories as their "own property" based on Pope Alexander VI's 1493 order that divided the non-Christian New World between Spain and Portugal. Although this papal division was not recognized by other imperial nations such as France and England, it is remarkable that what became known as the Line of Demarcation, which Portugal and Spain agreed on in the 1494 Treaty of Tordesillas, continued to be invoked by the Spanish three centuries later.

Territory was claimed by the Spanish through ceremonies notarized by a secular official and sometimes by a chaplain. These symbolic acts represented a legal process of a claim of sovereignty. From the time of the first Columbian encounters, most Spanish explorers and conquistadors in the Americas conducted ceremonies of possession by taking at least some of the actions described here. These rituals equated with claims of sovereignty, and the ceremonies were probably standardized and became rote by 1600. While some religious aspects of the rituals, such as the singing of hymns, were not always carried out, most accounts record the captain carrying a sword, walking the area, and so forth "without encountering opposition" and with a procession bearing a cross, marking trees, and leaving behind a buried bottle or other marker of the event. The importance of naming—or renaming—a place to create a culturally specific identity was clearly an important element of the ritual.

From the 16th through the 18th centuries, England and France also claimed sovereignty over territory in the Americas based on symbolic acts of possession taking. These nations also carried these practices to the Pacific Northwest, where British and Spanish diplomats late in the 18th century debated the idea of whether occupancy, not ceremony alone, should constitute title to territory.

Most notably, possession taking through ceremony was an entirely European story and completely disregarded indigenous peoples, the real possessors of the territory. European practices of mapping, naming and renaming, performing acts of possession, and citing European religious and so-called legal precedents for doing so legitimized one cultural group at the expense of another.

25. Bruno Hezeta's Map of the Columbia River, 1775

INTRODUCTION

North is oriented to the left on Bruno Hezeta's map of the entrance of the Columbia River, which was recorded on his chart as Bahia de la Asunción. The first European mariner to observe and chart the mouth of the river, Hezeta wrote in his diary on August 17, 1775, that "in the afternoon of this day I discovered a large bay. . . . Its latitude and extent are subject to the most exact determinations that theory and practice offer in this career. The latitudes of the most prominent capes of said bay, particularly the one on the N., are calculated from the observation of this

day. . . . These currents and the seething of the waters have led me to believe that it may be the mouth of some great river or some passage to another sea" (Beals 1985, 86). Hezeta's cartographic expertise is evident in the three maps of bays that he drew with the expectation that others might navigate by his charts. He carefully delineated shapes of coastlines and prominent headlands, included a central compass to plot direction, and added a scale of miles. Theoretically, any pilot with a small-scale map of the coastline could use dead reckoning to sail directly to any of the bays.

Hezeta's latitude recordings of the mouth of the river correspond closely with modern determinations. The origin of the name Bahia de la Asunción was probably the recent feast day of the Assumption of Our Lady, and the feast day of Saint Roch may have been the origin of Hezeta's northern cape Roque, present-day Cape Disappointment. Historians are less certain of the origin of the name Frondoso (implying luxuriant vegetation) and its modern location near the southern cape. Hezeta's map suggests that the two capes were separated by 10 to 11 miles, whereas today's Point Adams is only about 6 nautical miles from Cape Disappointment. Later in 1775, Juan Bodega drew the first map to record the observations of the two ships of the second Bucareli expedition. He labeled the area between the two capes Roque and Frondoso, "Entradade scubierta por Dn Bruno Ezeta" (Entrance Discovered by Bruno Hezeta).

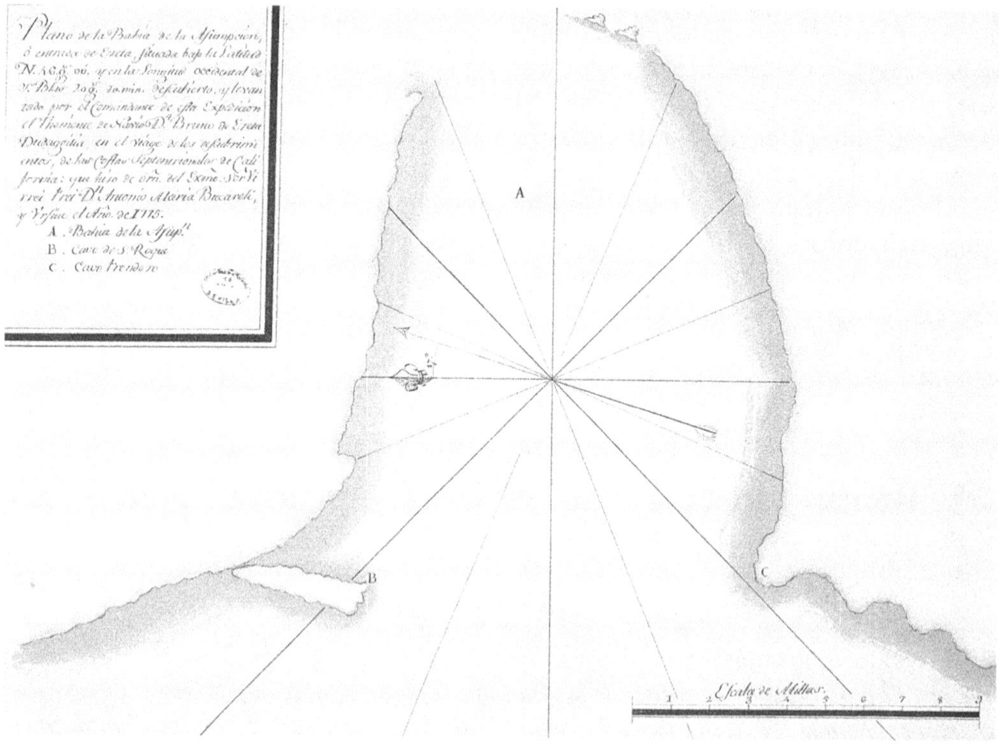

"Map of Assumption Bay or [H]ezeta's entrance." (Archivo General de Indias, Seville, Spain, Secretary of State and the Office of State, Reference Code ES.41091.AGI / 27.17 // MP-MEXICO, 306)

ANALYSIS

By August 17, 1775, Hezeta had been at sea for nearly four months without being able to adequately reprovision the expedition. In the absence of food that provided crew members with vitamin C, scurvy was a threat when voyages extended beyond four months and sometimes less. By mid-August many of the crew of the *Santiago* were ill, most of them suffering from scurvy. Despite Hezeta's desire to explore Bahia de la Ascunción, he did not have the manpower necessary to do it. He wrote that "I did not enter and anchor in the port that appears on the map, formed by what I suppose is an island, despite my ardent desire to do so. This was because, having taken the opinion of second captain and pilot Don Juan Pérez and that of pilot Don Cristóbal Revilla, they insisted that I should not attempt it, for in letting go the anchor we did not have men with which to get it up, nor to attend to the work that would thereby result. . . . I resolved to put about heading outward. Finding myself at a distance of three of four leagues, I lay to. That night I experienced intense currents to the SW that made it impossible for me to attempt to enter this bay on the following morning. . . . These currents likewise convinced me that at ebb tide a great quantity of water issues from this bay" (Beals 1985, 88–89).

Hezeta's clear recognition and charting of the Columbia River was documented on all Spanish manuscript maps throughout the rest of the century. The Spanish policy of secrecy, however, forbade publication of exploration results, even though multiple ceremonies of possession were performed during the last quarter of the 18th century to demonstrate to other imperial powers that Spain could claim sovereignty. King Carlos did allow exceptions to this policy, notably with the publication of two brief notices of Hezeta's voyage in Madrid in 1776 and with a small-scale coastal map published in Mexico City in 1781 that contained the first published reference to the Ensenda de Ezeta. This important map was not widely disseminated, and it may not have been known in Britain, since engravers of James Cook's charts in 1784 did not include Hezeta's discovery.

From these Spanish publications, however, word reached London about the Spanish voyages of 1774 and 1775, and it is clear from Cook's journal that he had some knowledge of Spanish discoveries before he left on his last voyage in July 1776. While exploring the Pacific Coast, he missed seeing the entrance to the Columbia River. In 1788, the British fur trader John Meares looked for Hezeta's Entrada and named the northern promontory Cape Disappointment. During initial reconnaissance of the coast in April 1792, George Vancouver also failed to find the Entrada de Ezeta. After the American Robert Gray charted the entrance of the Columbia River in May 1792, he subsequently shared his information with Bodega, who shared it with Vancouver. Three nations now had information about a possible major river system that extended far into the interior of the continent.

In March 1793 in a final attempt to maintain a position of strength through exploration and the discovery of a possible water passage across the continent, the Spanish viceroy Conde de Revillagigedo ordered another expedition to explore the coast north from San Francisco. He stressed the importance of a "lengthy and minute" examination of the Columbia River as far inland as possible. On that expedition, one ship managed to cross the dangerous bar of the Columbia and travel upriver 14 miles before running aground and eventually turning back, without making any new discoveries or carrying out any ceremonies of possession.

Any of Spain's claim to right of possession by prior discovery of Hezeta's river was trumped by Vancouver's charts that showed and named the Columbia River, which were published in 1798. No English, French, or American map ever identified the entrance of the Columbia River as the Entrada de Ezeta.

26. Ayala in San Francisco Bay, 1775

INTRODUCTION

These documents are from the accounts of the captain and the chaplain of the supply ship San Carlos, *one of three vessels that comprised the second Bucareli expedition of 1775. On orders from Viceroy Bucareli, 30-year-old Lieutenant Juan Manuel de Ayala sailed into San Francisco Bay, one of the first Europeans to enter the Golden Gate. From August to September, they explored, surveyed, and frequently interacted with people who lived in the region. The accounts and maps of Ayala, his sailing master José de Cañizares, and a Franciscan missionary, Father Vicente Santa Maria, constituted valuable information about the configuration and sailing conditions of the commodious harbor as well as an ethnographic record of first encounters with local indigenous people. After leaving San Francisco Bay, Ayala sailed to Monterey, where he met the* Santiago *and, later, the* Sonora. *He returned to San Blas on November 6.*

Ayala in San Francisco Bay

From Captain Juan Manuel de Ayala to Viceroy Bucareli, November 9, 1775

Sir: I have carried out the orders under which I embarked in the supply ship *San Carlos,* and have come in from my return voyage at this harbour of San Blas this 6th day of November after having been at the harbours of Monterey and San Francisco. . . . I can at least give briefly here a few particulars to convey more speedily the news that the harbour of San Francisco is one of the best that I have seen in these seas, from Cape Horn northward.

I then hoisted sail to seek out the harbour of San Francisco, which I reached on the night of the 5th of August. I stayed there forty-four days, carrying out sometimes myself, sometimes in the person of the sailing master mentioned [José de Cañizares], as faithfully as possible, the exploration of as much as could be brought under the methodical and attentive inspection the enterprise required.

This is certainly a fine harbour: it presents on sight a beautiful fitness, and it has no lack of good drinking water and plenty of firewood and ballast. Its climate, though cold, is altogether healthful and it is free from such troublesome daily fogs as there are at Monterey, since these scarcely come to its mouth and inside there are very clear days. To these many good things is added the best of all: the heathen all round this harbour are always so friendly and so docile that I had Indians aboard several times with great pleasure, and the crew as often visited them on land. In fact, from the first day to the last they were so constant in their behaviour that it behove me to make them presents of earrings, glass beads, and pilot bread, which last they learned to ask for in our language clearly.

Journal of Father Vicente Santa Maria, 1775

As we came near the shore, we wondered much to see Indians, lords of these coasts, quite weaponless and obedient to our least sign to them to sit down, doing just as they were bid. There remained standing only one of the eldest, who mutely made clear to us with what entire confidence we should come ashore to receive a new offering which they had prepared for us at the shore's edge. . . .

The time we were with them seemed to us short, but it was enjoyable, all the more when, upon my pronouncing the most sweet names of Jesus and Mary Most Holy, they repeated them clearly, a great satisfaction and pleasure to me and to my companions. . . .

When I saw there was so large a gathering, I began to speak to them for a short time though I knew they could not understand me unless God should work a miracle. All the time that I was speaking, these Indians, silent and attentive, were as if actually comprehending, showing by their faces much satisfaction and joy. . . .

Throughout the time the Indians were on board we tried to attract them to Christian practices, now having them cross themselves or getting them to repeat the "Pater Noster" and "Ave Maria," now chanting the "Alabado," which they followed so distinctly that it was astonishing with what facility they pronounced the Spanish.

Source: John Galvin, ed., *The First Spanish Entry into San Francisco Bay, 1775* (San Francisco: John Howell Books, 1971), 9, 27, 29, 33, 43, 59.

ANALYSIS

Lieutenant Ayala surveyed mainly from a single longboat while the *San Carlos* remained at anchor near Angel Island, which Ayala named. His two sailing masters, Cañizares and Juan Bautista Aguirre, explored and charted most of the bay. In terms of the extent of territory that these men covered in such a small craft, Cañizares's first large-scale map of San Francisco Bay was a remarkable accomplishment.

Ayala's account of the natural advantages of San Francisco Bay pleased Bucareli and officials in Madrid, who had sought a northern outpost in Alta California for some time to extend missionary work and counter potential threats of Russian and British encroachment. Although the actual harbor was not identified from the sea until Ayala's voyage, the European discovery of the bay had been anticipated for many years. As early as 1542, Juan Rodríguez Cabrillo probably had sighted Point Reyes just north of San Francisco Bay, and Francis Drake probably careened his ship in present-day Drake's Bay, just south of Point Reyes, in 1579. In 1595, Sebastián Rodríquez Cermeño anchored in Drake's Bay and named it La Bahia de San Francisco, after St. Francis of Assisi. A few years later in 1603, Sebastián Vizcaíno named Point Reyes.

It would be over 150 years before Gaspar de Portolá led an exploring party north from San Diego in 1769, intending to create an overland route to the port of La Bahia de San Francisco. By chance, they were the first Europeans to see the waters of San Francisco Bay. In 1775, an overland expedition led by Juan Bautista de Anza marched north with the intent of reaching the port of San Francisco that Portolá had described. Anza arrived in March 1776, and in late June a presidio and mission were established by Franciscans Junípero Serra and Francisco Palou, accompanied by a military escort, Native aides, and potential colonists. This important extension of Spanish dominion on the Pacific Coast occurred less than a week before the signing of the American Declaration of Independence. Within a year, Ayala and Anza had facilitated expansion of Spanish settlement to San Francisco. The Spanish would not establish any missions, presidios, or permanent settlements farther north.

In his account, Father Santa Maria included far less information about the geography of San Francisco Bay and far more about the nature of the people who lived there, a subgroup of the Miwok. Although the Spanish were wary of being attacked and the local residents were often described as bearing weapons, there were no armed conflicts such as those that occurred during Bodega's landing at Grenville Bay in July. The friar's repeated proselytizing to the Indians was an attempt to fulfill one of the primary purposes of Spanish expansion, to convert the so-called heathens to Christianity.

Finally, Father Santa Maria's references to language suggest a deeper theme that had pervaded much of Spanish colonial literature since Columbus: the idea that knowledge of an intelligible language, particularly the Spanish language, was an indicator of a civilized culture. The friar's "astonishment" that the Indians could master a few words of Spanish seemed to imply that the work of what they thought of as civilizing and Christianizing the Indians of San Francisco Bay might be less difficult than expected. The scene in which Indians were listening to his preaching "as if actually comprehending" might be better interpreted as an all too common instance of mutual incomprehension.

27. Human Cost and First Encounters, 1775

INTRODUCTION

Bodega's account of this violent encounter was the most riveting of the six surviving reports of the expedition. The senior officers of each ship—Bodega and Mourelle of the Sonora *and Hezeta and Pérez of the* Santiago*—kept journals of the voyage. They were aware that their entries would become part of the official record and would be thoroughly examined by senior Spanish administrators. In Bodega's case, this likely included Carlos III. Bodega recounted this event in the manner of a military commander who prepared for all contingencies and who, despite the imminent danger to his ship and the "treachery" that resulted in the loss of half of his working crew, managed to escape and repel another assault. Bodega's superiors would likely have been impressed that he was one of three men defending the ship while simultaneously directing the helmsman and remaining alert for the calls of the depth sounder and lookout.*

Journal of Juan Francisco de la Bodega y Quadra, July 14, 1775

[On the *Sonora,* south of the Quinault River, in present-day Washington State] in order to free my men of scurvy and other illnesses that are due to thirst, I decided to complete [the water supply], collect some firewood. . . . I designated six men who were most sensible and capable for such instances, armed for all contingencies. Each of them was carrying his musket, a sabre, and some [of them] two pistols. . . . Having given [the landing party] glass beads to offer as presents if any Indians got close, they departed. . . .

The sea being rather heavy, they were unable to reach the shore in perfect order. As soon as they began to alight, more than 300 Indians unexpectedly came out of the woods, attacking them in a defenseless moment. They were treacherously murdered, and finding myself without the means to help, I ordered the discharge of some muskets to see if it would frighten them. But as it could not do them harm, they paid little heed. . . . [F]inding myself isolated, without boat or men with which to go to their defence, I determined to weigh anchor . . . moving myself toward the frigate [the *Santiago*] in order that I be given the launch to take the most rigorous satisfaction; and to find out if any of the unfortunate ones had gotten free, because . . . two [sailors] were seen getting away, swimming to get aboard. But the coldness of the water prevented them from attaining their purpose, and it is not known if they turned back and died at the hands of the barbarians, or found refuge among the thickets of trees.

As soon as sail was set, nine canoes of Indians immediately came forth, each one of which had more than thirty in number. In a short time they assembled at a uniform distance, such that one of them . . . continued toward our side with all bows stretched taut and covered by some beautiful elk hides for defence. . . . But though we were few and under sail amidst shoals, we were prepared in good order. . . .

[T]he only ones able to fire were the master, I and my servant, having one [sailor] at the helm, another sounding, and another at the masthead on the lookout for shoals, one making cartridges, a boy to fetch them, and five sick with the seven unfortunate ones making up the remainder of the schooner's crew. . . . [T]hey determined to board us at the bow, seeing that there were no men at it. But at the same instant, we opened fire with two swivel guns and three muskets, spraying them in such a manner that it did not permit them much space between each shot; and in which they, with supreme agility, gave a stroke of the paddle in order to flee. But their dexterity did not protect them, because the first shots downed the majority of them. Taking the precaution of covering themselves with hides, the remaining ones finally found refuge in their agility, carrying six less than the nine who came. I might have captured or killed those who remained, but I found myself without men and amidst shoals to which we were forced to give the most attention.

Source: Herbert K. Beals, trans. and ed., "The Journal of Juan Francisco de la Bodega y Quadra," in *Four Travel Journals: The Americas, Antarctica and Africa, 1775–1874,* edited by Herbert K. Beals et al. (London: Hakluyt Society, 2007), 106–107. Used by permission of the Hakluyt Society.

ANALYSIS

Stories of the first encounters of Europeans and Native people during early explorations in the Americas are often bound together with tragic human costs. The deaths of the Quinaults and the sailors from the *Sonora* occurred in the lee of Cape Elizabeth on the coast of present-day Washington, probably just south of the Quinault River. A few hours earlier, Bruno Hezeta in the *Santiago* had taken possession of present-day Grenville Bay for Spain. These two landings were the first by Europeans along the coast of the Pacific Northwest south of Alaska, and they added substantial new cartographic knowledge of the region. While Bodega's and Hezeta's journal entries revealed a greater understanding of the local geography, they also reveal the extent of misunderstanding of one cultural group for another. Bodega was incredulous that the same people who had earlier negotiated trade with sailors would a few hours later attack with such vengeance. Nowhere in their journals did either man reflect on the possible motives for the attack.

Historians have suggested that Hezeta's landing and possession-taking ceremony early on July 14 may have been directly related to the assault on Bodega's men later in the day (Tovell 2008, 28–29). The territorial sovereignty of the Quinault might have been threatened by the rituals they observed—chaplains chanting while a cross was planted on the beach at Point Grenville, with the participation of armed men. The symbolic intent of the Spanish ceremony may have been clearly understood by the Indians who later defended their territory. As David Chance noted (Tovell 2008, 29), this may be the reason why current Quinault histories relate to the encounter on July 14 only at Point Grenville, not to the location of the armed assault. In addition, Point Grenville is a traditional sacred place on the Quinault reservation, which also may help explain the assault.

The area of shoals that trapped Bodega's ship is today called Sonora Reef. Present-day Grenville Bay was given the name Rada de Bucareli by Hezeta in honor of Viceroy Bucareli. Hezeta originally named point Grenville Punta de los Mártires (Place of the Martyrs) to honor the slain sailors.

28. Bruno Hezeta's Coastal Map, 1775

INTRODUCTION

Naval officers in San Blas were well trained in mapmaking. At the termination of the second Bucareli expedition of 1775, officers of the three ships prepared several detailed maps of bays they had explored from San Francisco north, as well as small-scale maps of the coast from Mexico to latitude ~58° north. Several of these maps were copied for wider distribution among officials and officers at San Blas and Mexico City and in Spain. Bodega and Hezeta shared astronomical data and unique observations, such as Hezeta's sighting of the entrance of the Bahia de la Asunción, and each produced at least two charts of the coast. The officers sent all journals and maps to Viceroy Bucareli, who was pleased with the "new intelligence" and the "highly useful information." Bucareli forwarded the original documents or copies of them to Spain, where Carlos III was impressed enough to publish two notices of the voyage in a royal-sponsored periodical. In 1781, a composite map of the discoveries of the expedition from San Blas to Prince of Wales Island was published in Mexico City. This rare map was not widely disseminated and apparently had little effect on European cartographers.

The historical record is incomplete about the disposition of the journals and maps produced by the officers and chaplains on the voyage. It is likely that copies of the journals, and possibly the maps, were made in San Blas and sent to Mexico City, where additional copies were sometimes made. The originals or copies were then sent to Spain, where officials placed them in the archives of the Casa de Contratación. In 1785, the council began transferring the vast storehouse of records of Spanish presence in the Americas to the Archivo General de Indias in Seville. Most of the manuscript journals and maps are now preserved at the Archivo General de la Nación in Mexico City, the Archivo General de Indias, or the Archivo Historico Nacional and the Museo Naval in Madrid.

When officials in Madrid decided to send another voyage of exploration to the north, the officers at San Blas studied the available cartography of the coast, and Bodega oversaw construction of a map in 1779 that he and the other senior officers used as a guide. The maps that Bodega consulted included a 1766 printed map of the North Pacific by French cartographer Jacques-Nicolas Bellin, two Russian maps, and Bodega's own map of 1775, which was similar in most respects to this map of Hezeta's. The title of the map can be translated as "Reduced Map of the Coasts and Seas of Northern California Drawn from the Observations and Notations of a Lieutenant of the Navy and Commander of the Expedition Don Bruno de Ezeta Dudagoitia: On the Voyage of Discovery of These Coasts and Seas by the Order of Lord Viceroy Don Antonio Maria Bucareli and Ursua Made in the Year 1775."

ANALYSIS

Hezeta's map illustrates the fundamental importance to Spain of mapmaking as a tool of empire building. Viceroy Bucareli initiated two northern voyages of exploration in 1774 and 1775, in large part to preempt the expansion of Russia and Britain into the Pacific

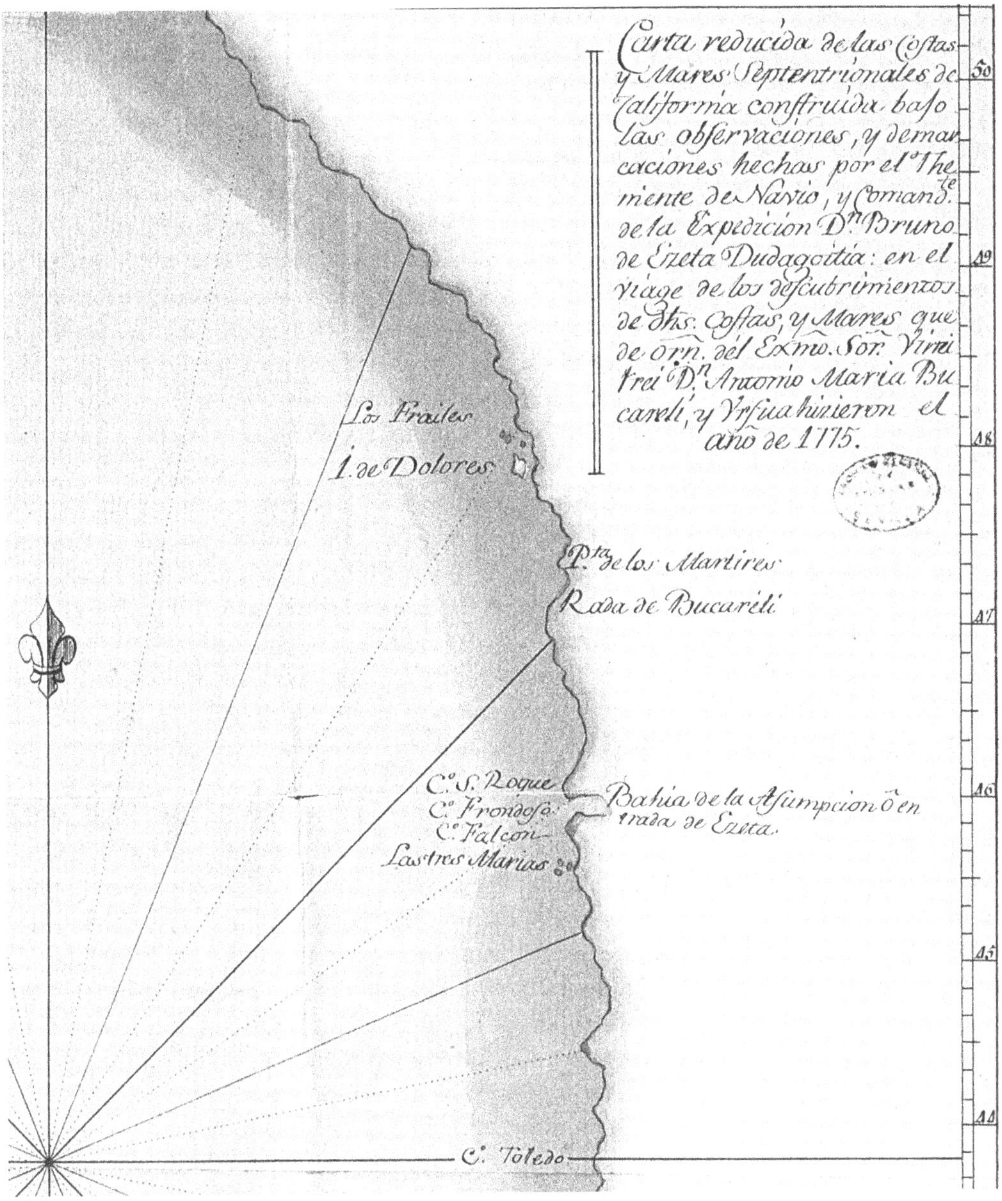

"Reduced Map of the Coasts and Seas of Northern California Drawn from the Observations and Notations of . . . Commander of the Expedition Don Bruno de Ezeta Dudagoitia," 1775. (Archivo General de Indias, Seville, Spain, Secretary of State and the Office of State, Reference Code: ES.41091.AGI / 27.17 // MP-MEXICO, 306)

Northwest. In the absence of an established foreign presence on the coast, possession-taking ceremonies established sovereignty at the location of the ceremony and contiguous regions. Although a notarized account of the ceremony "legalized" the activity, it was a map drawn from direct observation of the location and the nearby coastline that was the authoritative record of the extent of dominion.

Hezeta's full map ranged from latitudes ~36° to ~50° north. The section of the map shown here covered the coast from ~43° to ~50°, from just south of present-day Coos Bay,

Oregon, to north of Vancouver, British Columbia. Hezeta's names reflected only what was observed on the 1775 voyage. For example, he did not include Juan Pérez's anchorage in 1774 at Surgidero de San Lorenzo, which might have been in proximity to today's Nootka Sound on the west coast of Vancouver Island. Although Hezeta is the only mapmaker identified in the title, he may have had an assistant.

The purpose of the map was to record locations of events, such as anchorages and possession-taking ceremonies, and the direction of the observed coastline. The map illustrated the essential elements of a maritime voyage of reconnaissance and was not intended to impress a public audience with embellishments, descriptive legends, or conceptualized (not observed) information. Using only pen, ink, and watercolor, Hezeta produced a map that is distinguished by its elegant simplicity. His use of the word "California" in the title suggested that Spanish officials believed that the extent of that long-established region of New Spain extended indefinitely to the north. With rare exceptions, the names on Hezeta's map and in his journal have not survived.

Hezeta drew a small rectangular-shaped island, Isle de Dolores, at a calculated latitude of 47°58′. Historians have puzzled over the origin of the name and its modern equivalent, speculating that it is James Island or Destruction Island off the coast of Washington state. Nearby, Hezeta depicted three small islands on the map, which he named Los Frailes (The Friars). These islands were not mentioned in Bodega's or Hezeta's journals, and the origin of the name is unclear. Farther south, Pta. de los Martires (Martyr's Point) was the site of the deadly conflict between Bodega's crew and a group of Quinaults on July 14. This place is most likely today's Cape Elizabeth. Rada de Bucareli, named after the viceroy of New Spain, was most likely present-day Grenville Bay.

In the area of latitude 46°, the Bahia de la Asumpción o entrada de Ezeta (Bay of the Assumption or Entrance of Hezeta), Co. S. Roque (Cape of San Roque), and Co. Frondoso (Cape Frondoso) were the names Hezeta applied to the Columbia River estuary and points to the north and south. The location of Co. Falcon corresponds within 3′ of latitude to today's Cape Falcon, off the Oregon coast. The name probably originated from Saint Clare of Montefalco, whose feast day was August 16, two days before Hezeta recorded sighting the cape (Beals 1985, 89, 146n; Wagner 1968, 387). Three small islands, Las Tres Marias (The Three Marys), may correspond to present-day Three Arch Rocks, which lie off the Oregon coast at 45°28′. Both Bodega and Hezeta applied the name Co. Toledo to approximately the same location, but the name does not appear in the journal of either officer.

Spanish cartographers incorporated almost all of these place-names on many updated manuscript maps until the end of the 18th century. In so doing, mapmakers helped reinforce Spanish hegemony in the region and, concurrently, the context of encounter and conflict that frequently occurred. When these names were changed by non-Spanish cartographers, much of that important context was lost.

29. Rare Publicized Reports, 1776, 1781

INTRODUCTION

These documents are examples of information that became available beyond New Spain and Spain about the results of the second Bucareli expedition. While Spanish policy usually forbid the spread of exploration information beyond administrative control, officials occasionally made exceptions and released news or an abridged account of a recent voyage. Shortly after the accounts of the officers of the second Bucareli expedition reached Madrid, probably in early 1776, an announcement appeared in the Gaceta Oficial, *the official bulletin of the state. Two months later the same periodical declared the astonishing news that King Carlos planned to publish a map of the explored "coasts and seas of northern California." Carlos may have reconsidered that decision, since the engraved Spanish map described here was not published until 1781.*

Information about the voyages also reached people surreptitiously. Sometime between 1775 and 1781, Daines Barrington in London obtained and translated a copy of the diary of pilot Francisco Mourelle (sometimes Maurelle). When Barrington published extracts of the journal in 1781 he included a map, an unsophisticated product that likely had been constructed in London (probably not by Barrington). He was an intellectual, antiquarian, jurist, geographer, and member of the Council of the Royal Society, one of the oldest scientific societies in Europe, and he came to believe in the feasibility of polar navigation from Europe to the Pacific Ocean. He also exercised influence with the Royal Society and the British Admiralty and Parliament. The preface to Mourelle's diary reveals Barrington's "distrust" and "suspicions" of the Spanish and his promotion of British interests in recent discoveries as purely scientific.

Publications about the Expedition

Gaceta Oficial de Madrid

[March 14, 1776] As a consequence of these valuable events, the king ordered officers sent to the port of San Blas to extend these navigations and discoveries as much as possible. To this end at the beginning of 1775 *Teniente de Navío* Bruno de Hezeta in the frigate *Santiago* and *Teniente de Fragata* Juan Francisco de la Bodega y Quadra in the schooner *Sonora* and for the same purpose *Teniente de Fragata* Juan de Ayala in the packetboat *San Carlos* sailed to Monterey.

The first [*Santiago*] reached 50° of latitude, the second [*Sonora*] 58° and the third [*San Carlos*] 37°42′. They reconnoitered the coast in between, which is populated by savage Indians. The harbour of San Francisco was charted carefully, and the Indians in the neighbourhood were especially docile and tractable.

His Majesty, in consideration of the special and distinguished service which these naval officers and *pilotos* have rendered and to encourage others, has promoted Bruno de Eceta [*sic*] to *capitán de fragata,* Juan de Ayala and Juan Francisco de la Bodega y Quadra to *teniente de navio,* and Juan Pérez *to teniente de fragata.* The *pilotos* José de Cañizares and Francisco Maurelle [*sic*] have been promoted to *alférez de fragata.*

[May 19, 1776] [The viceroy] has also reported that the officers and *pilotos* have made very accurate maps and charts of the coasts and seas of northern California, which His Majesty has decided to have engraved and so publicize the results of the aforementioned navigation from Monterey.

Francisco Antonio Mourelle, Journal of a Voyage

The following journal having been placed in my hands for perusal, I conceived it to be so interesting for the improvement of Geography, that I desired permission to translate and publish it.

I was principally induced to take this trouble, because I supposed, that the Spaniards, from their most peculiar jealousy with regard to their American dominions, would never permit that navigators of other countries (particularly the English) should know the excellent ports of the Western part of America in high Northern Latitudes, which are here laid down with such accuracy and precision, together with the abundant supply of masts, fire wood, and water which may be procured in most of them. . . .

Nothing however can be more groundless than these suspicions, for whenever a N. W. or any other Northern communication is found between the Atlantic and Pacific Oceans, it may be boldly pronounced that such passage will be so very precarious, as never to answer the purpose of expeditions in time of war, or commerce during peace.

The Spaniards should, after our late voyages of discovery (which reflect so much honour upon his Majesty's reign), be convinced that the English Nation is actuated merely by desiring to know as much as possible with regard to the planet which we inhabit, and to which our geographical inquiries are necessarily bounded.

This distrust on the part of Spain would more wisely be directed against the Russians, who from Camíkatiíka [Kamchatka] might easily establish themselves on the W. coast of America, and from thence perhaps in time shake their unwieldy, and tottering empire.

Sources: Freeman M. Tovell, *At the Far Reaches of Empire: The Life of Juan Francisco de la Bodega y Quadra* (Vancouver: UBC Press, 2008), 44, 46, reprinted with permission of the publisher from *At the Far Reaches of Empire* by Freeman Tovell © University of British Columbia Press 2008, all rights reserved by the publisher; Francisco Antonio Mourelle, *Journal of a Voyage in 1775* (London, 1781), 471–472.

ANALYSIS

By the mid-18th century, several maritime nations were increasingly focused on obtaining direct information on the geography, natural resources, and inhabitants of the North Pacific Coast of North America. In this international context, it is not surprising that the results of the two Spanish explorations of 1774 and 1775 would be of interest to ministers and naval officials beyond Madrid and Mexico City. In a rare departure from Spanish policy, King Carlos III promoted the publication of two notices of the second Bucareli expedition a few months after the ships returned to San Blas in November 1775. He may have been thinking strategically by giving international notice of Spanish primacy of discovery.

The notices, published in Madrid, were picked up and reprinted by British newspapers, the *London Evening Post* on May 29, 1776, and the *London Annual Register* on June 28. They also appeared in brief in an anonymous work, *Summary Observations and Facts Collected from Late and Authentic Accounts of Russian and Other Navigators,* published in London that same year. Although news from the uprising in the American colonies dominated the headlines, the accounts of Spanish movement on the other side of the continent were influential. Just before departing England on his voyage of discovery in July 1776, Captain James Cook learned about the Spanish voyage of 1775, apparently from these newspaper accounts.

Perhaps the most surprising Spanish publication of this period was a map of the coast from latitude 19°58′ north that was published in Mexico City in 1781. The cartographers, probably Bodega and José de Cañizares, noted in the title that the map incorporated Spanish discoveries from the years 1769 through 1775. This composite map contained abundant details about place-names, ships' tracks, and legends. The map apparently had little impact among European geographers and cartographers, however, perhaps because it was not published in Europe and was soon overshadowed by the release in 1784 of Cook's atlas and the multivolume account of his voyage.

Although Daines Barrington was not a diplomat, a member of the British Parliament, or a naval officer, his influence in the intellectual community provided him a significant audience to express his views. His personal interest in the possibility of a Northwest Passage and his apparent anti-Spanish and anti-Russian sentiment suggest that his primary interest in publishing Mourelle's account had less to do with revealing the latest knowledge of geographical exploration than with promoting British interests on the Northwest Coast.

Further Reading

Beals, Herbert K., ed. and trans. "The 1775 Journal of Juan Francisco de la Bodega y Quadra." In *Four Travel Journals: The Americas, Antarctica and Africa, 1775–1874,* edited by Herbert K. Beals et al., 1–139. London: Hakluyt Society, 2007.

Beals, Herbert K., trans. *For Honor and Country: The Diary of Bruno de Hezeta.* Portland: Oregon Historical Society Press, 1985.

Tovell, Freeman M. *At the Far Reaches of Empire: The Life of Juan Francisco de la Bodega y Quadra.* Vancouver: University of British Columbia Press, 2008.

Wagner, Henry R. *The Cartography of the Northwest Coast of America to the Year 1800.* Amsterdam: N. Israel, 1968.

Third Expedition of James Cook

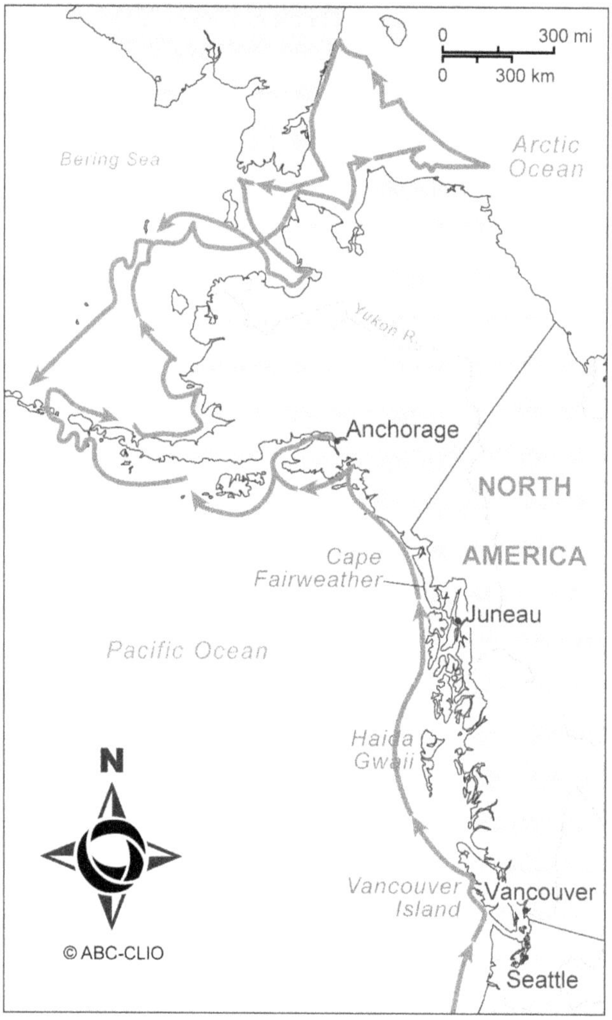

CHAPTER 4

James Cook, First British Foray to the Northwest Coast, 1778

James Cook is perhaps the best-known mariner from Britain during the Age of Sail, "one of the world's great explorers," in the words of scholar Anne Salmond (2003, xix). His three voyages exploring the Pacific Ocean—1768–1771, 1772–1775, and 1776–1780—had great influence on mariners and contributed significantly to Britain becoming the world's dominant seafaring nation. Cook's accomplishments mark him as a singular force in British naval history and the history of human endeavor in the Pacific Ocean. He set a high standard in cartography, especially his 1778 map of the Northwest Coast, published in 1784. He was the first to use chronometers and to confirm their utility in determining accurate longitude, and he curbed the incidence of scurvy in his crew, which changed shipboard practices for British and other mariners. Cook earned a reputation—perhaps exaggerated—for humane and just relationships with Native people, especially in the South Pacific. And finally, his published journals in 1784 created a sensation in the discovery of a rich pelagic fur trade between the Northwest Coast and China. Even if we do not lionize Cook, it is hard to exaggerate his importance to the maritime exploration of the Pacific Northwest. Jean-François La Pérouse, Alejandro Malaspina, and George Vancouver—all important mariner explorers of the Pacific Northwest—sailed in Cook's wake from 1785 to 1792, following his lead in their voyages and mimicking his successful strategies.

Nathaniel Dance painted this portrait in May 1776, showing Cook as the master mariner of the Pacific, a rendering surgeon David Samwell regarded as the most accurate depiction of the great explorer. (Library of Congress)

Born in 1728 in Marton, Yorkshire, Cook grew up in a middling family; his father worked as a day laborer for area farmers. The young Cook worked for a grocer before apprenticing with a Quaker ship owner in 1746. Cook sailed on collier ships hauling coal and learned the rudiments of seamanship. In 1755 in the run-up to war with France, he joined the Royal Navy as a hand on a 60-gun frigate. During the war, he served in North America and made a mark as an accomplished chartmaker in Newfoundland and American waters. He also received attention and promotion from Captain Hugh Palliser, who was well connected to the Admiralty. In 1762 Cook married Elizabeth Batts, 14 years his junior, and settled in East London. The couple had six children, but none survived the 18th century or had children themselves. Elizabeth Cook died in 1835.

During the 1760s, the Royal Society—Britain's premier scientific organization, founded in 1660—planned an expedition to observe the transit of Venus across the sun to gather important information to aid navigation. The Admiralty chose Tahiti as the site for the observation, an island that Samuel Wallis had discovered and claimed for Britain in 1767. Alexander Dalrymple, a cartography expert and later the first hydrographer to the British government, had expected to lead the expedition because of his comprehensive knowledge of the South Pacific. He had personally chosen the ship *Endeavour* for the voyage, and the Royal Society "had recommended Mr Dalrymple to Lords of Admiralty for the command of a vessel" (Fry 1970, 120). But the Admiralty could not allow a nonnaval commander. Instead, it tapped James Cook, a master seaman, to command the expedition. The Royal Society based its choice on Cook's chartmaking experience and his publication of a scientific paper on a solar eclipse in the Royal Society's *Transactions* in 1767, which marked him as an able mathematician and astronomer.

In 1768 James Cook received commission as a lieutenant and commander of the *Endeavour* to explore the Pacific. A full-rigged 368-ton bark, with a length of 106 feet, a beam of 29 feet, and a deep hold, the *Endeavour* left Plymouth in 1768 with 94 people aboard, including a crew of 73 officers and sailors, 12 marines, and 9 scientists. The expedition had two primary goals: to observe the transit of Venus in 1769 and, in a secret order from the Admiralty, to search for Terra Australis Incognita, an imagined great continent in the South Pacific that had been featured on European maps since the early 16th century. It was an avowedly scientific mission. On the *Endeavour* was botanist Joseph Banks, a close friend of Lord Sandwich of the Admiralty and the Royal Society's onboard representative. Banks had inherited estates in 1764, when he was 21 years old, that brought him £6,000 a year, considerably more than Cook's annual wage of just under £100. Banks's wealth allowed him to finance the passage of 7 other scientists, including Swedish naturalist Daniel Solander, a protégé of the great botanist Carl Linnaeus.

Cook sailed around Cape Horn and 4,000 miles across the Pacific to Tahiti, where he set up an observatory to view the transit of Venus—Point Venus—under the direction of astronomer Charles Green. From Tahiti, Cook sailed to New Zealand and the eastern coast of Australia, where he found no great continent, which Dalrymple believed surely existed. The *Endeavour* sailed around the Cape of Good Hope to England, arriving on July 12, 1771. Cook returned home to learn of the deaths of two of his youngest children. Nonetheless, he had two remaining boys, whom he included on his muster rolls to gain additional pay for his household.

On his return to England, Cook met with King George III, who authorized his promotion to captain in August 1771. Cook also conferred with members of the Royal Society, although it was Banks and Solander who received most of the public recognition for the voyage, especially their scientific collecting. One newspaper referred to Cook as a naval officer "who sailed round the Globe with Messrs Solander, Banks etc.," while another predicted that Banks would lead a second voyage in 1772 (O'Brian 1987, 149).

In November 1771 the Admiralty gave command of a second Pacific voyage to Cook, with the goal of testing chronometers and continuing the search for Terra Australis Incognita in the South Pacific. He captained the *Resolution,* a Whitby-built 110-foot full-rigged ship that displaced 462 tons and carried 92 crew and officers plus 20 marines. Tobias Furneaux, a veteran of Samuel Wallis's circumnavigation in 1767, captained the 336-ton *Adventure,* Cook's companion vessel. Onboard scientists included astronomer William Wales and naturalists Johann Forster and his son George. Cook carried three chronometers for testing, including K-1, Larcum Kendall's exact copy of John Harrison's H-4 timepiece. By the end of the journey, which lasted just over four years, the K-1 had lost only 18 minutes.

The *Resolution* and the *Adventure* left England in July 1772 and headed around the Cape of Good Hope on a nearly direct path for New Zealand, where the *Resolution* anchored in March 1773. The *Adventure* made the passage early and investigated Tasmania before joining Cook in May. Cook and Furneaux explored the Pacific in high latitudes, crossing the Antarctic Circle (latitude 66°33′ south) in January 1773 in their search for the imaginary southern continent. By August they had sailed north to Tahiti, where they stayed for two months. In October they headed back to New Zealand, but they became separated once again. From New Zealand, where his crew engaged in hostilities with Maoris, Furneaux headed east around Cape Horn on his return to England, where he arrived in July 1774. Cook continued searching in Antarctic waters, crossing the Antarctic Circle twice in 1774 before expanding his survey of South Pacific regions from Easter Island to Tahiti. He returned to New Zealand before heading home by way of the Cape of Good Hope, returning in July 1775 to a heroic welcome and honors.

DID YOU KNOW?

Captain James Cook thoroughly disliked a diet of boiled penguins, which he and officers ate on his second Pacific voyage when they sailed near Antarctica, but he swallowed them willingly rather than eat salt pork, a shipboard food he had grown to detest.

Cook became a member of the Royal Society and in January 1776 published "The Method for Preserving the Health of His Majesty's Crew" in *Philosophical Transactions,* which won him the Copley gold medal from the Royal Society in July. In February, the 48-year-old mariner agreed to lead a third voyage to the Pacific in response to a request from the Admiralty to find the Northwest Passage. Although Cook had the opportunity to take a hospital administrative position, he could not shake his passion for a mariner's life. "A few Months ago," he wrote to a friend, "the whole Southern hemisphere was hardly big enough for me and now I am going to be confined within the limits of Greenwich Hospital, which are far too small for an active mind like mine" (qtd. in Salmond 2003, 301). There was an additional lure: Parliament offered £20,000 (about $5 million today) to the British discoverer of the Northwest Passage and members of his crew, who would receive up to two years' pay as their reward.

The Admiralty refitted the *Resolution* and purchased a companion ship, the *Discovery,* which was put under the command of Charles Clerke. The *Discovery* was just over 90 feet long, weighed 300 tons, and was fully rigged; it carried 70 crew and officers. Cook selected his crew from trusted veterans of earlier voyages and added young masters such as George Vancouver and William Bligh, both expert navigators.

The two ships left Plymouth in mid-July 1776 and sailed to the Pacific by way of the Cape of Good Hope, but delays put the expedition behind schedule. In February 1777 Cook landed in New Zealand, where he learned more about the conflict that Tobias Furneaux had with some Maoris in 1774, knowledge that put Cook on alert. Cook's relations with South Sea islanders, which had been positive during the first two voyages, began to sour, and crew members noted his anger at petty theft and other conflicts. He was also less patient with his crew and officers and used punishment more often than he had on earlier voyages. He had administered 354 lashes on the first voyage and 288 on the second; during the third voyage, he applied 624 stripes on his crew with a cat-o'-nine-tails.

Cook sailed in the South Sea throughout 1777, leaving the Society Islands for the Northwest Coast in January 1778. On the way he was the first mariner to visit Hawaii, landing at Waimea Bay on the island of Kuai on January 18, 1778. By March, the *Resolution* and the *Discovery* had coursed north and sighted North America at 47° north, the beginning of a six-month search for a Northwest Passage in Alaskan waters. Although they spotted several inviting bays and openings, each one turned out to be a dead end, and a disappointed Cook found no Northwest Passage. He continued northwest along the coast to Bering Strait, where he sailed as far north as 70° in the present-day Chukchi Sea before a wall of sea ice blocked his progress. While at Unalaska, he met with Russian mariners and sorted out the complex maritime geography of the Aleutian Islands, which stretched from Alaska to the Kamchatka Peninsula in Asia. In late October with his ships in need of repair, Cook headed south to winter in Hawaii.

DID YOU KNOW?

Nuu-chah-nulth Indians at Nootka Sound on Vancouver Island charged Cook a substantial fee to gather grasses for feeding his ship-bound milk goats, but he willingly paid, commenting that the Indians had "high notions of everything the Country produced being their exclusive property."

Cook spent a month repairing his ships and preparing for the remainder of the voyage. He and his crew had harmonious relations with Hawaiians, and the ships' men left in February 1778 in good cheer. But tempestuous weather damaged a foremast, and they beat a hasty return to Hawaii, where they set about making repairs. As the time for leaving grew near, relations between the British and Hawaiians began to sour. Some locals stole the ship's cutter, an important vessel that Cook could not abandon. To retrieve the boat, the captain went ashore on February 14 to secure the local chief's help, but Cook and his armed guard soon became targets for Native hostility, especially when they learned of the death of another chief at the hands of British seamen who had a fracas with Hawaiians in another part of the bay. Friction developed quickly, putting Cook, his attending marines, and two boats in jeopardy. Shots were fired, and the Hawaiians threw stones, escalating the conflict. Several marines died in the conflict, and Cook was stabbed and beaten to death.

Cook's death devastated the crew of the *Resolution* and the *Discovery*, but naval discipline held firm, and the expedition sailed north to continue surveys of Kamchatka and the Bering Sea. Charles Clerke took over as commander, but by August he had succumbed to tuberculosis in Kamchatka. The expedition returned to England in October 1780, with John Gore in command. In their interactions in Kamchatka and in their last pursuit of a Northwest Passage, they made a discovery that would change the history of the Pacific Northwest. While in Kamchatka, they sold furs acquired through trade with Mowachaht at Nootka Sound. What they received for the furs shocked them. As John Ledyard reported, skins "sold at Kamchatka . . . for sixty rubles, which is near 12 guineas, and had it been sold in China it would have been worth 30 guineas. We purchased while here [Nootka] about 1500

This is the most accurate portrait of Cook's death in Hawaii, drawn by expedition artist John Webber, a 24-year-old French-trained painter who had joined the *Resolution*'s crew at Cook's request. (Bettmann/Corbis)

beaver, besides other skins, but took none but the best, having no thoughts at that time of using them to any other advantage than converting to the purposes of cloathing, but it afterwards happened that skins which not cost the purchaser sixpence sterling sold in China for 100 dollars" (Munford 1963, 70). That revelation, which was published in London in 1783, helped stimulate a rush to the Northwest Coast that set in motion an international fur trade that spanned the North Pacific. More important perhaps, Cook's last voyage produced exceptional maps and charts of a region that had been dimly understood, and his investigations made it much less likely that a Northwest Passage existed.

In 1781, the first account of Cook's third voyage appeared in *Journal of Captain Cook's Last Voyage,* in which 2nd Lieutenant John Rickman portrayed the captain as an aggressive, imperial mariner. Heinrich Zimmerman published a brief account of the journey in Mannheim, Germany, in 1781, and the *Discovery*'s surgeon mate, William Ellis, wrote a memoir of the voyage in 1782. Ledyard's account appeared in 1783, and the authorized Admiralty account appeared in 1784. Compiled and edited by John Douglas, the three-volume *A Voyage to the Pacific Ocean* included Cook's journal and Lieutenant James King's post-Hawaii journal. Cook's image in this official rendition portrayed him as a hero and described his death as the demise of an Enlightenment idol and a man-god among Hawaiians.

30. Secret Instructions from the Admiralty, 1776

INTRODUCTION

James Cook's third voyage followed two earlier and highly successful expeditions that had pursued geographical discoveries in the Pacific Ocean and filled in details of the South Pacific and the Antarctic. Cook had mastered methods of protecting his seamen from the privations and illnesses of long voyages, and he had used new technologies, especially chronometers, for determining longitude. His instructions for the third voyage, marked as "Secret Instructions for Capt. James Cook, Commander of His Majesty's Sloop the Resolution," dated July 6, 1776, were executed by the "Commissioners for executing the Office of Lord High Admiral of Great Britain & Ireland &ca." Rear Admiral Hugh Palliser and the Earl of Sandwich, Charles Spencer, signed the document.

The instructions laid out the route that Cook should take on this circumnavigation, with stops in Tahiti and New Zealand before striking the coast of New Albion, the name Sir Francis Drake had given the Northwest Coast of North America in 1579 when he surveyed the area in the Golden Hind *as part of the second European circumnavigation of the world. Cook's principal goal was to investigate potential Northwest Passage inlets in the North Pacific and the shape and content of the Northwest Coast, especially places of advantage for Britain as a seafaring nation.*

Secret Instructions for Captain James Cook, Commander of His Majesty's Sloop the *Resolution*

[Y]ou are to leave those Islands [Tahiti], in the beginning of February, or sooner if you shall judge it necessary, and then proceed in as direct a course as you can to the coast of New Albion, endeavouring to fall in with it in the latitude of 45°0′ North; and taking care, in your way thither, not to lose any time in search of new lands, or to stop at any you may fall in with, unless you find it necessary to recruit your wood and water.

You are also, in your way thither, strictly enjoined not to touch upon any part of the Spanish dominions on the Western continent of America, unless driven thither by some unavoidable accident; in which case you are to stay no longer there than shall be absolutely necessary, and to be very careful not to give any umbrage or offence to any of the inhabitants or subjects of his Catholic Majesty. And if, in your farther progress to the Northward, as hereafter directed, you find any subjects of any European Prince or State upon any part of the coast you may think proper to visit, you are not to disturb them, or give them any just cause of offence, but, on the contrary, to treat them with civility and friendship.

Upon your arrival on the coast of New Albion, you are to put into the first convenient port to recruit your wood and water and procure refreshments, and then to proceed Northward along the coast, as far as the latitude of 65°, or farther, if you are not obstructed by lands or ice; taking care not to lose any time in exploring rivers or inlets, or upon any other account, until you get into the before-mentioned latitude of 65°, where we could wish you to arrive in the month of June next. When you get that length, you are very carefully to search for, and to explore, such rivers or inlets as may appear to be of a considerable extent and pointing towards Hudson's or Baffin's Bays; and if from your own observations, or from any information you may receive from the natives (who, there is reason to believe, are the same race of people, and speak the same language, of which you are furnished with a Vocabulary, as the Esquimaux) there shall appear to be a certainty, or even a probability, of a water passage into the afore-mentioned bays, or either of them, you are, in such case, to use your utmost endeavours to pass through with one or both of the sloops, unless you shall be of opinion that the passage may be effected with more certainty, or with greater probability, by smaller vessels; in which case you are to set up the frames of one or both [of] the small vessels with which you are provided, and when they are put together, and are properly fitted, stored, and victualled, you are to dispatch one or both of them, under the care of proper officers, with a sufficient number of petty officers, men, and boats, in order to attempt the same passage. . . . [I]f you shall find it more eligible to pursue any other measures than those above pointed out, in order to make a discovery of the before-mentioned passage (if any such there be), you are at liberty, and we leave it to your discretion, to pursue such measures accordingly.

Source: *A Voyage to the Pacific Ocean Undertaken, by the Command of His Majesty, for Making Discoveries in the Northern Hemisphere . . . Performed under the Direction of Captains Cook, Clerke, and Gore, in His Majesty's Ships the* Resolution *and* Discovery, 3 vols., published by Order of the Lords Commissioners of the Admiralty (Dublin, 1784), 1:xxxiii–xxxv.

ANALYSIS

The Admiralty's instructions to Cook left little to happenstance or potential distraction. His mission on the third voyage pointed him to the Northwest Coast, an important unexplored portion of the Pacific that had drawn intense attention from the Admiralty. In April 1778 Cook stopped at Nootka Sound for three weeks in what was to be a wood- and water-gathering stop, because poor preparations of the *Resolution* before sailing had left the masts vulnerable to the storms the ship had encountered on its route to North America. The time Cook had to spend repairing his masts forced him to speed to Alaskan waters as far north as 65°, or farther, in pursuit of the Strait of Anian or any inlet that would signal the western opening of the so-called Northwest Passage.

The promise of the Northwest Passage through the continent transfixed European seafaring nations, holding out the promise of an efficient and economical path to new sources of agricultural, mineral, and horticultural wealth that could be acquired through slave or low-cost labor. Cook's instructions included direct references to these goals. The search for the Strait of Anian—the name was a misreading of Marco Polo's label for a waterway between Asia and America—had been approached first from the Atlantic by early explorers, including Martin Frobisher in the 16th century and William Baffin in the 17th century. The first cartographic representation of the Strait of Anian appeared on maps by Italian mapmakers Giacomo Gastaldi and Paolo Forlani in the 1560s. Other 16th-century maps also showed a Northwest Passage, stretching from east to west across North America. By the early 18th century, French geographer Guillaume Delisle and others had conceived of watercourse linkages across the continent that included rivers, lakes, and bays (Williams 2002, 21). In the Pacific Northwest, these imagined geographies prominently located a major inlet or river that penetrated the continent in the far northern latitudes. The report of Bartholomew de Fonte's apocryphal voyage of 1640 had put the passage as far south as 53° north. The British Parliament offered a "£20,000 reward for the discovery of a Northwest Passage, and forced the Hudson's Bay Company to indulge in maritime and land exploration in the 1760s and 1770s beyond the extent demanded by trade" (Norris

1977–1978, 13). Those explorations in present-day Canada eliminated potential waterways south of about 60° north.

The direction to bypass all inlets south of latitude 65° north alerted Cook to the area that the Admiralty believed harbored the Strait of Anian. Cook's investigations, however, began well south of 65° north in Prince William Sound, at 60.1°, disclosing a general ignorance of the North Pacific among European geographers and cartographers. Cook would not reach latitudes above 64° until he penetrated the Bering Strait north of the Aleutian Island archipelago in August 1778, months after leaving Nootka Sound. Ironically, Cook did not eliminate the desire to search for the Northwest Passage, because he did not investigate all of the coastline from 45° to 55°, an effort that would be taken up by later mariners.

31. On the Oregon Coast, 1778

INTRODUCTION

The Resolution *left Plymouth, England, on July 12, 1776; the* Discovery *followed on August 1. The ships reached the Cape of Good Hope in November and headed east to Kerguelen Island in the Indian Ocean. By January 1777, they had reached Tasmania and New Zealand. They left there in late February, considerably behind schedule. Cook's instructions had proposed that he leave the South Pacific islands by early February, a destination that was months away when he left New Zealand. He knew the seas and winds in the South Pacific from his second voyage in 1772–1773, but he was sailing earlier in the year on this voyage and encountered much different wind patterns and storms, which delayed his progress. The* Discovery *and the* Resolution *did not leave Tahiti until late December. Cook sailed to Christmas Island in early January 1778 and anchored in Hawaii on January 18. "I never saw Indians so much astonished at the entering a ship before," he wrote in his journal, "their eyes were continually flying from object to object, the wildness of thier* [sic] *looks and actions fully express'd their surprise" (Beaglehole 1967, 3:265). Cook remained in the Hawaiian Islands for nearly three weeks visiting several of the islands, taking careful notes of discovery, and establishing relations with the local people. He left for the Northwest Coast on February 2, 1778.*

Journal of Captain Cook, March 13–22, 1778

The wind continued at West, and North West; storms, moderate weather, and calms, succeeding each other by turns, till the morning of the 21st; when, after a few hours calm, a breeze sprung up at South West. This bringing with it fair weather, I steered North Easterly, in order to fall in with the land, beyond that part of it where we had already so unprofitably been tossed about for the last fortnight. In the evening, the wind veered to the Westward; and, at eight o'clock, the next morning, we saw the land, extending from North East to East, nine leagues distant. At this time we were in the latitude of 47°5′ North, and in the longitude of 235°10′ East.

I continued to stand to the North with a fine breeze at West, and West North West, till near seven o'clock in the evening, when I tacked to wait for day-light. At this time we were in forty-eight fathoms water, and about four leagues from the land, which extended from North to South East half East, and a small round hill, which had the appearance of being an island, bore North three quarters East, distant six or seven leagues, as I guessed; it appears to be of a tolerable height, and was but just to be seen from the deck. Between this island or rock, and the Northern extreme of the land, there appeared to be a small opening, which flattered us with the hopes of finding an harbour. These hopes lessened as we drew nearer; and, at last, we had some reason to think, that the opening was closed by low land. On this account I called the point of land to the North of it Cape Flattery. It lies in the latitude of 48°15′ North, and in the longitude of 235°3′ East. There is a round hill of a moderate height over it; and all the land upon this part of the coast is of a moderate and pretty equal height, well covered with wood, and had a very pleasant and fertile appearance. It is in this very latitude where we now were, that geographers have placed the pretended strait of Juan de Fuca. We saw nothing like it; nor is there the least probability that ever any such thing existed.

Source: *A Voyage to the Pacific Ocean Undertaken, by the Command of His Majesty, for Making Discoveries in the Northern Hemisphere . . . Performed under the Direction of Captains Cook, Clerke, and Gore, in His Majesty's Ships the* Resolution *and* Discovery, 3 vols., published by Order of the Lords Commissioners of the Admiralty (Dublin, 1784), 2:262–263.

ANALYSIS

The crew left Hawaii's climate reluctantly, knowing that the North Pacific destinations would bring much colder temperatures and likely stronger winds. "We have been so long Inhabitants of the torrid Zone," Captain Charles Clerke recorded on February 11, "that we are all shaking with Cold here with the Thermometer at 60" (Beaglehole 1967, 3:288n1). Two weeks later, Cook commented that if he had not known that North America was close to their location at sea he would have thought they were "some thousand leagues" (3,000 miles) from land, for they had seen no birds since leaving Hawaii (Beaglehole 1967, 3:288). On March 6 they saw two seals and a few whales, and the following day they saw North America, which Cook estimated at 44°55′ north and about eight leagues (24 nautical miles) distant from their ships. He noted that they sailed no closer to shore than 70 fathoms (125 feet) of water and that the landscape was unremarkable except for a striking headland that he named Cape Foulweather, a reference to the poor weather that engulfed the ships after sighting the cape.

Unable to make much headway near the shore, Cook tacked away from the coast and was soon sailing westward to find clear weather and the opportunity to take lunar sightings. Gale-force winds, squalls, and heavy rain forced the ships to keep off the coast for three days, but by March 10 Cook had been able to sail to within eight leagues of land. He could find no likely harbor, but the following day they saw a large promontory, which he named Cape Perpetua, at 44°06′, 30-some miles south of Cape Foulweather. He also named a smaller nearby promontory Cape Gregory (present-day Cape Arago).

Thirty to 40 nautical miles to the south, Cook spotted another major headland that he thought was Cape Blanco, which Spanish mariner Martín Aguilar had named in 1603. Aguilar's exploration had been incorporated into a map by Jacques-Nicolas Bellin in the 1760s, a map that Cook probably had in hand. Cook also noted that he was near the area where "Geographers have placed a large entrance, or Strait, the discovery of which they ascribe to the same Captain [Aguilar]" (Beaglehole 1967, 3:292), but he was too distant to verify a large river, and it is more likely that he had mistaken interior mountains for Cape Blanco. This notation documents his attention to earlier named landscape features and the potential for navigable interior rivers. Nonetheless, Cook seemed skeptical of early maps of the Northwest Coast, especially the mythical river openings that suggested passages through the continent. As he noted in his journal, the Strait of Juan de Fuca, some 6 degrees in latitude north of Cape Blanco, was likely mythical. In any event, the storms, the lack of easily found harbors, and his anxious need to sail to 60° north before the season ended compelled him to keep his investigations of the present-day Oregon and Washington coastlines minimal.

32. Ship Repairs at Nootka Sound, 1778

INTRODUCTION

All mariners on long voyages faced the threat of damage to their ships. They had to be prepared to conduct all manner of repairs on their vessels, from replacing key components to caulking and recaulking the bottoms. Deepwater ships were complex. The Resolution *and the* Discovery *were flat-bottomed, which allowed seaman to inspect and make repairs on a beach, where they used shore anchors and cables to careen, or lean, a ship on its side. At times, captains had to warp their ship to shore by taking an anchor by boat and dropping it near the shore, then pulling the ship toward the anchor, using the ship's capstan to reel in the anchor line.*

Built in the Whitby yards on the River Esk in Yorkshire on the North Sea, both ships were rigged as sloop-of-war vessels. The Resolution, *built in 1770, displaced 462 tons and was 110 feet long and 30 feet wide. Its armament included 12 six-pounder guns and 12 swivel guns, and it carried 112 men, including 20 marines. The* Discovery, *built in 1774, displaced 300 tons. Stretching 91 feet and spanning 27 feet, the ship carried 8 six-pound guns and 70 seamen and marines. The essential components of sailing ships were made of wood—the masts, yards, decking, larboards, and cabins and compartments—and weakness or failure in any of them posed risks to crews and often the ships themselves. The hulls of both the* Discovery *and the* Resolution *were sheathed to protect against* teredo, *or shipworm. Long-distance voyages required experienced sailmakers, carpenters, and blacksmiths who were outfitted with requisite tools and had work space, a forge, and an assemblage of wood and metal materials to make repairs at sea. Major repairs required a harbor or good anchorage and a ready supply of timbers and wood.*

Journal of Captain Cook, April 4–7, 1778, at Ship Cove, Nootka Sound

We resumed our work in the afternoon, and, the next day, rigged the fore-mast; the head of which being rather too small for the cap, the carpenter went to work, to fix a piece on one side, to fill up the vacant space. In cutting into the mast-head for this purpose, and examining state of it, both cheeks were found to be so rotten, that there was no possibility of repairing them; and it became necessary to get the mast out, and to fix new ones upon it. It was evident, that one of the cheeks had been defective at the first, and that the unsound part had been cut out, and a piece put in, which had not only weakened the mast-head, but had, in a great measure, been the occasion of rotting every other part of both cheeks. Thus, when we were almost ready to put to sea, we had all our work to do over again; and, what was still more provoking, an additional repair was to be undertaken, which would require some time to be completed. But, as there was no remedy, we immediately set about it. It was fortunate for the voyage, that these defects were discovered, when we were in a place, where the materials requisite were to be procured. For, amongst the drift-wood, in the cove where the ships lay, were some small seasoned trees very fit for our purpose. One of these was pitched upon; and the carpenters began, without loss of time, to make out of it two new cheeks.

In the morning of the 7th, we got the fore-mast out, and hauled it ashore; and the carpenters of the ships were set to work upon it. Some parts of the lower standing rigging having been found to be very much decayed, as we had time now to put them in order, while the carpenters were repairing the fore-mast, I ordered a new set of main-rigging to be fitted, and a more perfect set of fore-rigging to be selected out of the best parts of the old. . . .

[April 8] According to the old proverb, Misfortunes seldom come single, the mizen was now the only mast on board the Resolution that remained rigged, with its top-mast up. The former was so defective, that it could not support the latter during the violence of the squalls, but gave way at the head under the rigging. About eight o'clock the gale abated; but the rain continued with very little intermission for several days; and, that the carpenters might be enabled to proceed in their labours, while it prevailed, a tent was erected over the fore-mast, where they could work with some degree of convenience. . . .

On the 11th, notwithstanding the rainy weather, the main-rigging was sixed and got over head; and our employment, the day after, was to take down the mizen-mast, the head of which proved to be so rotten, that it dropped off while in the slings. . . .

In the afternoon of the next day, I went into the woods with a party of our men, and cut down a tree for a mizen-mast. On the day following, it was brought to the place where the carpenters were employed upon the fore-mast. . . .

The fore-mast being, by this time, finished, we hauled it along-side; but the bad weather prevented our getting it in till the afternoon; and we set about rigging it with the greatest expedition, while the carpenters were going on with the mizen-mast on shore. They had made very considerable progress in it on the 16th; when they discovered, that the stick upon which they were at work was sprung, or wounded; owing, as supposed, to some accident in cutting it down. So that all their labour was thrown away; and it became necessary to get another tree out of the woods, which employed all hands above half a day.

During these various operations, several of the natives, who were about the ships, looked on with an expressive silent surprize, which we did not expect, from their general indifference and inattention.

Source: *A Voyage to the Pacific Ocean Undertaken, by the Command of His Majesty, for Making Discoveries in the Northern Hemisphere . . . Performed under the Direction of Captains Cook, Clerke, and Gore, in His Majesty's Ships the* Resolution *and* Discovery, 3 vols., published by Order of the Lords Commissioners of the Admiralty (Dublin, 1784), 2:275–278.

ANALYSIS

By the time Cook reached Nootka Sound, his ships had suffered from the storms they encountered as they sailed north up the coastline of North America. The refitting of the *Resolution* at Whitby yards had been slipshod—the vessel leaked even before clearing the English Channel—and significant repairs were mandatory by the time they reached Nootka. Cook selected a small harbor, Ship Cove (present-day Resolution Cove), on the ocean side of a large island that dominated the center of the sound, present-day Bligh Island. David Samwell described the anchorage on March 31, 1778: "This morning we warped the

Ship into a snug Cove and made her fast with Hawsers [stout lines] to the Trees on shore" (Beaglehole 1967, 3:1089). It was a relatively protected harbor, but more important, the dense forest near the shore included cedar, hemlock, and fir, good selections for timber to repair ships. In addition, there was ample space for an observatory, where Cook could establish position and his astronomers could make observations. Foremost in Cook's mind, however, was repair of the *Resolution*'s foremast and the replacement of the mizzenmast.

Cook's ships were full-rigged, which meant there were three masts and at least three sets of sails—course, topsail, and topgallant sail. The foremast was set closest to the bow; the mainmast, the tallest, rose from the midsection of the ship; and the mizzenmast, the shortest, stood aft of the mainmast. Each mast was composed of three sections: the base rested on the keel (or bottom deck of the ship), and the lower mast and topmast sections rose above the top deck. At Nootka, the *Resolution*'s foremast proved the most difficult. Several components had become rotten or, as Cook noted, had been defective from the time the ship left the Whitby shipyard. The "cap" is a fitting with two holes, used to fix together two sections of the mast. The "cheeks" are square blocks attached to the lower mast to support "trestle trees," or timber braces, that support the topmast.

More difficult was finding and shaping a proper timber for the mizzenmast. As Cook described, the mizzen nearly fell apart as the carpenters removed it, and Cook led a party to search for a stout fir to replace it. On April 14, however, as the carpenters prepared the new timber, they discovered that it had a pronounced split, evidently the result of wrestling it over some large rocks to the ship. The damage made it untrustworthy and forced a new search for a mast. It took Cook's crew more than 10 days to complete the repairs and make the *Resolution* seaworthy.

During the work, as David Samwell recorded in his journal, there was constant activity around the ship, with men brewing pine-needle beer and Native traders visiting daily to exchange furs for nearly any metal object. Cook noted that brass was a favorite of Native traders, "so that before we left the place, hardly a bit of brass was left in the Ship, except what was in the necessary instruments" (Beaglehole 1967, 3:302). Nonetheless, Cook kept his men diligently repairing both ships, while he and Captain Clerke surveyed the sound and visited villages. The *Resolution* and the *Discovery* left Nootka on April 26.

33. Buying Nature at Nootka Sound, 1778

INTRODUCTION

In two previous circumnavigations, James Cook had encountered many Native people in the Pacific. He had laid out his interactions in a narrative of discovery, and it was through the published accounts of his voyages that readers learned and saw the world through his eyes. In British lending libraries, the accounts of Cook's voyages easily topped the list of the most borrowed volumes in the late 18th and early 19th centuries. His account—and those of other explorers—made reality of the romantic idea of the noble savage, an image of people who had not been corrupted by modern civilization.

In his account of the third voyage, Cook portrayed himself and his crew as commonsense modern men who approached Native people with dignity and fairness, even though they did not hesitate to use force and violence in response to actions deemed dangerous or deceitful. In effect,

Cook offered an amendment to the noble savage prototype, describing Native people as capable of what he characterized as "savage" and "civilized" behavior. He offered detailed descriptions of Native groups, including observations and explanations of their responses to a range of environmental and social conditions.

Journal of Captain Cook, April 22, 1778, at Nootka Sound

Next morning, about eight o'clock, we were visited by a number of strangers, in twelve or fourteen canoes. They came into the cove from the Southward; and as soon as they had turned the point of it, they stopped, and lay drawn up in a body above half an hour, about two or three hundred yards from the ships. At first, we thought, that they were afraid to come nearer; but we were mistaken in this, and they were only preparing an introductory ceremony. On advancing toward the ships, they all stood up in their canoes, and began to sing. Some of their songs, in which the whole body joined, were in a slow, and others in quicker time; and they accompanied their notes with the most regular motions of their hands; or beating in concert, with their paddles, on the sides of the canoes; and making other very expressive gestures. At the end of each song, they remained silent a few seconds, and then began again, sometimes pronouncing the word *booee!* forcibly, as a chorus. After entertaining us with this specimen of their music, which we listened to with admiration, for above half an hour, they came along-side the ships, and bartered what they had to dispose of. Some of our old friends of the Sound, were now found to be amongst them; and they took the whole management of the traffic between us and the strangers, much to the advantage of the latter.

Our attendance on these visitors being finished, Captain Clerke and I went, in the forenoon, with two boats, to the village at the West point of the Sound. When I was there the day before, I had observed, that plenty of grass grew near it; and it was necessary to lay in a quantity of this, as food for the few goats and sheep which were still left on board. The inhabitants received us with the same demonstrations of friendship which I had experienced before; and the moment we landed, I ordered some of my people to begin their operation of cutting. I had not the least imagination, that the natives could make any objection to our furnishing ourselves with what seemed to be of no use to them, but was necessary for us. However, I was mistaken; for the moment that our men began to cut, some of the inhabitants interposed, and would not permit them to proceed, saying they must "*makook;*" that is, must first buy it. I was now in one of the houses; but as soon as I heard of this, I went to the field, where I found about a dozen of the natives, each of whom laid claim to some part of the grass that grew in this place. I bargained with them for it, and having completed the purchase, thought we were now at liberty to cut wherever we pleased. But here, again, it appeared, that I was under a mistake; for the liberal manner in which I had paid the first pretended proprietors, brought fresh demands upon me from others; so that there did not seem to be a single blade of grass, that had not a separate owner; and so many of them were to be satisfied, that I very soon emptied my pockets. When they found, that I really had nothing more to give, their importunities ceased, and we were permitted to cut wherever we pleased, and as much as we chose to carry away.

Here I must observe, that I have no where, in my several voyages, met with any uncivilized nation, or tribe, who had such strict notions of their having a right to the exclusive property of everything that their country produces, as the in habitants of this Sound. At first, they wanted our people to pay for the wood and water that they carried on board; and had I been upon the spot, when these demands were made, I should certainly have complied with them. Our workmen, in my absence, thought differently; for they took but little notice of such claims; and the natives, when they found that we were determined to pay nothing, at last ceased to apply. But they made a merit of necessity; and frequently afterward, took occasion to remind us, that they had given us wood and water out of friendship.

Source: *A Voyage to the Pacific Ocean Undertaken, by the Command of His Majesty, for Making Discoveries in the Northern Hemisphere . . . Performed under the Direction of Captains Cook, Clerke, and Gore, in His Majesty's Ships the* Resolution *and* Discovery, 3 vols., published by Order of the Lords Commissioners of the Admiralty (Dublin, 1784), 2:283–285.

ANALYSIS

Cook's encounters with Native groups on the Northwest Coast were distinct from his experiences in the South Pacific. More than their physical appearance, clothing, and houses—which stood in sharp contrast to appearances in the warm Pacific islands—Cook learned quickly that Northwest Indians had much different ways of dealing with conflict, addressing friends and enemies, and possessing land. Oration presaged most encounters among the Mowachaht at Nootka, for example. Cook recorded the canoes maneuvering near the *Resolution,* concluding that it was entertainment, with the seamen playing the role of spectators, but for the Mowachaht it was more a preparation for trading.

The Natives dealt with potential conflict in oratory as well. Ship's surgeon David Samwell described their behavior as they faced a potential enemy. As he described the scene, canoes full of angry-appearing armed men threatened other Natives onshore. An "Engagement of Tongues" ensued, with oratory and a corresponding reaction onshore whereby a "Man started up in a violent fury . . . [and] made a long Speech, sputtering his words out with the utmost Vehemence & throwing out Saliva from his Mouth. . . . In short his whole actions were those of a man out of his Senses. After continuing reviling his Enemies in this manner for some minutes or till he was tired he went back to his party & sat down among them as composed as if nothing had happened" (Beaglehole 1967, 3:1093). The incident, a product of a quarrel over theft of some goods, ended with no blows struck and an engagement in trade.

Cook's experience gathering fresh fodder for the ship's animals dramatically underscored the Natives' sense of ownership and approach to trade. The Mowachaht at Nootka had a practical and political concept of land and landownership. John Ledyard recorded that the Natives at Nootka made the British understand their control and possession of the area. "They intimated to us," Ledyard reported, "that the country all round further than we could see was theirs. Water and wood they charged us nothing for." When Cook tried to ensure that his crew could take wood, Ledyard recorded that "one of the Indians took him by the arm and thrust him from him, pointing the way for him to go about his business," which meant go ahead and collect. "Cook was struck with astonishment," Ledyard recorded, "and turning to his people with a smile mixed with admiration exclaimed, 'This is an American indeed!'" (Munford 1963, 70). The Mowachaht held property and possession dear but in ways that were surprising to the Europeans. As cultural anthropologist Peter Drucker explained, they "carried the concept of ownership to an incredible extreme. Not only rivers and fishing places close at hand, but the waters of the sea for miles offshore, the land, the houses, carvings on a house post, the right to marry in a certain way or the right to omit part of an ordinary marriage ceremony, names, songs, dances, medicines, and rituals, all were privately owned property" (qtd. in Clayton 2000, 101).

34. Native Art, 1778

INTRODUCTION

David Samwell, surgeon's mate on the Resolution, *had gone to Greenland at age 20 and had experience in maritime exploration before joining Cook on the third voyage. Samwell's nonsurgeon interests verged toward art, culture, and social conditions. He was a keen observer of Native life, and he recorded the details of local living conditions and people's skills. An outgoing Welshman, Samwell embraced seafaring life but mainly as preparation for life as a surgeon on land. "For my part," he wrote to a friend on boarding the* Resolution, *"I live as happy as I cou'd wish only that one's cut off from the Society of the Dear Girls" (Beaglehole 1967, 3:lxxxv). Like his crewmates, Samwell criticized Native men and women for their "dirtiness," but he confided to his journal that "it must be confessed we sometimes found some Jewels that rewarded our trouble, Namely two sparkling black Eyes accompanied with a beautiful Face, & when such was our fortune we never regretted the time & trouble" (Beaglehole 1967, 3:1100).*

This excerpt from Samwell's detailed journal includes his descriptions of Native implements of war, clothing, and manufacturing technologies. He took particular notice of their skill in rendering metal, notably fashioning knives and daggers, and he noted that they carried on trade with the Spanish for metal goods. It was the Native art, however, that evoked his praise and wonder.

Some Account of a Voyage to South Seas, Written by David Samwell, Surgeon of the *Discovery*, April 25, 1778

Their Canoes are very well constructed & made out of one Piece of Wood; they are of different sizes from 10 feet long to 40, being of a good breadth they require no outrigger and are altogether much like a Thames wherry only that they rise up a little more at the Head & stern; the large ones are painted on the outside with various Figures of Beasts and fish, & some have broad waving white Lines along the Gunwale much like the greenland Fishing Boats; they are by much the best Canoes we have seen this Voyage & many of them capable of holding 30 or 40 Men; their Paddles are long and narrow & terminate in a sharp point, which makes it probable that they use them as Weapons of offence as well as for navigating their Canoes. They know nothing of the use of sails having none among them. After their skill in ship building may be mentioned their Proficiency in the Art of Carving, in which they very much excel, there is hardly any thing in the Heavens above or the Earth below or in the Water under the Earth that they have not the Image of; their Masks of the Human face have been already mentioned, the Execution of which is much better than could be expected from a rude & savage race of men as these are.

They are likewise no mean Artists at making large Boxes in which they keep their Furs & other things, some of these are curiously ornamented; when they came to trade with us they always brought these in their Canoes full of Skins & other articles of Traffic; they appear so much like our Chests at first view that we took them for the work of Europeans, being made with Lid[s], one half of the Box is made to inclose the other. . . .

Their Food consists chiefly of Fish & whale Fat. . . . We bought much [whale] oil of them for our Lamps, which they brought to us in bladders & sold very cheap; however they more than once took us in by giving us a bladder full of water instead of oil, however we soon put a stop to this by always afterwards emptying the bladder before we paid them &

when they found us use this Caution they never attempted to impose upon us. They have large Combs made of wood & are mostly carved. We found Iron, Copper and Brass among them tho' in no great Quantities, they sold us two Silver Spoons of an old fashioned make which we judged to be spanish. . . . The Spaniards we know have been further to the Northward than this part of the Coast & it is not unlikely but they might have touched here; however this is far from being certain, as the Indians might have got the Spoons & the Iron by way of Traffic from other Tribes to the Southward, as the Coast is probably inhabited all the Way from this Sound to California.

Source: David Samwell, "Some Account of a Voyage to South Seas," 1778–1779, MS EG 2591, British Museum, London, accessible in UW Microform A1687, Suzzallo Library, University of Washington, Seattle.

ANALYSIS

All Europeans who encountered the Indians on the Northwest Coast commented on their artistic creations, from massive totems and house structures to domestic utensils, war clubs, and wooden boxes. Artisans worked primarily in wood, especially cedar. Nootka and Salish used western red cedar, while Haida and Tlingit in the north often used sitka spruce and western hemlock. The northern groups—the Kwakiutl, Haida, and Tlingit—and the southern peoples—Nuu-chah-nulth/Mowachaht and Salish—created different art forms, but a common motif used two-dimensional painted and incised images on flat surfaces. Many images were stylized representations of animals—for example, Eagle, Raven, Frog, Bear, Orca, Grizzly—that were connected to Native clans. Other images, often of supernaturally powerful animals and figures, played important roles in shamanistic rituals or dances. Samwell also saw three-dimensional sculpture forms in masks, house posts and totems, and as integral components of canoes. As Samwell noted, masks and headdresses often had stylized human faces.

The canoes on the Northwest Coast looked quite different from those seen in the South Pacific. Long, heavy, and flat-bottomed, the canoes easily handled waves, sea turbulence, and cargo. Their shape reminded Samwell of "wherrys," boats used on the Thames River for cargo and passenger travel that featured a long upturned prow. Wherry craft were clinker-sided boats, while Native canoes at Nootka were made from solid cedar logs with a raised bow and stern that featured carved and painted images. Nootkans made canoes from mature cedar, straight grained and without boles or any trace of rot. They could only find such trees—usually 150 to 200 feet tall—in mature stands. To fashion a large canoe, Nootkans used stone and shell adzes and hot rocks to hollow out interior space; stretcher braces forced the sides apart and helped shape the craft. The archaeological record documents canoe building in the region at least 3,500 years ago.

Cook collected art objects during his stay at Nootka Sound, and expedition artist John Webber sketched Mowachaht people, structures, and artistic creations. In fact, Webber's drawings may have communicated more to British readers about conditions on the Northwest Coast than Cook's journals did. Among the most impressive objects were the cedar boxes Samwell mentions in his journal. The boxes were storage units where Indians put treasured items, often ceremonial artwork. Artisans began with a cedar plank, which they notched or cut and then steamed to bend the plank into a box shape. They then sewed on a bottom, fitted a top cap piece, and adorned the top and sides with designs composed of integrated images, many of them abstract and representing clan identifications.

35. Looking for the Northwest Passage, 1778

INTRODUCTION

In mid-May 1778, the Resolution *and the* Discovery *entered Prince William Sound, a sprawling bay with prominent islands and inlets that penetrate the interior. "We had often tried to make the Indians understand," Lieutenant James King noted in his journal, "that we wanted to know if the Ships coud proceed to the N or nearly in that direction" (Beaglehole 1967, 3:1419). Those queries reflected Cook's continued pursuit of an opening that might signal a Northwest Passage. An inlet, Valdez Arm, extended east-northeast from the Sound and seemed promising, but investigations by Lieutenant John Gore convinced them that it was not a passage, even though they did not pursue it to its end. Native informants had told Gore that the way to the sea was south—the way they had come—or north, but not east. Cook recorded his decision on May 17: "I resolved to spend no more time in searching for a passage in a place that promised so little success. . . . We were now upwards of 520 leagues to the Westward of any part of Baffins or Hudsons bays, and what ever passage there may be it . . . must lie North of 72° latitude; who could expect to find a passage or strait of such extent?" (Beaglehole 1967, 3:353). With that conclusion, Cook sailed northwest and came upon a much more promising opening, the waterway now called Cook Inlet at latitude 60.33° north.*

Journal of Captain Cook, June 1, 1778, at Cook Inlet

At two o'clock next morning, being the 1st of June, the master returned, and reported that he found the inlet, or rather, river, contracted to the breadth of one league, by low land on each side, through which it took a Northerly direction. He proceeded three leagues through this narrow part, which he found navigable for the largest ships, being from twenty to seventeen fathoms deep. The least water, at a proper distance from the shore and shoals, was ten fathoms; and this was before he entered the narrow part. While the ebb or stream ran down, the water was perfectly fresh; but, after the flood made, it became brackish; and, toward high water, very much so, even as high up as he went. He landed upon an island, which lies between this branch and the Eastern one; and upon it saw some currant bushes, with the fruit already set; and some other fruit-trees and bushes, unknown to him. The soil appeared to be clay, mixed with sand. About three leagues beyond the extent of his search, or to the Northward of it, he observed there was another separation in the Eastern chain of mountains, through which he supposed the river took a North East direction; but it seemed rather more probable that this was only another branch, and that the main channel kept its Northern direction, between the two ridges or chains of mountains before mentioned. He found that these two ridges, as they extended to the North, inclined more and more to each other, but never appeared to close; nor was any elevated land seen between them, only low land, part woody, and part clear. All hopes of finding a passage were now given up. But as the ebb was almost spent, and we could not return against the flood, I thought I might as well take the advantage of the latter, to get a nearer view of the Eastern branch; and, by that means, finally to determine whether the low land on the East side of the river was an island, as we had supposed, or not. With this purpose in view, we weighed with the first of the flood, and, having a faint breeze at North East, stood over for the Eastern shore, with boats ahead, sounding. Our depth was from twelve to five fathoms; the bottom a hard gravel, though the water was exceedingly muddy. . . . But thinking that what the ships could not do, might be done with boats, I dispatched two, under the command of Lieutenant King, to examine the tides, and to make such observations as might give us some insight into the nature of the river. At ten o'clock, finding the ebb begun, I anchored in nine fathoms water, over a gravelly bottom. Observing the tide to be too strong for the boats to make head against it, I made a signal for them to return on board, before they had got half way to the entrance of the river they were sent to examine, which bore from us South 80° East, three leagues distant. . . . It was a satisfaction to me, however, to reflect, that, if I had not examined this very considerable inlet, it would have been assumed, by speculative fabricators of geography, as a fact, that it communicated with the sea to the North, or with Baffin's or Hudson's Bay to the East; and been marked, perhaps, on future maps of the world, with greater precision, and more certain signs of reality, than the invisible, because imaginary, Straits of de Fuca, and de Fonte. In the afternoon, I sent Mr. King again, with two armed boats, with orders to land, on the Northern point of the low land, on the South East side of the river; there to display the flag; and to take possession of the country and river, in his Majesty's name; and to bury in the ground a bottle, containing some pieces of English coin, of the year 1772, and a paper, on which were inscribed the names of our ships, and the date of our discovery.

Source: *A Voyage to the Pacific Ocean Undertaken, by the Command of His Majesty, for Making Discoveries in the Northern Hemisphere . . . Performed under the Direction of Captains Cook, Clerke, and Gore, in His Majesty's Ships the* Resolution *and* Discovery, 3 vols., published by Order of the Lords Commissioners of the Admiralty (Dublin, 1784), 2:394–397.

ANALYSIS

John Ledyard described Cook Inlet on May 27, 1778: "conjectured to be a vast river, having a strong southerly current—sounded 40 fathoms. This gave us hopes again of a Passage" (Munford 1963, 81). On May 31, a Sunday, Cook recorded that the *Resolution* lay at anchor in 16 fathoms of water some 12 miles from the west shore of Cook Inlet, looking east through an opening in the mountains that invited inspection—and potential as the Northwest Passage. He noted an ebb tide of water running three knots toward the sea, spilling freshwater over the deeper saltwater. In addition, Cook recorded that "we had many other & but too evident proofs of being in a great River; such as low shores, very thick and Muddy water, large trees and all manner of dirt and rubbish floating up and down with the tide" (Beaglehole 1967, 3:365).

Cook sent out two boats to explore the inlet, heading northeast. William Bligh, the *Resolution*'s master and an experienced surveyor and chartmaker, commanded the party. Taking soundings until they reached three fathoms, they returned with news that present-day Knik Arm was not a passage. This left a larger opening that headed directly east, present-day Turnagain Arm, which had promise even though its conformity suggested a river, not a continental passage. Nonetheless, Cook could not ignore the possibility, and it required investigation.

Cook's choice to command the second reconnoitering party exposed a minor but not unimportant conflict among his officers. To lead the Turnagain Arm survey, he selected 23-year-old Lieutenant James King, a brainy and well-educated man. King later would command the *Discovery* on the voyage home in 1780 and would coauthor the official account of the third voyage. He had been chosen for the voyage because of his training in astronomy at Oxford University, and he became close to Cook during the expedition.

Nonetheless, the lieutenant irritated some of the crew, especially Bligh, who later commented that King feigned illness whenever conditions grew rough. In short, Bligh thought that King was too refined, even genteel, to handle the pressures of seafaring.

King held no animosity toward Bligh, however, who many thought was more than brusque, even vulgar in his manner. Bligh had struggled in lower positions on shipboard in his teens before gaining the master's position on the *Resolution* at age 22. His temperament inclined him to perfectionism and a fierce defense of his rank and responsibilities. Bligh later complained that Cook had too often intruded into his work on navigation charts. Of the veterans of Cook's third voyage, Bligh had to wait the longest to gain the rank of captain, and his future included the disastrous mutiny on the *Bounty*, which he commanded in 1787–1789.

King's report on his survey of Turnagain Arm filled in some details about the waterway. "As all prospects of finding a passage to an open sea by this Inlet is now given over, & which I believe everyone is satisfy'd," he wrote on June 2, "to be only the Entrance of a large & possibly a very extensive river, the Captn before his return out of it sent me to the North point of the Eastern low land, there to take Possesion of all the Country, in the Name & for the use of his Majesty & his Successors" (Beaglehole 1967, 3:1421).

36. At the Bering Strait on the Edge of the Continent, 1778

INTRODUCTION

The three textual volumes and accompanying atlas of James Cook's A Voyage to the Pacific Ocean *included two copper-engraved small-scale folding maps and some 85 plates, views, and illustrations. Several were large-scale charts of harbors and discrete areas of ocean that Cook had surveyed, including this chart of Bering Strait. Many scholars believe that it was the most significant of all of Cook's published maps, because it was the first to depict so accurately the relationship of Asia and America. A smaller-scale version of this region of the North Pacific appeared on the larger chart, which also included the most northern tracks of Cook's ships. Curiously, the name Bherings Strait on the smaller-scale map is clearly placed between the two continents, whereas that toponym appears only in the title on the larger-scale version. Nonetheless, it is likely that this was the first separately published European map of this area of the Pacific, a map that a contemporary audience, mindful of the extent of Cook's surveying, would have quickly accepted, including Cook's choice of place-names. Cartographers surely felt that way, because many included Cook's name, Bering's Strait, on their subsequent maps of the North Pacific.*

ANALYSIS

The "Chart of Norton Sound and of Bherings Strait" was neatly framed by a border conspicuously marked out in degrees of longitude (measured east of Greenwich) and latitude, a testimony to the scientific precision of calculating position with advanced instruments. The coasts of Asia and America are separated by the strait, which is marked and crisscrossed by the tracks of the two ships, the *Resolution* and the *Discovery*, but at different times. The dotted track lines represent Cook's command of the expedition in these waters, from

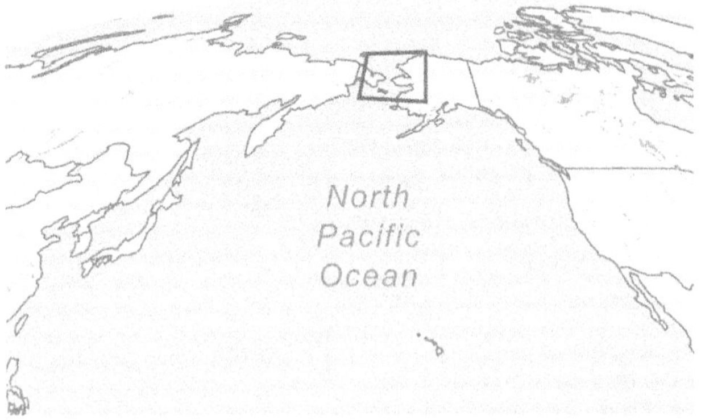

approximately early August 1778 to late October, when he sailed for the Hawaiian Islands. The continuous track lines were those of the two ships, now commanded by Captain Charles Clerke, who took command after Cook's death in February 1779. In July 1779, Clerke again attempted to negotiate the northern reaches of Bering Strait but failed and returned south in August. The engravers of this chart carefully entered as many data points along the ships' tracks as possible, including dates, sounding depths in fathoms, and compass variations.

Anderson's Island, a solitary island in the lower middle part of the chart, was named for William Anderson, the surgeon on the *Resolution* who had died and been buried at sea on August 3, 1778. The island appeared on the large chart in the atlas as well, but apparently no such island was recorded on the manuscript charts that served as the templates for the published engraved charts. Later William Bligh, a skilled draftsman, lamented the inclusion of the nonexistent Anderson's Island on the published maps: "Here is a gross mistake for Anderson's Id

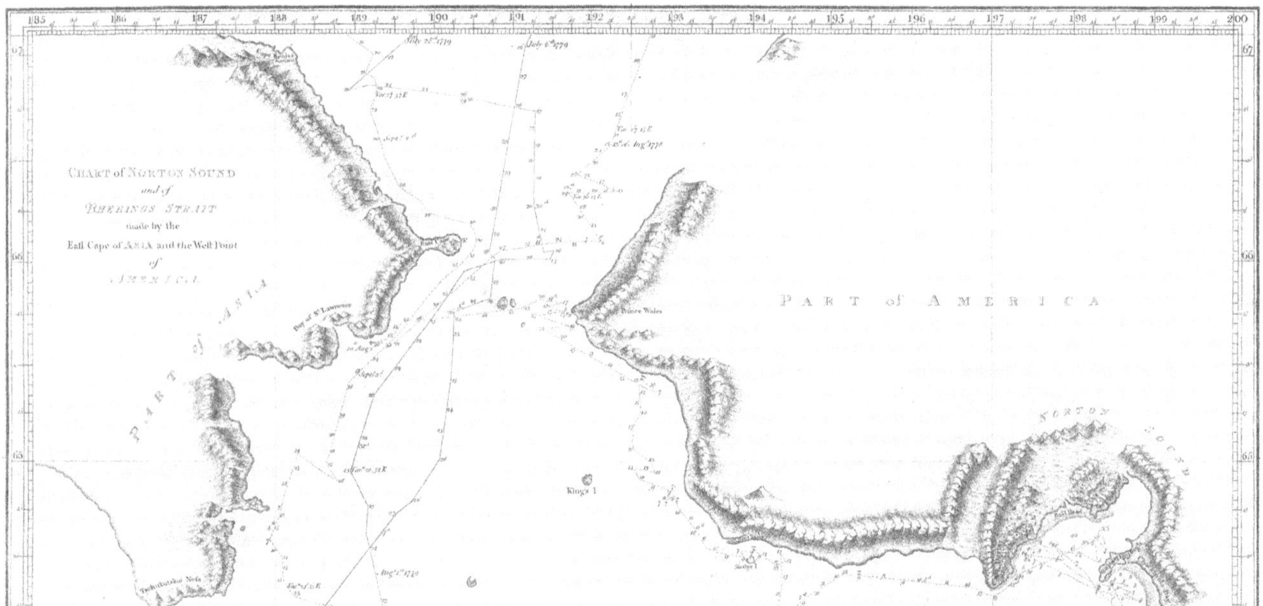

"Chart of Norton Sound and of Bhering's Strait Made by the East Cape of Asia and the West Point of America." (Copyright © 2000 by Cartography Associates)

& the East end of Clerk's Id is one and the same land & how they have blundered to lay them down as two I cannot conceive" (David 1988, liv).

In mid-August, Cook reached and named the westernmost point of North America, Cape Prince of Wales. Storms had caused the ships to cross the narrow 50-mile-wide Bering Strait to their first contact and landing on Asian shores at the Bay of St. Lawrence. On August 11, 1778, as the ships returned to the midpoint of Bering Strait on a clear day, Cook saw East Cape, the most eastern point of Siberia, and the American coast in the distance. For the first time, a European saw the geographical relationship between the continents.

In early September after returning from a disappointing attempt to penetrate the ice barrier at 70°45′ north, Cook's ships passed by the vicinity of present-day Nome, Alaska, and arrived in Norton Sound, which Cook named in honor of Sir Fletcher Norton, Speaker of the House of Commons. They charted the sound, named several more features, and departed after less than two weeks in the area to return to the Aleutians and then on to winter in the Hawaiian Islands. En route, Cook sighted another island or islands in the southern reaches of the strait and applied the name Clerke's Islands (present-day St. Lawrence Island) for his fellow officer Charles Clerke. Cook's tracks after this sighting extended beyond this map.

We can appreciate Captain Cook's legacy through the permanency of many of the place-names on this map as well as on the smaller-scale map of the Pacific Coast of North America. Among the many posthumous accolades Cook received was one from Captain James King, the author of the third volume of Cook's *Voyages:* "He . . . ascertained the proximity of the two great continents of Asia and America; passed the straits between them, and surveyed the coast, on each side, to such a height of Northern latitude, as to demonstrate the impracticability of a passage in that hemisphere, from the Atlantic into the Pacific Ocean, either by an Eastern or Western course" (Lada-Mocarski 1969, 142).

37. Meeting the Russians, 1778

INTRODUCTION

Cook's investigation of the Northwest Coast and the potential Northwest Passage ended as the Resolution *and the* Discovery *sailed southwest along the Aleutian Islands. Russia, Kamchatka, and the undetermined coastline of the far western edge of North America became Cook's prime focus. He had researched Asian North Pacific regions in preparing for the third voyage, including a 1729 chart of Vitus Bering's travels from the first Kamchatka expedition, a history of Kamchatka written in 1764, and a 1774 map of Kamchatka by Russian geographer Yakov Stählin. In addition, Cook had a copy of a map of the Northwest Coast of North America drawn by cartographer Gerhard Müller, based on the discoveries of Bering's second Kamchatka expedition. Importantly, the Stählin and Müller maps disagreed on their depictions of North America, presenting Cook with an opportunity to confirm or deny either of their renditions of the relationships between Asia and North America in the North Pacific. Once Cook entered the Gulf of Alaska, he used Müller's charts and noted each landmark that Bering had seen in 1741, from Mt. St. Elias to Unalaska Island in the Aleutians. Cook sailed north to explore Bristol Bay and Norton Sound and to investigate Bering Strait, which he named. Blocked by pack ice north of the strait, he had to turn back south and return to Unalaska, anchoring at Samgoonoodha on the island's north coast on October 2, 1778.*

John Ledyard Journal, October 1778, at Unalaska

It was beginning to be dark when the canoe came to us. It was a skin canoe after the Equimaux plan with two holes to accommodate two setters. The Indians that came in the canoe talked a little with my two guides, and then came to me and desired I would get into the canoe . . . and went head foremost very swift through the water about an hour, when I felt the canoe strike a beach, and afterwards lifted up and carried some distance, and then set down again. . . . I was drawn out by the shoulders by three or four men, for it was now so dark I could not tell who they were, though I was conscious I heard a language that was new. . . . [I] concluded from their appearance they were Russians, which I soon after found to be true. As we entered the hut which was particularly long I saw arranged on each side a platform of plank a number of Indians, who all bowed to me, and as I advanced to the farther end of the hut the arrangement was composed of other Russians. . . . I had a change of garments brought me, consisting of a blue silk shirt and drawers, a furr-cap, boots and gown. . . . [A]ll the Russians in the house set down round me, and the bottles of spirits, tobacco, snuff, and whatever Perpheela [Ledyard's Native guide] had was brought . . . ; these I presented to the company, intimating that they were presents from Commodore Cook, who was an Englishman. One of the company then gave me to understand, that all the white people I saw there were subjects of the Empress Catherine of Russia, and rose and kissed my hand the rest uncovering their heads. I then informed them as well as I could that Commodore Cook want to see some of them, and had sent me there to conduct them to our ships. . . . After I had lain down the Russians assembled the Indians in a very silent manner, and said prayers after the manner of the Greek Church, which is much like the Roman. . . .

The number of Russians were about 30, and they had with them about 70 Kamchadales, or Indians from Kamchatka, these with some of the American Indians whom they had entered into friendship with occupied the village, enjoyed every benefit in common with the Russians, and were converts to their Religion. . . .

The next day I set off. . . . We embarked at the village in a large skin boat much like our large whale-boats rowing with 12 oars . . . [and] we arrived by sundown at the bay where the ships lay, and before dark I got on board with our new acquaintance. The satisfaction this discovery gave Cook, and the honor that redounded to me may be easily imagined, and the several conjectures respecting the appearance of a foreign intercourse rectified and confirmed.

Source: John Ledyard, *Travels and Adventures of John Ledyard,* 2nd ed. (London: Henry Colburn, 1834), 111–118.

ANALYSIS

Cook's maps of the Alaskan coastline and the Bering Strait represented an important expansion of geographic knowledge. For the first time, European cartographers could include accurate representations of the relationships between two great continents and the entrance to the Arctic from the Pacific. Cook's expeditionary instructions, however, specified how he should proceed if he encountered "Subjects of any European Prince or State" in his North Pacific investigations. If he met Europeans, the instructions ordered, he was "not to disturb them or give them any just cause of offence, but on the contrary to treat them with civility & friendship" (Beaglehole 1967, 3:ccxxi).

At Samgoonoodha, Cook tended to his battered ships, repairing the sheathing and stopping leaks, and he discovered that Russians had established a fur-trading post on the opposite side of the island. An "Indian named *Derramoushk*" offered Cook "a rye loaf or rather a pie made in the form of a loaf," a present from "some Russians in our Neighbourhood." Cook dispatched "Corpl Ledyard of the Marines, an inteligent man[,] in order to gain some further information, with orders, if he met with any Russians, or others, to endeavour to make them understand that we were English, Friends and Allies" (Beaglehole 1967, 3:448–449).

The romantic son of a sea captain, Ledyard was born in Connecticut in 1751 and had been a student at Dartmouth College before he fled what he considered a restrictive life to pursue adventure travel in 1773. By 1774 he was in Gibraltar, where he enlisted in the British Army, only to return to America in 1775 as part of the occupation of Boston. He chafed at his role in an occupying force and joined the British Royal Navy as a marine, which led to his posting on the *Resolution* in 1776. Throughout the third voyage Ledyard volunteered several times to accompany Natives, so his willingness to accompany Indians to the Russian outpost was not unusual.

Ledyard returned to the *Resolution* on October 10 with three Russians: Yakov Ivanov Sapozhnikov, chief of the Russian factory on the island of Umnak, west of Unalaska; Gerasim Gregoriev Izmailov, a navigator; and Peter Natrubin, a translator from one of the Russian vessels. Cook quizzed the Russians on their trade and their relations with Natives and discovered that the Unalaska post, founded in 1772–1774, was the earliest Russian permanent settlement in North America. He learned that 75 Russians lived on Unalaska and 97 lived on Umnak, with upwards of 500 Russians and Kamchatkans spread out on other islands in the Aleutians. Izmailov let Cook copy Russian charts of Kamchatka and the Sea of Okhotsk, which discredited the Stählin map and verified Müller's rendition of Bering's discoveries. Cook credited Bering and honored him by naming the strait that separated the two continents for him. Cook also learned from Sapozhnikov that the harbor identified in his instructions as his destination in Kamchatka, Petropavlosk, had few or no fresh provisions, so he decided to return to Hawaii for the winter.

38. Death of Cook, 1779

INTRODUCTION

The death of James Cook on February 14, 1779, on the beach at Kealakekua Bay on the east coast of the island of Hawaii dramatically changed the course of the expedition. He had repaired to Hawaii in January 1779 to stay for several weeks before returning to the North Pacific. On arrival he went ashore, where Natives honored him, and as John Ledyard recorded, "the multitude on the beach fell prostrate with their faces to the ground" (Munford 1963, 105). Relations between the British and Hawaiians became close, with some chiefs spending days aboard the ships.

On leaving Hawaii in early February, however, Cook had to abruptly return because heavy winds damaged the foremast and required repair. He anchored again in Kealakekua Bay, but this time his reception was cool. The mood had changed by February 14, when Natives stole the Resolution*'s cutter boat, a transgression that Cook determined to remedy. With Lieutenant*

Molesworth Phillips and nine marines, Cook took two boats to shore, intending to bring King Kariopoo (Kalei'oup'u) to the Resolution *and work out the boat's return. Cook went to Kaawaloa village to get Kariopoo while the marines stayed on the beach, some 30 yards from the bay. According to David Samwell, they were "entirely surrounded by the Indians to the Number of two or three Thousand, which in the present hostile appearance was by no means an eligible Situation" (Beaglehole 1967, 3:1196). Samwell's journal takes up the events of February 14.*

Some Account of a Voyage to South Seas, written by David Samwell, Surgeon of the Discovery, February 14, 1779

The Indians opened and made a Lane for the marines to pass. . . . Captain Cook followed them having hold of Kariopoo's Hand who came with him very willingly . . . accompanied by his two Sons. . . . When they had come near the Beach an old Woman came crying to the King & throwing her arms round his neck. . . . They wou'd not suffer him to go any further, telling him that if he went on board the Ship he would be killed. . . . A Chief well known to us whose name is Coho was seen skulking about them concealing a long iron dagger under his Cloak. . . . [T]he Man still closing upon them the Officer struck him with the butt end of his Musket. . . . Captn Cook seeing the Indians resolute & armed all round, said 'if we take him [the chief] by force I see we must kill a great number of them,' and as he was pausing . . . an Indian was seen threatening to hit him with a Stone, upon which he fired at him with Small shot with which one of the barrels of his double Piece was loaded; but as the man had a thick matt before him He received no hurt. . . . All Thoughts of getting Kariopoo on board were now given up & Captn Cook acted entirely on the defensive. . . . The Indians were now grown very daring & troublesome, they threw several Stones at our people. . . . One was seen . . . threatening to dart his Spear at Captain Cook, who on being shewn the man shot at him, but it seems killed another who stood by him; the Serjeant told him that he had shot the wrong Person upon which he ordered him [the sergeant] to shoot the right one, which he did and killed him. . . . [A] Volley of Stones now came among our People on which the Marines gave a general fire & left themselves without a Reserve, this was instantly followed by a fire from the Boats [the lieutenant's launch], on which Capt. Cook expressed his Astonishment. . . . [T]he Indians immediately rushed down upon them, dragged those [who] could not swim upon the Rocks where they dashed their brains out. . . . [T]he Stones flew as thick as hail which knocked the Lieut. [Phillips] down & as he was rising a fellow stuck him in the back with a Spear, however he recovered himself shot the Indian dead and escaped into the Water. Captain Cook was now the only Man on the Rock, he was seen walking down towards the Pinnace, holding his left hand against the Back of his head to guard it. . . . An Indian came running behind him. . . . [T]hen taking him unaware he sprung to him, knocked him on the back of his head with a large Club. . . . [T]he blow made Captain Cook stagger. . . . [A]nother Indian came running to him & before he could recover himself . . . drew out an iron Dagger . . . & stuck it with all his force into the back of his Neck, which made Capt. Cook tumble into the Water in a kind of a bite by the side of the rock . . . ; here he was followed by a croud of people who endeavoured to keep him under water. . . . [N]ot being able to swim he endeavoured to scramble on the Rock, when a fellow gave him a blow on the head with a large Club and he was seen alive no more.

Source: David Samwell, "Some Account of a Voyage to South Seas," 1778–1779, MS EG 2591, British Museum, London, accessible in UW Microform A1687, Suzzallo Library, University of Washington, Seattle.

ANALYSIS

Cook's death changed the third voyage by robbing the expedition of its commander and then saddling the crew, the Admiralty, and eventually all interested in Cook with an unresolved description of the events on the beach at Kealakekua Bay. First, the accounts from crewmen and officers told different stories, with some renditions coming from men who could not have witnessed the death. Second, the events exacerbated the existing divisions among officers and crew, which extended for years. Finally, and to make matters worse, several crewmen wanted Lieutenant John Williamson, who was in charge of one of the boats, investigated for failure of duty for not going to Cook's aid during the fatal moments on the beach. They pressed Charles Clerke to find evidence of Williamson's guilt, but he could get no single verifiable account.

The official account, written by James King, cast Cook as a martyred Enlightenment hero, retaining his image as a master of seamanship and an honest representative of Britain to the world, especially to Native people. But that view was not held by all. William Bligh, for example, called King's account "a most infamous lie" (qtd. in Ashley 2007, 114) and complained that King had eliminated reference to Cook's aggressive behavior during the encounter on the beach. Further, several crew reported that Cook had initially reacted strongly and vengefully when he went ashore to secure return of the cutter and that he had intended to make Kariopoo a hostage on board the *Resolution.* Once he faced Kariopoo, however, Cook's mood had softened, and he had become enraptured, as crewman Heinrich Zimmerman described it, "by an infatuation that is altogether unaccountable [and] trifled away his time" (qtd. in Salmond 2003, 425). King's account ignored contrary viewpoints and created favored officers, who would fare better in their naval careers than those who were not favored. Lieutenant John Rickman, who likely ordered the firing on Natives, which probably precipitated the attack on Cook, was the only officer to never receive a promotion after the expedition.

Captain Clerke, now commander of the voyage and captain of the *Resolution,* considered destroying the village with the ship's big guns, but instead he sent King ashore to negotiate the release of Cook's body. King, now captain of the *Discovery,* "talk'd with different People," Clerke recorded, from whom "he clearly understood . . . that the Body was cut up" (Beaglehole 1967, 3:541). On February 16, a local man came on board the *Resolution* with a part of Cook's thigh. Two days later Clerke had some village houses burned down, as some Hawaiians continued to throw stones while members of the crew gathered wood and water, even though Native priests consoled the British with gifts of "2 hogs and a large quantity of roots" that they said were "a present from Torre'oboo [Kariopoo]." On February 21, Clerke received some of Cook's bones, which he "committed to the deep" on the following evening (Beaglehole 1967, 3:546–548). By March 1, the *Resolution* and the *Discovery* had left the islands behind, following Cook's plan to return to the North Pacific.

The crew carried the events at Kealakekua Bay with them all the way back to Britain in 1780. "Everyone on the ships was stricken dumb, crushed, and felt as though he had lost his father," crew member Heinrich Zimmerman wrote in his journal (qtd. in Salmond 2003, 424). Cook's death also lived on in Hawaii, where his memory was kept alive in honor. The fur trader John Meares, on visiting Hawaii in 1787, remarked on local honorific references to Cook. Peter Puget, an officer on George Vancouver's *Discovery,* recorded in 1793 that Hawaiians held Cook in high veneration, and Edward Bell, also on Vancouver's expedition, reported that Hawaiians called Cook "Orono [Lono], or Great God" (qtd. in Sahlins

1995, 87). British mariners erected a monument in 1825, a staff with a plaque, and a second one in 1845, a plaque on the stump of a coconut palm.

The longer image of Cook's death has become the property of historians and anthropologists, who have disagreed on why events transpired as they did in February 1779 and what Cook's death can tell us about interactions between the mariners and Natives. There are no certainties in their interpretations except that there are no easy explanations and, further, that relationships between 18th-century European explorers and Pacific peoples were complex and cannot be rendered simply as meetings of humans incapable of mutual understanding.

39. Cook's Great Chart, 1784

INTRODUCTION

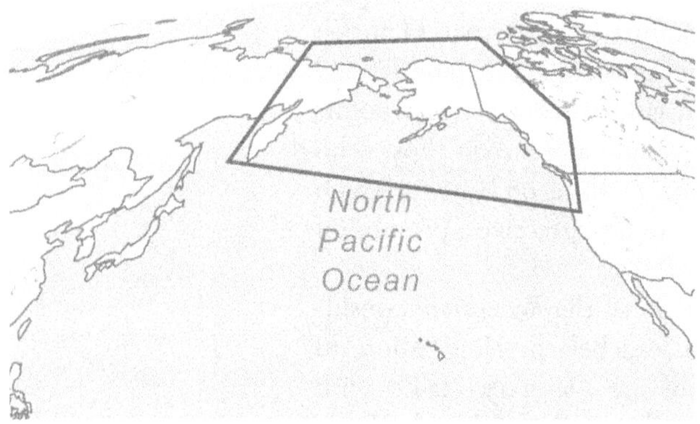

In December 1779 as the returning ships from James Cook's third voyage entered Macao, all journals kept by officers and crews were taken on prior orders of the Admiralty, apparently to prevent dissemination of information until the official account was published. Nevertheless, within a year after the expedition returned to England in August 1780 several anonymous accounts of the voyage were published, and a few maps appeared with fragmentary geographical information from the voyage along the Northwest Coast. In 1784 the Lords Commissioners of the Admiralty ordered that an official account by Cook and James King be published, accompanied by an atlas volume. This first edition, which included 85 plates, views, and illustrations and two maps, sold out within three days. Popular response to Cook's A Voyage to the Pacific Ocean *remained robust, and several editions followed, including ones published in London and Dublin the same year and another in London in 1785. Subsequently, many European and American mapmakers incorporated onto their maps information from Cook's charts on the topography of the Pacific Northwest. Until the 1798 publication of the account of George Vancouver's voyage, with its more detailed maps, Cook's map was the definitive rendition of the entire littoral of the Pacific Northwest, from latitudes ~43° to 70° north.*

ANALYSIS

As indicated by the locator map, this small-scale chart covers a very large area, and place-names are difficult to read without magnification. Nonetheless, the chart makes us appreciate the expanse of ocean that Cook's ships covered during their reconnaissance. The tracks of the expedition begin at the lower right corner of the map at approximately 43°, just above the current state line of Oregon and California. The tracks extend northwestward to reach 70°45′ before turning to Asian shores and then in a roughly southwesterly course. Notably, Cook carefully shaded coastal areas that he directly observed, and he filled in coastal areas that he

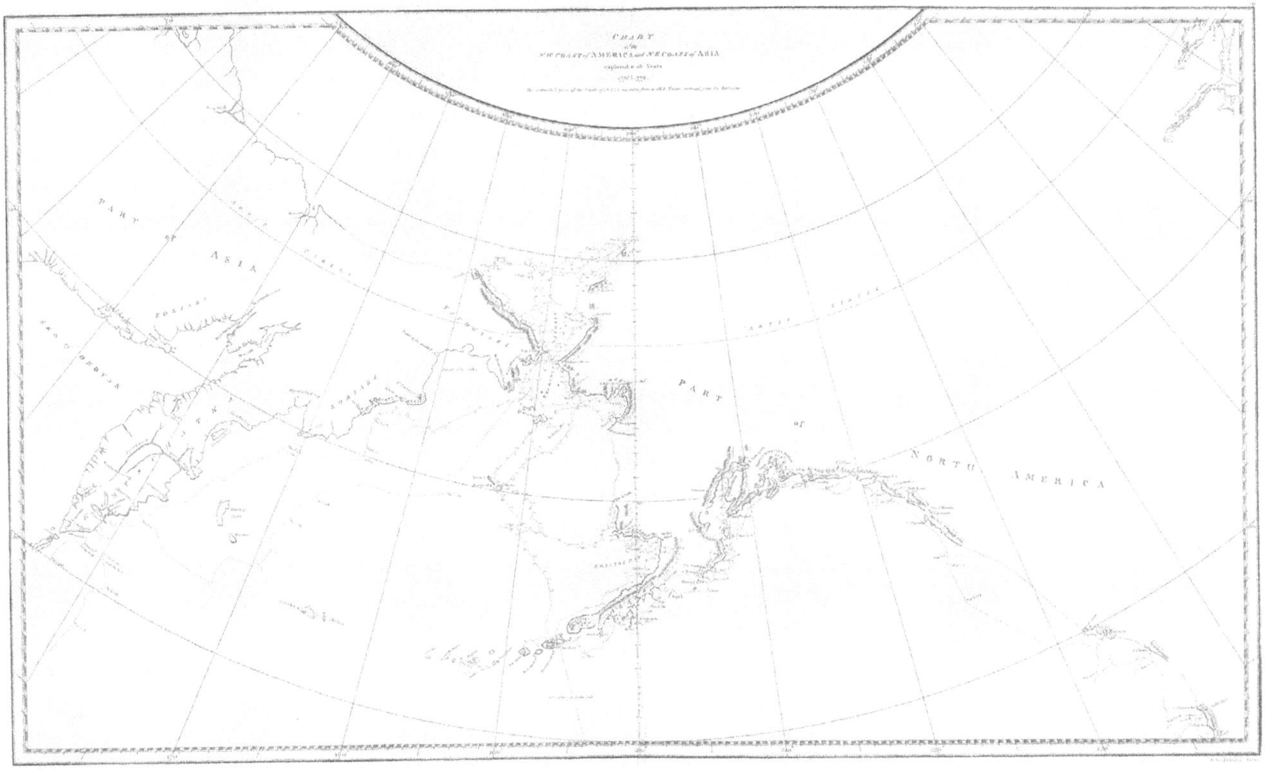

"Chart of the NW Coast of America and NE Coast of Asia explored in the Years 1778 & 1779." (Copyright © 2000 by Cartography Associates)

did not actually see, leaving them unshaded. For a great part of the Asian coast, Cook used (and credited) manuscript maps that the commander of a Russian settlement shared with him in October 1788 during a brief stop on Unalaska Island in the Aleutian Islands.

The unshaded coastal areas on the North American coasts are particularly relevant. As indicated by the ships' tracks, poor weather conditions required Cook to sail farther out to sea and out of sight of land along extensive stretches from ~45° to 48° and then again from ~50° to 56°. Consequently, he missed seeing the entrance to the Columbia River at ~46°27′. Farther north at ~48°, on March 22 Cook observed a cape that he named Flattery because he anticipated that he would soon come upon the legendary Strait of Juan de Fuca, but he passed the entrance in darkness. These failed sightings of potential major waterways into the interior of the continent dismayed geographers who believed in the likelihood of such a passage (Hough 1979, 30). Subsequent observations by fur-trading expeditions, however, rekindled hopes that Cook had missed the entrance. In addition, the unshaded areas on Cook's chart suggested to some fur-trading captains that new, as yet undiscovered, harbors existed along the coast from ~50° north, where they might encounter Native settlements with furs to trade.

Cook left England in July 1776 and first sighted the shores of the Pacific Northwest on March 7, 1778, at ~44°55′, noted on his map in the lower right corner as "7th March 1778 made the Coast." By August 18, he had reached his most northern latitude near Icy Cape. After negotiating much of the Bering Strait and returning to the Aleutians, he attempted to depart for the Hawaiian Islands in late October. After nearly 29 weeks, Cook had sailed along roughly 3,500 miles of coastline. Although his great map was replaced within 20 years, his legacy was in the place-names he applied to a great many locations along this extensive course.

Starting in March in latitudes from ~43° to ~45°, Cook's first-named site was Cape Foulweather, a testimony to prevailing weather conditions. Cape Gregory (present-day Cape Arago) was named close to the day that honors St. Gregory. Cook also named Cape Perpetua after a saint. North of Cape Flattery, the expedition entered a harbor that Cook christened King George's Sound, although the Admiralty apparently changed the name to Nootka before this map was drawn. Cook was not aware that he was anchored on the southwest side of an island (Vancouver Island), but it was here that his crew initiated trade with local inhabitants for sea otter skins that subsequently proved to be an exceedingly valuable commodity.

During May, Cook sailed closely along the coast from ~57° to the region of Cook's Inlet, and many of his place-names survive today (much of this information on Cook's names is from Wagner 1968). Both Cape Edgecumbe and Mt. Edgecumbe may have been derived from a mountain of the same name in Plymouth Harbor, England. Cook named Cross Sound after Holy Cross Day on the church calendar, and Cape and Mt. Fairweather likely reflected current sailing conditions. Cape Suckling and Comptrollers Bay were named after Maurice Suckling, the comptroller of the navy.

On May 12 Cook landed on Kaye's Island (now Kayak Island), where he buried a bottle containing silver coins given by the ship's chaplain, Dr. Kaye. Cook named Prince William Sound after that regent, and the name Cooks River was applied to the map in Cook's honor by the publisher of the journal. When Cook entered this body of water in late May he did not recognize it as an inlet, and George Vancouver later changed the name to its current form, Cook's Inlet. Many of these locations were sighted and given names by Spanish explorers, but adaptation of Cook's place-names and common usage by fur traders, George Vancouver, and British commercial cartographers ensured that his names survived.

Further Reading

Ashley, Scott. "How Navigators Think: The Death of Captain Cook Revisited." *Past & Present* 194 (February 2007): 107–137.

Beaglehole, J. C., ed. *The Journals of Captain James Cook on His Voyages of Discovery.* Vol. 3, *The Voyage of the* Resolution *and* Discovery, *1776–1780.* 3 vols. Cambridge: Cambridge University Press, 1967.

Clayton, Daniel. *Islands of Truth: The Imperial Fashioning of Vancouver Island.* Vancouver: University of British Columbia Press, 2000.

David, Andrew, ed. *The Charts and Coastal Views of Captain Cook's Voyages,* Vol. 1, *The Voyage of the* Endeavour, *1768–1771.* Canberra: Australian Academy of the Humanities, 1988.

Fry, Howard T. *Alexander Dalrymple and the Expansion of British Trade.* London: Frank Cass, 1970.

Hough, Richard A. *The Last Voyage of Captain James Cook.* New York: William Morrow, 1979.

Lada-Mocarski, Valerian. *Bibliography of Books on Alaska Published before 1868.* New Haven, CT: Yale University Press, 1969.

Ledyard, John. *Travels and Adventures of John Ledyard.* 2nd ed. London: Henry Colburn, 1834.

Munford, James Kenneth, ed. *John Ledyard's Journal of Captain Cook's Last Voyage.* Corvallis: Oregon State University Press, 1963.

Norris, John. "The Strait of Anian and British Northwest America: Cook's Third Voyage in Perspective." *BC Studies,* no. 36 (Winter 1977–1978): 1–22.

O'Brian, Patrick. *Joseph Banks: A Life.* Chicago: University of Chicago Press, 1987.
Sahlins, Marshall David. *How "Natives" Think: About Captain Cook, for Example.* Chicago: University of Chicago Press, 1995. ACLS Humanities e-book edition.
Salmond, Anne. *The Trial of the Cannibal Dog: The Remarkable Story of Captain Cook's Encounters in the South Seas.* New Haven, CT: Yale University Press, 2003.
Samwell, David. "Some Account of a Voyage to South Seas." 1778–1779, MSS EG 2591. British Museum, London. Accessible in UW Microform A1687, Suzzallo Library, University of Washington, Seattle.
Wagner, Henry R. *The Cartography of the Northwest Coast of America to the Year 1800.* Amsterdam: N. Israel, 1968.
Williams, Glyn. *Voyages of Delusion: The Quest for the Northwest Passage.* New Haven, CT: Yale University Press, 2002.

Expeditions of George Dixon, John Kendrick, and John Meares

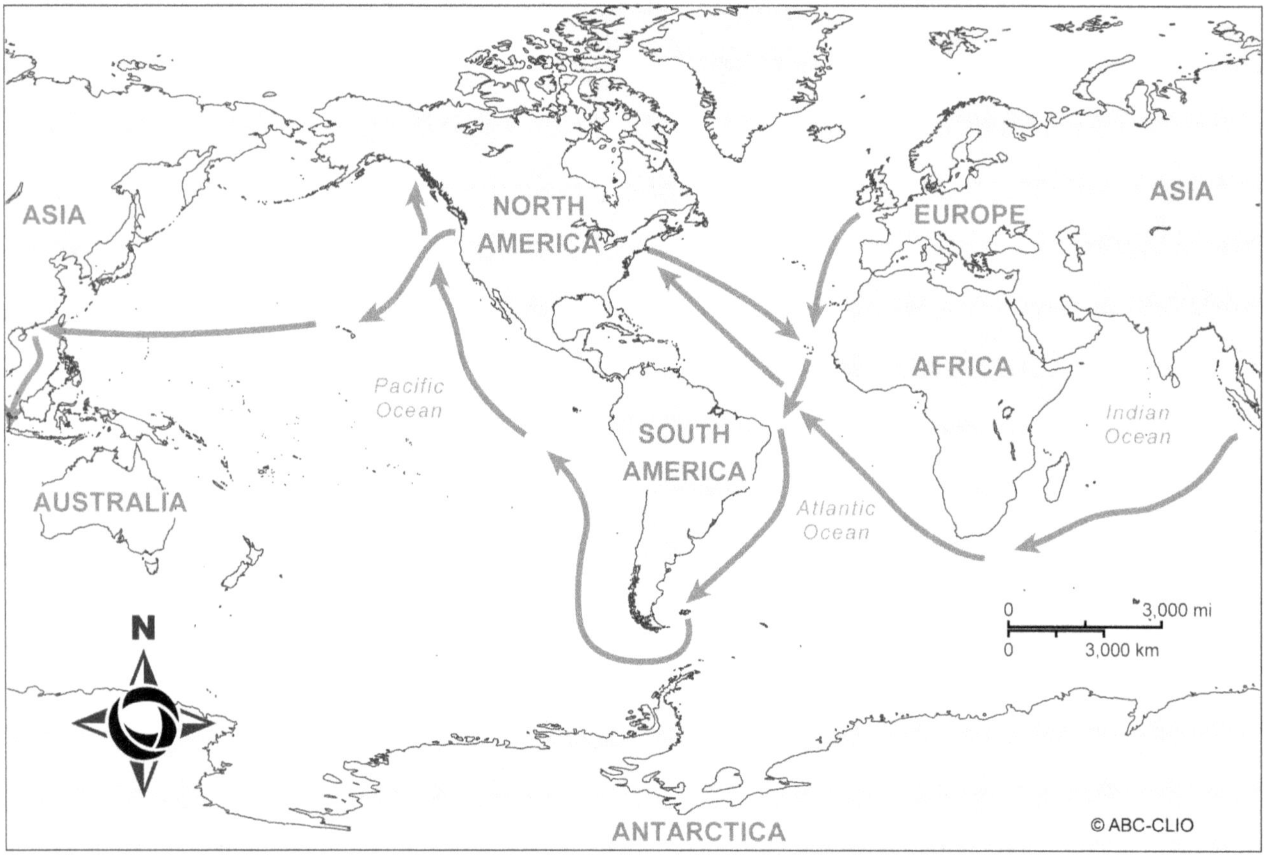

CHAPTER 5

Fur Traders, Commerce, and Cartography, 1785–1792

During the 18th century, fur traders became aware that northern sea otters thrived along the North Pacific Coast, from California to the northern extent of the Bering Strait between Asia and America. Active trade for these and other pelagic fur-bearing animals began with the Russians after the second Kamchatka expedition in 1733–1743. By the time the British entered the maritime trade in the 1780s, Russian traders had spent over 40 years extending their reach to the Aleutian Islands, Kodiak Island, and Prince William Sound in present-day Alaska. This commercial expansion meant that the Russians understood the geography of an extensive area of the Pacific Northwest before British fur traders explored the area and before any of the 18th-century circumnavigations by Spanish, French, British, and American explorers and traders.

Merchants from England, America, Spain, and other countries entered the market later, which contributed significantly to geographical knowledge and also created geopolitical conflict in the Pacific Northwest. In late 1780, with the return of the ships on Captain James Cook's third voyage, the news arrived in Europe of the great profit that might be realized from the sale of sea otter furs to China and Japan. Cook's crews had obtained nearly 1,500 skins, principally from the area of Friendly Cove in Nootka Sound, and crew members had sold many in Canton for substantial profit. Word of mouth and early unofficial published accounts of the voyage incited interest in sea otter furs, and the official account of Cook's voyage, published in 1784, included a scheme for exploring and trading in the Pacific Northwest. By 1785 merchants in London, Bombay, Calcutta, and Canton had capitalized ventures and employed experienced mariners to engage in the maritime fur trade.

In 1785, Englishman James Hanna became the first trader on the coast. During a short trading season, he realized substantial profit by selling over 550 skins for more than 20,000 Spanish dollars. British trading vessels descended on the Northwest Coast in 1786. For many years British and American traders explored an enormous area. Their investigations included Cook's Inlet in the western part of the Gulf of Alaska, the southeast coast of Alaska, the mainland of British Columbia and Vancouver Island, and the Strait of Juan de Fuca—a vast region stretching across roughly 13 degrees of latitude and 28 degrees of longitude.

The dense, dark, and glossy fur of the sea otter (*Enhydra lutris*) was the most valuable fur on the international market. Coastal Natives also valued otter coats. Only chiefs and highly esteemed members of Nootkan groups wore prestgious and beautiful coats of sea otter fur. (Porntep Kawjun/Dreamstime.com)

A few statistics illustrate the extent of this activity (data are from Cook 1973, 551; Howay 1930; and Woollen 1924, 229). From 1774 to 1779, 7 ships, 5 of them Spanish, were on the coast, none principally engaged in the fur trade, and no European ships visited the coast between 1779 and 1784. From 1785 to 1792, however, 103 ships from six nations were in Pacific Northwest waters, including approximately 38 British and 20 American vessels. Most of them competed in the fur trade. During the 10-year period from 1786 to 1797, more than 114,000 sea otter skins were harvested, and 170 American and 18 British ships visited the area from 1797 to 1814. The predictable outcome of this unregulated extraction of a marginally renewable resource, was the virtual disappearance of sea otters by the 1830s.

British and American enterprises operated under different models. American traders could profit from three points in the venture: through the favorable exchange of furs for items of minor value in Native villages, by selling furs directly to merchants in Canton and Macao, and through the sale in Boston of merchandise obtained in China, such as silk and tea. British regulations required that traders operate under license by government-approved monopolies such as the East India Company, which required that furs brought to China be turned over to the company's intermediaries for subsequent sale and limited the profit to a ship's owner. Furthermore, traders could not export Chinese goods to Europe for resale.

Some British traders, such as John Meares, occasionally resorted to subterfuge to evade these restrictions.

In all cases, the allure of great profit in the early years of the trade quickly gave way to the realities of a fiercely competitive industry. The first one or two ships arriving in a harbor might accumulate most of the hides for a full season before the next ships arrived. Because of this advantage, the desire to push the limits of a trading season occasionally had a significant human cost, such as the loss of crew members during the region's hard winters. Eventually the number of suitable furs in a particular location dwindled to such a degree that traders abandoned it. Nootka Sound, for example, was the center of the fur trade beginning in 1785, but within five years a decline in the number of fur-bearing animals sent traders farther north. Native chiefs, who mostly controlled the trade exchanges regionally, quickly adapted to European demand for furs by raising their prices and their expectations of trade goods, which eventually included arms and ammunition. In addition, the increased influx of furs into China depressed prices. Finally, questions of sovereignty in the Pacific Northwest, particularly by Spain, interfered with British access to the Native trade.

DID YOU KNOW?

English fur trader John Meares was a notable character who stirred trouble and interest, including operating his ship under the Portuguese flag to avoid paying British licensing fees, losing half of his crew to scurvy, publishing a map that reinvigorated interest in discovery of the Northwest Passage, and inciting anti-Spanish sentiment among British officials and mariners during the Nootka Sound Controversy.

The consequences of these pressures on a global capitalist system of trade helped shape the way Europeans understood the geography and geopolitics of the Pacific Northwest. Although the meticulous and extensive surveys by George Vancouver in 1792 resulted in relatively modern maps of the region (published in 1798), fur traders had searched for new markets along the coast and mapped many new locations. This was especially true in the stretches of the coast north of Vancouver Island that James Cook had not explored. These trader-explorers included James Hanna, James Strange, Charles William Barkley, and George Dixon. Hanna described Fitzhugh Sound in British Columbia in 1786. Strange observed Queen Charlotte Sound and the Scott Islands north of Vancouver Island in 1786. Barkley first documented the entrance to the Strait of Juan de Fuca in 1787. Dixon identified Dixon's Entrance between Haida Gwaii (the Queen Charlotte Islands in British Columbia) and the Alexander Archipelago in Alaska, and in 1787 he was the first European to suggest that the Queen Charlotte landmasses were islands. Traders usually kept accounts of their voyages, but only a small number were published in the first years of the trade. In 1789–1790 Nathaniel Portlock, George Dixon, and John Meares published their accounts with accompanying maps. Commercial cartographers such as London-based Aaron Arrowsmith and others drew from these sources and disseminated the accumulated knowledge.

These early published discoveries resulted in the reemergence of a centuries-old belief in a transcontinental water route across North America. Some influential geographers, including Alexander Dalrymple, reintroduced these conjectural schemes on their maps, a reminder that mapping was not always a linear process. As fur traders accumulated new information about the complexity of the island groups in the Pacific Northwest, some geographers concluded that the discovery of an entrance to a Northwest Passage was still possible.

Within the fur-trading industry, geographical knowledge was linked to Native villages that could supply furs. The principal commodities were the skins of sea otters and fur seals, but traders also sought the pelts of land mammals such as the beavers, foxes, and sables.

During his lifetime, Alexander Dalrymple (1737–1805) served many influential roles, such as civil servant, surveyor and navigator with extensive experience in exploration in the China Sea and Philippines, historian of maritime exploration, geographer, and writer and publisher of many pamphlets and charts. He enjoyed a solid reputation in London intellectual circles and among the establishment of the East India Company and British Parliament. He strongly advocated for linking the worldwide expansion of British commerce with imperial strategy. (Private Collection/Bridgeman Images)

British and American companies occasionally ordered their ships' captains to explore the poorly understood archipelagoes of the Pacific Northwest for new markets, including those in the Strait of Juan de Fuca. The close association between location and profit played out on a global scale as well. As geographer Daniel Clayton (2000, 168–189) explained, the map-to-ledger link in the development of British imperial planning took place within a few years of James Hanna's venture in 1785 and before news of the Spanish impoundment of British fur-trading vessels at Nootka Sound reached London in early 1790. In Clayton's view, the British-authored map replete with British names, many bestowed by fur traders, legitimized British interests in the Pacific Northwest.

The maps of George Vancouver, published in 1798, became the ultimate cartographic manifestations of the process. Even earlier, maps by fur traders and a 1790 map by Aaron Arrowsmith helped politicians use "the ledger and the map to fashion a geopolitical space that Native peoples could not manipulate and to claim the right to hold it for a future generation of British settlers" (Clayton 2000, 182). It was common that Europeans used maps to appropriate newly explored territory before any European settlement was established. As mapmakers incorporated the geographies of fur traders and Royal Naval officers onto maps of North America, an imperial British presence in the Pacific Northwest preceded colonization by many years. Since the time of Christopher Columbus, this process appeared many times in the cartographic history of the Americas.

During 1785–1792, ministers and mapmakers in London understood the maritime fur trade principally in terms of place and power—that is, how commerce related to British imperial strategies in the Pacific Northwest. Written accounts and geographical knowledge acquired by fur traders provided the foundation for textual and cartographic narratives of the expansion of the British Empire. Some traders, such as John Meares, played a more direct role in promoting these policies. At Nootka Sound on Vancouver Island in 1789–1790, the Spanish and British almost waged war over trading rights in the region. During the international confrontation over what became known as the Nootka Sound Controversy, for example, Meares was an important catalyst in stirring up anti-Spanish sentiment in Parliament and among the general public.

In communities along the Pacific Northwest Coast, distinctions about power and place also operated among Native groups before and after the years of the European fur trade. Unlike in London, however, Europeans and Natives on the coast judged their encounters in human terms. The single-minded, competitive efforts by both groups to generate profit resulted in achievement, including monetary gain and new geographical discoveries, and at times considerable loss of life. Violence often broke out, and some accounts describe a gratuitous ferocity carried out against Natives by heavily armed ships' crews. Intense competition among fur-trading enterprises encouraged deceitful practices meant to circumvent licensure regulations, and ship captains were often deeply suspicious of information passed on by rival merchants. The rapid increase in wealth and prestige among some Native chiefs and groups and the introduction of guns as trade items disrupted traditional practices in complex ways.

DID YOU KNOW?

Northwest Coast Natives demanded a great variety of trade goods over the years, from metals to textiles, foodstuffs, arms and ammunition, and even unusual objects. One Boston trader sold the ship's cat for a top-quality sea otter skin and two otter tails.

40. George Dixon's Discoveries in the Pacific Northwest, 1787

INTRODUCTION

Fur traders recorded many observations about the geography of the Northwest Coast, particularly in areas that James Cook had not charted on his voyage. In June 1787 during George Dixon's second trading season in the Pacific Northwest, he left Port Mulgrave on the Queen Charlotte *at latitude ~59°34′ north and sailed south and southeast for Nootka Sound. Dixon had been an armorer on the* Discovery *during Cook's third voyage, and he knew that he was sailing and trading in waters that had not been well charted. In July and August, he traded with villages along a broken coastline from ~54° to ~52°. The cargo officer aboard ship, William Beresford, subsequently wrote an account of Dixon's voyage that was published in London in 1789. Beresford described the voyage in a series of letters, and Dixon contributed the introduction, charts, and appendices. These excerpts from Beresford's narrative record some of Dixon's observations about the commercial potential of his trading activity in the area and the nature of the landmasses the ship encountered.*

William Beresford, *A Voyage Round the World*

As the islands we have just left proved so uncommonly fortunate to us, a few remarks concerning them will, perhaps, not be unacceptable to thee. There is every reason to suppose, not only from the number of inlets we met with in coasting along the shore, but from our meeting the same inhabitants on the opposite sides of the coast, that this is not one continued land, but rather forms a group of islands; and as such, we distinguished them by the name of *Queen Charlotte's Islands* [present-day Haida Gwaii]. They are situated from 51 deg. 42 min. to 54 deg. 24 min. North latitude; and from 130 deg. to 133 deg. 30 min. West longitude. The land, in some places, is considerable elevated, but not mountainous, and is totally covered with pines, which in many places afford a pleasing contrast to the snow that perpetually covers the higher grounds. . . .

The number of people we saw during the whole of our traffic, was about eight hundred and fifty; and if we suppose an equal number to be left on shore, it will amount to one thousand seven hundred inhabitants, which, I have reason to think, will be found the extreme number of people inhabiting these islands, including women and children. The great plenty of furs we met with here, sufficiently indicated that these people have had no intercourse whatever with any civilized nation; and I doubt not, but we may justly claim the honour of adding these islands to the geography of this part of the coast. . . .

[T]hey [other fur traders], not expecting to meet with furs in any other places than those mentioned by Cook, quietly sat down with what they could get in those harbours, and consequently we had to seek for other resources, or return home empty handed. By this means we fell in with Queen Charlotte's Islands, which . . . surpassed our most sanguine expectations, and afforded a greater quantity of furs than, perhaps, any place hitherto known.

That we have made considerable additions to the geography of this coast, cannot be denied, yet much remains to be done; indeed, so imperfectly do we still know it, that it is in some measure to be doubted whether we have yet seen the main land; certain it is that the coast abounds with islands, but *whether any land we have been near is really the continent,* remains to be determined by future navigators: Thus much we can venture to affirm, (and which is of the *first* consequence to this undertaking) that the fur trade is inexhaustible wherever there are inhabitants, and they . . . are not confined to any particular situation, but are scattered in tribes all along the coast, which . . . extends from 40 to 61 deg. North latitude; and from 126 to 155 deg. West longitude.

Source: William Beresford, *A Voyage Round the World,* 2nd ed. (London: Geo. Goulding, 1789), 223–224, 235–236.

ANALYSIS

The Haida Gwaii archipelago lies between 54°14′ at present-day Langara Island (Dixon's North Island) and 51°56′ at Cape St. James (named by Dixon for that saint's day). Dixon's coordinates closely approximated these, and Beresford's account and a map drawn by Dixon, both published in 1789, credibly added to knowledge of the geography of the islands, which appeared to be separated from the mainland by Dixon's Straits (present-day Hecate Strait). John Meares's published map (1790) and a map by Dixon's partner, Captain Nathaniel Portlock (1789), also illustrated what were then labeled the Queen Charlotte's Islands. In 1790, commercial cartographer Aaron Arrowsmith credited Dixon with discovering the island group.

Some influential geographers quickly tried to link Dixon's firsthand observations with long-standing conceptual information about the geography of the region. In 1789 the influential hydrographer to the East India Company, Alexander Dalrymple, published a pamphlet using Dixon's observations as support for his theory that the supposed transcontinental water passage of Admiral Bartholomew de Fonte must exist in the vicinity of Haida Gwaii (Queen Charlotte's Islands). In 1708 in London, a story was published that an Admiral de Fonte had entered a river along the coast at 53° north and sailed practically across the continent until he encountered a ship that was sailing west from Boston. Although apocryphal, the story gained credibility among those who believed in the possibility of a Northwest Passage.

However important Dixon's geographical discoveries were to expanding knowledge for the sake of science, Beresford made it clear that finding previously untapped and "inexhaustible" sources of furs was the issue "of first consequence to this undertaking." Mapping new geographical discoveries and realizing commercial profit, however, were inexorably linked. In the introduction to *A Voyage Round the World,* Dixon made this point explicitly: "it is presumed that the advantages which both Geography and Commerce will derive from this Voyage, are neither uninteresting or unimportant; so that there perhaps will need no apology for laying the result of it before the world" (Beresford 1789, xi). In England, Beresford's narrative and Dixon's maps became part of the textual and visual discourse that framed the Pacific Northwest as a region that was important because it was rich in furs and was populated by people who willingly supplied them. As geographer Daniel Clayton notes, this kind of construct was an early step in the process of extending British sovereignty to the region, one of the "modes of appropriation that prepared the ground for colonists" (Clayton 2000, 234).

41. John Kendrick's Land Purchases, 1791

INTRODUCTION

Spanish and English naval officers attempted to extend the dominion of their governments to the Pacific Northwest under the centuries-old practice of claiming territory through acts of possession taking. The results of the Nootka Convention in October 1790 reached Nootka Sound in March 1791 and ensured that British fur traders had access to the waters around Vancouver Island and to the north. In July, American fur trader John Kendrick took an alternative approach. Within three weeks of arriving at Nootka, he had negotiated what he considered outright purchases of extensive tracts of land from Mowachaht chiefs who claimed authority in the area. This document is a deed that conveys land to Kendrick from Chief Maquinna and others on July 20, 1791.

Record of Land Purchase by John Kendrick, July 20, 1791

To all persons to whom these presents shall come: I, Macquinnah, the chief, and with my other chiefs, do send greeting: Know ye that I, Macquinnah, of Nootka sound, on the north-west coast of America, for and inconsideration of ten muskets, do grant and sell unto John Kendrick, of Boston . . . a certain harbor in said Nootka sound, called Chastacktoos, in which the brigantine Lady Washington lay at anchor on the twentieth day of July, 1791, with all the land, rivers, creeks, harbors, islands, &c, within nine miles north, east, west and south of said harbor, with all the produce of both sea and land appertaining thereto; only the said John Kendrick does grant and allow the said Maquinnah to live and fish on the said territory as usual. And by these presents does grant and sell to the said John Kendrick, his heirs, executors and administrators, all the abovementioned territory, known by the Indian name Chastacktoos, but now by the name of the Safe Retreat harbor; and also do grant and sell to the said John Kendrick, his heirs, executors and administrators, a free passage through all the rivers and passages, with all the outlets which lead to and from the said Nootka sound. . . . Signed with my own hand and the other chiefs' . . . to have and to hold the said premises, &c., to him, the said John Kendrick, his heirs, executors, and administrators, from henceforth and forever, as his property absolutely, without any other consideration whatever. . . .

MACQUINNAH, his × mark [followed by the marks of 5 other chiefs and 10 witnesses]

Source: Report to the Senate of the United States, Senate Doc. 335, 32d Cong., 1st sess., August 11, 1852, 20.

ANALYSIS

This deed describes a large portion of Nootka Sound and environs, a square 18 miles on a side, or 324 square miles. Whether through prior intent or spurred on by his initial success, Kendrick proceeded to other Native communities with similar offers to purchase land in exchange for cloth, guns, and powder. At Taskes village he purchased a 9-square-mile track at an inlet north of Nootka Sound, and his fourth and fifth purchases were similarly sized parcels at nearby Ahasset and Clayoquot Sounds. By August 11 he had signed deeds for nearly 1,000 square miles of prime land, which included the entirety of Nootka Sound and harbors to the north and south. In late 1792 when interviewed by Juan Francisco de la Bodega y Quadra, representing Spain, and George Vancouver, representing England, during the negotiations over the terms of the Nootka Convention, Maquinna affirmed the land sales to Kendrick. Some scholars have interpreted Kendrick's brash actions as an inspired defensive move to block claims of the Spanish and British and to create an American-Native alliance to oppose them.

On March 1, 1793, Kendrick sent copies of the deeds to Secretary of State Thomas Jefferson with a thinly veiled request for assistance in obtaining congressional approval of his purchases. He pointed out to Jefferson that "the future commercial advantages which may arise from the fur trade, besides many other branches which are daily opening to the view of those who visit the northwest American coast, may perhaps render a settlement there worthy the attention of some associated company under the protection of government. Should this be the case, the possession of lands previously and so fairly acquired, would much assist in carrying the plan into effect" (U.S. Senate 1852, 19). Kendrick's true motives, however, may not have been so patriotic. On November 15, 1787, the principal sponsor of his first voyage to the Pacific Northwest, Joseph Barrell, instructed Kendrick that "if you make any fort or improvement of land upon the coast, be sure you purchase the soil of the natives, and it would not be amiss if you purchased some advantageous tract of land in the name of the owners, if you should let the instrument of conveyance bear every authentic mark the circumstances will admit of" (Howay 1990, 111). By 1791 Kendrick apparently felt no obligation to his former employer, as all five deeds stated that the beneficiaries of the sold property were "John Kendrick, his heirs, executors and administrators." Jefferson took no action in response to Kendrick's letter.

Kendrick never had a chance to develop whatever plans he had devised. On December 12, 1794, he was killed by an accidental cannon discharge. John Howell, who was employed by Barrell, assumed control of the deeds and attempted to register them in Canton or Macao. In the meantime, Barrell published a remarkable notice in August 1795 that was circulated in Europe in four languages, offering to sell "American lands . . . that such as may be inclined to associate for settling a commonwealth on their own code of laws, on a spot of the globe no where surpassed in delightful situation" (U.S. Senate 1852, 29–30). On December 23, 1796, Howell wrote to Barrell that "I have had an opportunity of seeing most parts of Capt. Kendrick's purchases on the No. Wt. Coast of America; and cannot flatter you with any hopes of profit from them even to your great, great grand children" (Howay 1990, 493).

Apparently, Barrell did not receive the promised documents, and for many years it was not clear where they were located. In December 1817, Secretary of State John Quincy Adams wrote to Charles Bulfinch, a promoter of Kendrick's first voyage, about the whereabouts of the deeds. Adams may have considered that American claims to sovereignty in the Pacific Northwest might be stronger if Kendrick had made his purchases as a representative of the United States. But the deeds did not resurface until 1840, when they were submitted as evidence by the heirs of the promoters of the voyage in a petition to Congress for compensation. Congress took no action until August 11, 1852, when the Committee of Public Lands concluded that "the right to extinguish the possessory title of the natives does not rest in the individual discoverer, but in his government, and a purchase thus made by him of the natives would . . . give no title as against his government" (U.S. Senate 1852, 13). Kendrick had never rightfully acquired title to the land, which practically ended any claims for compensation.

Mapping the Pacific Northwest had many consequences. Fur traders who were motivated by profit bargained that the maps would allow them to identify and then control prime resource-rich land. It is unlikely that Maquinna or the other chiefs who signed the deeds interpreted ownership in the same way that Kendrick and Barrell and their heirs did, and it is true that the people who lived in and around Nootka Sound did not participate in the subsequent debate over the purchases. Nevertheless, Kendrick's actions became part of the historical record of the Nootka Sound Controversy, occupied the attention of American secretaries of state, and briefly resulted in land speculation in a European market.

42. John Meares and the Nootka Sound Controversy, 1790

INTRODUCTION

In 1789 at Nootka Sound, British fur traders were charged with trespassing on territory that Spain considered its sovereign possession. The two parties to the conflict used geographical knowledge differently. The Spanish looked to their maps for ways to defend their established colonies, while fur traders used maps to find ways to achieve commercial gain. Although the traders were privately funded, their mercantile sponsors, including the East India Company and the King George's Sound Company, had close ties with highly placed government officials and members of Parliament. In early 1790 London and Madrid exchanged letters of formal protest about the events at Nootka the year before, but it was the testimony of John Meares—navigator, fur trader, explorer, cartographer, and author—that helped catalyze the British Parliament and the Admiralty to challenge Spanish claims of dominion in the Pacific Northwest. These extracts were part of Meares's testimony before the British cabinet and Parliament.

Meares's Testimony before the British Cabinet and Parliament

Minutes of Cabinet Meeting, April 13, 1790

Upon consideration of the information which has been received from Mr. Meares of the detention and capture of several British vessels at Nootka Sound, on the coast of America, and of the circumstances of that transaction . . . Your Majesty's servants have agreed humbly to submit to Your Majesty their opinion that Your Majesty's minister at the Court of Madrid should be instructed to present a memorial demanding an immediate and adequate satisfaction for the outrages committed by Monsieur de Martinez; and that it would be proper, in order to support that demand and to be prepared for such events as may arise, that Your Majesty should give orders for fitting out a squadron of ships of the line.

John Meares's Memorial to Parliament, May 13, 1790

That your Memorialist, immediately on his arrival in Nootka Sound, purchased from Maquilla, the chief of the district contiguous to, and surrounding that place, a spot of ground, whereon he [Meares] built a house for his occasional residence, as well as for the more convenient pursuit of his trade with the natives, and hoisted the British colours thereon; that he also erected a breast-work, which surrounded the house, and mounted one three-pounder in front; that having so done, your Memorialist proceeded to trade on the coast. . . .

That, during the absence of your Memorialist from Nootka Sound, he obtained from Wicananish, the chief of the district surrounding Port Cox and Port Effingham, situated in the latitude of 48° and 49°, in consequence of considerable presents, the promise of a *free and exclusive trade with the natives of the district,* and also his permission to build any store-houses, or other edifices, which he might judge necessary; that he also acquired the same privilege of exclusive trade from Tatootche, the chief of the country bordering on the Straits of John De Fuca, and purchased from him a tract of land within the said strait, which one of your Memorialist's officers took possession of in the King's name, calling the same Tatootche, in honour of the chief. . . .

He [Meares] also has subjoined a statement of the actual as well as the probable losses which he and his associates have sustained from the unwarrantable and unjustifiable proceedings of Don Martinez, in open violation of the treaty of peace subsisting between this country and the Court of Spain. . . .

Your Memorialist therefore most humbly begs leave to submit the case of himself and his associates to the consideration of Government, in full confidence that the proper and necessary measures will be taken to obtain that redress, which he and his associates have, as British subjects, a right to expect.

Source: J. Richard Nokes, *Almost a Hero: The Voyages of John Meares, R.N., to China, Hawaii and the Northwest Coast* (Pullman: Washington State University Press, 1998), 188, 196, 206.

ANALYSIS

John Meares was in Macao when Esteban José Martínez seized the British ships at Nootka and arrested the crew of Meares's recently established Company of Associated Merchants of London and India. Meares received the news in November and December by word of mouth and through smuggled letters written by one of the arrested officers. In London, the cabinet first learned of the seizures of British registered vessels in January 1790. By the time Meares returned to England in early April, Spanish and British ministers of state had already widened the geopolitical scope of the crisis by attempting to enlist the support of other European powers, including France, Austria, and Sweden.

Meares was not an eyewitness at Nootka, but his memorial of April 30 and the accompanying affidavits provided detailed information of the seizures and the arrests, the nature of the sea otter trade in the Pacific Northwest, and his claims of having legally purchased and occupied a site at Nootka (which was later emphatically denied by Chief Maquinna). George III acted promptly on the recommendations of the cabinet, and in the ensuing months the Admiralty began mobilizing for war. After Parliament received Meares's memorial on May 13, it was published and disseminated for public consumption. As diplomatic maneuvering carried on over the next few months, Meares capitalized on his notoriety and prepared a narrative of his journeys for publication, *Voyages Made in the Years 1788 and 1789 from China to the North West Coast of America.* "Meares' connections at this time to individuals at the highest levels of British society and government," historian Richard Nokes wrote, "are evident in the list of Subscribers who supported the publication of *Voyages.* Nearly 350 names are listed, a veritable who's who in British administration, commerce, banking, finance and the military, in England, Ireland and China" (1998, 158). The list included the Prince of Wales and Sir Joseph Banks, head of the Royal Society.

On October 28, 1790, British and Spanish diplomats in Madrid reached an agreement on a pact that defused the specter of war in the Pacific Northwest. This First Nootka Convention contained several ambiguously worded articles, and both governments subsequently arranged for direct negotiations between naval representatives of their countries at Nootka Sound. By the time George Vancouver, representing Britain, and Juan Francisco de la Bodega y Quadra, representing Spain, met at Nootka in the fall of 1792 to negotiate the terms of the convention, there was no evidence of Meares's "house for his occasional residence." The location and nature of his "spot of ground" was never clearly determined, and it never appeared on any map. But British diplomats and "custodians of geographical knowledge," such as the influential Alexander Dalrymple (Clayton 2000, 179), used Meares's claim to transform issues of commercial enterprise into broader questions of territorial claims in the region. Diplomats, investors, and pundits turned to cartographic references when considering British interests in the Pacific Northwest. The Committee of Trade and Plantations, responsible for British overseas commerce, interviewed Meares with questions about the "highest and lowest latitude of successful commerce" in the Pacific Northwest and about the current availability of furs and their future marketability in China. As geographer Daniel Clayton noted, British diplomats "viewed Nootka Sound through the lexicons of the ledger and the map" (2000, 178). Certainly one of Meares's legacies was his role in highlighting the inexorable links among commerce, cartography, and geopolitical ambition in Britain.

43. John Meares's Orders to Explore New Markets, 1788

INTRODUCTION

The Company of Associated Merchants of London and India was established in Macao on January 23, 1789, and the owners, including John Meares, assumed control of four ships: the Argonaut, Ephigenia Nubiana (Iphigenia), Princess Royal, *and* North-West America *(the first vessel constructed on the Northwest Coast). The* Argonaut *had recently been purchased in India, while the others were already engaged in commerce in the North Pacific. Before leaving Nootka Sound for Macao and after launching the* North West America *on September 20, 1788, Meares issued orders to countryman William Douglas, captain of the* Iphigenia, *on how to proceed during the remaining months of 1788 and during the next trading season. After January 1789, Douglas remained captain of the ship under the same orders. Meares became a proprietor and placed James Colnett, captain of the* Argonaut, *in command of the four-ship squadron. His orders to Douglas were similar in substance to those given to Colnett and captains of the other ships.*

Orders of Captain Meares to Captain Douglas Commanding the *Ephigenia Nubiana*

[In the remaining time of the season of 1788] I recommend you to steer to the Lat. 46° and trace as much of the Coast from 46° to 40° as you can; as this month is esteemed tolerably good. You may fortunately Light on a new tribe by exploring this part of the Coast which I have the Strongest reason to imagine inhabit those parts. I have knowledge that the Spaniards have obtained a Prodigious number of Otter skins which were brought to Manilla in the Galleons and from thence to Macao where I saw them. I am morally sensible they were not purchased to the Northward of 46° as nowhere did I meet with one Item of the Manufacture of Spain, either Iron or Copper. This part of America is now all that remains to be known to us and so fortunate a moment be seized it will either determine us in future to give up this Southern pretension as unproductive of Mercantile advantages or seek sea Otters in those parts. It becomes a Question where did the Spaniards procure those Skins and this point I hope you will be able to clear up so as next year to make us able to take advantage of any discovery you may be able to make. . . .

The knowledge we have now obtained of the Coast of America, of the Periodical Winds and Seasons give us the advantage of all competitors. The Year 1789 must be viewed by both of us as the most productive of any we may expect. Having these views before us our exertions must be doubled to effectually sweep the Coast before any Ship from Europe can arrive. . . .

As you steer along the N.W. side of the great Island [Haida Gwaii] you will explore it minutely to the height of 54° where resides a Chief whose district is Large. . . . Between this Chief's residence and Cape S. James lies a part of the Coast unexplored and we may not flatter ourselves vainly that in such an extent of Land there reside many Chiefs and numerous Inhabitants.

There is one point which I most particularly recommend to your attention. In your Journal mention is made of prodigious Sounds which Inclines you to imagine a N.W. passage may exist. Such a thing is not impossible. . . . Strange as it may appear I believe it nevertheless true that Captain Cook never saw any part of the Continent of America until he reached the 60° of Latitude. Nor any other navigator either preceding or after till our time . . . [this belief] throws a light on the esteem'd Fabulous Voyage of Admiral Defont. It brings the Archipelago of S. Lazarus within the Line of probability and without much Vanity or assuming much credit I have ascertained the imagin'd Straights of Juan de Fuca to be a reality. . . . Though my expectations may rise with yours and my wishes keep pace with you which may render us anxious to make such a discovery yet to such an attempt I can by no means consent; Commerce is our view and should we effect the intention of our employers by wasting time to effect that which at best is precarious, we should lose that Character of Prudence which I trust we shall reap from bringing the Voyage to a happy conclusion. I therefore recommend to you in the most serious manner not to entangle yourself too far in those immense Sounds or Lakes so that a waste of time may ensue and consequently the Loss of the most valuable Season that we are likely to have except [if] you can make Discovery and commerce go hand in hand. From this Prudence I trust nothing will permit you to deviate.

Source: F. W. Howay, ed., *The Journal of Captain James Colnett aboard the* Argonaut (Toronto: Champlain Society, 1940), 32–33, 35–36.

ANALYSIS

Meares's full set of explicit orders occupy several pages and probably would have impressed the merchant owners who were not seamen by training. He expressed knowledge of coastal geography and weather patterns that he predicted could be a help or a hindrance, and he was familiar with the location of Russian settlements and established Native villages. He also suggested likely locations of villages that traders had not yet visited. During his time in Macao, Meares had apparently obtained confidential information about fur purchases by the Spanish "not . . . to the Northward of 46°." For him, knowledge of geography meant likelihood of profit.

In his last few paragraphs, Meares sounded optimistic that the "Archipelago of S. Lazarus" of the "Fabulous Voyage of Admiral Defont" or the "Straights of Juan de Fuca" might still be discovered and lead to the fabled Northwest Passage. But although Meares said he sympathized with Douglas's desire to make such a discovery, he invoked the tone of his new position as a merchant proprietor and reminded Douglas of the primary purpose of the voyage: commerce.

44. A Conniving Letter to Joseph Banks, 1788

INTRODUCTION

In 1785 Richard Cadman Etches, a wealthy London merchant, was one of the first in England to venture into the maritime fur trade. He and eight others founded the King George's Sound Company and obtained licenses from the South Sea Company to collect furs in the Pacific Northwest. To sell the furs in Canton, they were granted a license from the East India Company that included stringent economic terms favoring the company at the expense of the merchant. Although Etches also applied to the British government for support, hoping to reap the privileges of a nationally chartered organization, he did not receive that charter. He apparently considered that his first expedition under Captain Nathaniel Portlock and Captain George Dixon—from 1785 to 1788—was only a marginal financial success. By mid-1788, Etches began to strategize how he might reorganize his company, expand its operations, and circumvent the oppressive license obligations of the East India Company. He tried to solicit the British government by appealing to one of the most influential scientists in England, Joseph Banks.

Richard Cadman Etches to Sir Joseph Banks, July 17, 1788

The repeated acts of Friendship that I have experienced from You, particularly in our North West Expedition enduces me to presume to solicit your assistance in *maturing* it, tho' our first expedition hath by no means answered our expectation—yet it hath fully satisfied us of the real prospect of it proving that valuable commercial acquisition which we had originally reason to hope it wou'd—the present system must be very materially chang'd; I am sensible Sir Joseph [that you] need not be told the superiority *they* [the East India and South Sea companies] have over us. . . . But I am perfectly satisfied that a system *may be adopted* which will give us every advantage to be requir'd—satisfied that You have the Prosperity of the undertaking much at heart. I presume to trouble You with my sentiments for the future conducting of it.

Our Ships in their late Voyage discover'd a New Island [Haida Gwaii] which lays between Nootka and Prince Williams Sound, and some distance off the Coast . . . they met with abundance of Furs. . . . [A] small Factory shou'd be established on this Island as an Annual Rendezvous. . . .

All we wou'd wish is to have the forming the foundation (on any system that may be adopted) and a *short* and *partial* claim to its Commercial Part—an Establishment is what I have long had at Heart, not only as an Individual, but a National attempt—Your Patronage I have experience'd—Your Assistance I have found of the utmost consequence. The past enduces me to Solicit your future endeavour to complete our Views—Protection from Government is what we presume to hope for—You, Sir, I am sure can adopt an Idea in what manner this may be done with propriety. Such an Idea hinted to me, I wou'd adopt it and present a memorial to Government, and shou'd hope for your *kind assistance* in carrying it *thro'*. There is one Idea, which perhaps may not be immediately thought of, which with all possible defference I will beg leave to suggest to You, Government having adopted the measure of Colonizing with Convicts, I shou'd presume the same wou'd hold good with forming an establishment on the N. W. Continent of America or Islands, equally with New Holland. If that Government wou'd adopt the same Idea . . . and giving us such power *over the Commercial part,* for a *limited time* as *they* shou'd approve, a plan form'd on such an Idea, I am persuaded wou'd not only secure the complete discovery of that extensive and unexplor'd part of the World, but wou'd open, and secure a source of commerce of the most extensive magnitude to this Country. . . .

I am sensible I had your good wishes for the Success of the past, and my [any?] future attempt I may make, I have therefore only to entreat Your pardon for any Inaccuracy.

Source: F. W. Howay, "Four Letters from Richard Cadman Etches to Sir Joseph Banks, 1788–92," *British Columbia Historical Quarterly* 6(2) (1942): 130–132.

ANALYSIS

Etches's principal complaint was that the monopoly held by the East India Company resulted in great economic disadvantage to a competing, privately supported English company that wanted to sell sea otter furs in China. If he wanted to challenge the monopoly or create one of his own, he knew that he had to appeal to influential nonmerchants and make a case for pursuing common interests. In a few paragraphs in this letter Etches proposed an alliance of merchants, ministers of state, and scientists. He suggested that if his company was permitted unrestrained commerce on the Northwest Coast, he could establish a British settlement by building a "Factory," initially using convict labor. His ships would eventually "secure the complete discovery of that extensive and unexplor'd part of the World," and Britain would reap the financial benefits. Etches proposed a conjunction of science, private enterprise, and British colonial presence in a part of the world whose uncharted regions would be made known through commercial geography.

In 1788, Etches might have thought that he had good reason to expect the attention of the government. Three years earlier, he had received government sanction when he helped organize the King George's Sound Company, and he now hoped that Banks would be his ally as a conduit to high government officials. Banks was an internationally respected scientist, president of the Royal Society, and an adviser to George III. He had also sailed on James Cook's first voyage, was keenly interested in exploration, and had participated during the launching of Etches's first voyage in 1785 by naming one of the vessels the *Queen Charlotte.* Etches knew that in 1779 Banks had advocated before the House of Commons the establishment of a British colony at Botany Bay populated by convicts. Banks apparently did not respond to Etches's flattering requests, however, so in January 1789 he moved ahead and merged the King George's Sound Company and the Company of Merchants in Macao to create the Associated Merchants Trading to the Northwest Coast of America.

Etches's attempt was one of several by merchants who sought assistance from the government to support commercial enterprise in the Pacific Northwest. They had reason to be optimistic. During 1789, for example, the Admiralty and the administration of Prime Minister William Pitt initiated preparations for a voyage of exploration to the Pacific Ocean. The draft instructions stated that "one of the objects of the expedition being to form a settlement on the nor.-west coast of America, it is his Majesty's pleasure that you should select from among the people with you a proper number of persons to compose it" (Harlow 1964, 2:439).

45. Alexander Dalrymple's Fur Trade Plan, 1789

INTRODUCTION

In 1789, geographer Alexander Dalrymple published a pamphlet in London, Plan for Promoting the Fur-Trade, Securing It to This Country, by Uniting the Operations of the East-India and Hudson's-Bay Company. *Highly educated and articulate, he was hydrographer to the East India Company and a consultant on James Cook's third voyage. Dalrymple was an advocate for locating the Northwest Passage, despite Cook's failure to find evidence of that fabled waterway. "When diplomatic negotiations turned to territorial questions," geographer Daniel Clayton noted, "the Foreign Office turned to Alexander Dalrymple." He was "one of the principal custodians of geographical knowledge about North America and the Pacific" (2000, 179).*

Dalrymple's position with the East India Company allowed him to gather the latest geographical information from all British commercial expeditions to the Northwest Coast, beginning with the first fur-trading voyage of James Hanna in 1785. For two years beginning in 1789, before news of the Spanish seizure of British ships at Nootka Sound reached London, Dalrymple published at least six pamphlets that framed British mercantile and geopolitical interests in the Pacific Northwest in geographical terms.

Plan for Promoting the Fur Trade and Securing It to This Country

The late Voyages, particularly those of Captains *Cook, Hanna, Portlock, Dixon,* and *Barkley,* have ascertained with competent precision, the Longitudes of the *Lands* lying upon the *Ocean,* commonly called the *South Sea,* or *Pacifick Ocean.* . . .

It is the opinion of Capt. *Dixon,* that all the *Lands,* yet discovered, of the *West Side,* at least from *Nootka* inclusively Northward, are *Islands,* and *not* the *Continent:* If this opinion be well grounded, which every particular Examination of *parts in detail,* as well by *English* as *Spanish Navigators,* seems strongly to confirm; Then, the distance between the *two Seas* will be still more reduced; and the *Navigation,* being carried on through *Sounds* covered by *Islands,* abounding with excellent *Harbours,* will be exposed to little danger when the Coasts are properly surveyed.

But Capt. *Dixon,* and other late *Navigators,* have found an *Archipelago of Islands,* and the strongest indications of a *large River,* in the place where such are described by *De Fonta;* This gives some countenance to that too hastily exploded Narration! . . .

This opinion is more strongly confirmed by Capt. *Hanna* in the *Sea-Otter* 1786, who entered *Fitzhugh Sound,* about 50° beyond the furthest Land that Capt. *Dixon* saw. . . .

Capt. *Barkley* has made a *Chart* of the *Coast,* from *Nootka* S. Eastward, for which, as well as for his *Journal,* I am indebted to Mr. *Cox.* . . .

This long discussion was necessary, to introduce the proposed Union of the Operations of The *East-India* and *Hudson's-Bay Companys,* because the further-in the *Sea Communication* extends, on the West-side of AMERICA, with the greater facility will their joint Operations be connected. . . .

But it is probable that the Communication between *Hudson's-Bay* and the *West Coast* of AMERICA would . . . be made in a higher Latitude; by means of the *Chesterfield Inlet,* or some of the *Inlets* and *Rivers* from *Hudson's Bay,* connecting with the *Arathapescow, Dobaunt,* and other *Lakes.* . . .

A bare inspection of the Map, with the slightest attention to what has been said, will evince, that the Co-operation of the Two Companys, would effectually secure to this Country the Command of the *Fur-Trade;* for by no other way, than Hudson's-Bay, can the Communication be made with so much facility, nor with so little expence as by a Ship from China: and having thus the option of the *Chinese,* as well as the *European* Markets, no other Traders can stand in competition. . . .

It appears, from the Indian Maps, that the *Arathapescow Lake* communicates with *Hudson's-Bay;* it is therefore highly expedient to examine what obstructions there are to navigate thither; for this Lake is reported by Mr. Hearne, from the information of the Indians, to be about 400 miles long: The most effectual manner of making this examination, would be from the *Arathapescow Lake,* which, by the Observation of the Longitude at *Hudson's House,* appears to be much nearer *Hudson's-Bay* than Mr. Hearne's Map represents.

Source: Alexander Dalrymple, *Plan for Promoting the Fur-Trade and Securing It to This Country* (London: George Bigg, 1789), 7, 9, 10, 21, 25, 26, 31, 32.

ANALYSIS

Dalrymple's *Plan for Promoting the Fur-Trade and Securing It to This Country* was a masterful reworking of newly acquired geographical knowledge from fur traders and scientists into a British imperial strategy. He envisioned a union between the Hudson's Bay Company, which controlled much of the fur trade in the interior of North America, and the East India Company, whose monopoly in China effectively controlled the maritime fur trade in the Pacific Northwest. If a navigable water passage existed between Hudson Bay and the Pacific Ocean, then merchandise from the interior and the coast could flow to Canton and Macao as well as to Europe without having to be shipped around Cape Horn.

Dalrymple's construction was influential because it was based on three sources of authority. First was his own standing as an expert on geographical matters. He was respected in both scientific and commercial communities, and his publications were widely disseminated to the public. Second were fur traders' accounts, which included firsthand observations of Natives living in discrete harbors among the archipelagoes in the region. In this 1789 pamphlet, Dalrymple was careful to cite information from all the traders to date—James Hanna, Nathaniel Portlock, George Dixon, Charles William Barkley, and James Strange. Although their geographical knowledge was incomplete, he extrapolated it to fit his belief that the entrance to a transcontinental navigable water passage probably lay somewhere above latitude 50° north.

The third source of authority was the map itself. In 1790 in a separate pamphlet, Dalrymple emphatically stated that "the Proper Authorities to consult are *Geographers;* for although a *Map* may not be considered as *conclusive,* it must be admitted as *presumptive, testimony* of a *fact*" (qtd. in Clayton 2000, 179). Most readers likely shared this belief. Dalrymple published his own map to illustrate his geographical theory, and his ideas were also reflected in the maps of others, including those by John Meares.

Dalrymple's concerns were commercial and imperial. His antagonists were the Spanish, Russians, and Americans and also the Canadians, British subjects who were competitors in the North American fur trade and not aligned with the Hudson's Bay Company. Neither Dalrymple nor his audience were concerned about Natives, whose presence was nowhere to be seen on contemporary maps. In earlier documents, private correspondence and accounts of voyages at least acknowledged Native chiefs and the importance of locating Native trading villages to ensure the commercial success of their endeavors. In Dalrymple's publications, Natives were missing altogether.

46. Tragic Human Costs of Pursuing Furs, 1787

INTRODUCTION

Accounts of encounters between European traders and coastal Natives contain many descriptions of conflict and loss of life, but human costs were also incurred as a consequence of errors in judgment. In March 1786 John Meares, operating without the required license from the South Sea Company, left Bengal for the Pacific Northwest on the Nootka. *By May, the ship's crew of 50 men was apparently experiencing signs of scurvy. This would have been a very early onset of the disease unless some of the sailors had been nutritionally depleted at the time of departure. Over the next few months, Meares made his way to the Aleutian Islands and Kodiak Island and arrived in Prince William Sound in September. Presumably, he was able to obtain provisions containing enough vitamin C to forestall the development of the disease among his crew. He had not had a very successful commercial season, but he was encouraged by the promise from a local chief that he could obtain a regular supply of furs. Contrary to usual practice, Meares decided to overwinter in the sound instead of returning to China or the Hawaiian Islands to wait for spring. The ship soon became icebound, and few fresh provisions were available. By February 1787, 11 months after their departure, 8 men had died of scurvy. A month later, 30 were ill with the disease.*

In May, the British traders Nathaniel Portlock on the King George *and George Dixon on the* Queen Charlotte *came upon Meares and the* Nootka, *which was still partially icebound. Twenty-three men had died, and the remaining crew members were too ill to bury their dead. Meares's and Dixon's accounts of their voyages, both published in 1789, told the story.*

Accounts of the Tragedy at Nootka

Captain George Dixon,* Voyage Round the World but More Particularly to the North-West Coast of America, *May 1787

[Meares] had wintered in the creek where I now found him, and his vessel was still fast in the ice. The scurvy had made sad havock amongst his people, he having lost his Second and Third

Mates, the Surgeon, Boatswain, Carpenter, Cooper, Sail-maker, and a great number of the foremast men, by that dreadful disorder; and the remaining part of his crew were so enfeebled at one time, that Captain Meares himself was the only person on board able to walk the deck.

It gave him very great pleasure to find two vessels so near, who could assist him in some measure in his distress; and I had no less satisfaction in assuring him, that he should be furnished with every necessary we could possible spare. . . .

They told us various stories respecting their success in trade, but on these much reliance was not to be placed, as the Captain and Mate seldom agreed in their accounts: at one time, they had purchased two thousand sea-otter skins; at another, not more than seven hundred. This, however, is pretty certain, that they had procured a considerable quantity of valuable furs, and these chiefly, if not altogether, in Prince William's Sound. . . .

On our enquiring whether the scurvy, unattended with any other complaint, had made such fatal havock amongst the people belonging to the Nootka, we were informed, that a free and unrestrained use of spirits had been indiscriminately allowed them during the extreme cold weather, which they had drank to such excess about Christmas, that numbers of them kept their hammocks for a fortnight together; add to this, their liquor was of a very pernicious kind, so that there is reason to suppose its effects, when drank to such an extreme, were not less fatal than the scurvy itself . . . and I am afraid that a scanty supply of the various antiscorbutics absolutely necessary for these voyages, put it out of their power to check this cruel disorder, so often fatal to sea-faring people. . . .

Nathaniel Portlock's Agreement with John Meares to Render Assistance to the **Nootka**

[A]nd as your [Meares's] remaining in the sound and trading must, of course, stop a considerable part of the trade that I might get, I find myself in duty bound, on account of my employers' interests, to propose the following conditions . . . that you bind yourself in a bond of five hundred pounds that no trade be carried on for skins of any kind by yourself or any of your crew, during your stay in the Sound this season, and that you let me have twenty bars of iron and some beads.

Sources: William Beresford, *A Voyage Round the World,* 2nd ed. (London: Geo. Goulding, 1789), 155, 157; John Meares, *Voyages Made in the Years 1788 and 1789 from China to the N.W. Coast of America* (London, 1790), xxvii.

ANALYSIS

The death of 23 men from scurvy on the *Nootka* may have been the greatest loss of life from that disease among all the European voyages to the Pacific Northwest from 1774 to 1800. The disaster on Meares's voyage, however, was not inevitable. While scurvy was still a major problem for the British Royal Navy and for crews on extended voyages at sea during the last two decades of the 18th century, experienced mariners had accumulated a great deal of information about how to keep their crews healthy. Reprovisioning with citrus fruits or some leafy plants and even concoctions of pine needles not only could reverse symptoms but also could prevent the onset of the crippling disease, which was invariably fatal if not treated. Some of the men on the *Nootka* apparently did treat themselves by chewing young pine branches. With the knowledge that members of the crew might already have marginal nutrition, Meares's ill-advised decision to overwinter in Prince William Sound appeared to sacrifice good judgment for potential profit.

With only half of his original crew remaining, Meares left the Northwest Coast for Hawaii and returned to Macao in October. After selling his furs in early 1788, he quickly formed a new trading company, outfitted two ships, and sailed back to the Pacific Northwest, this time under the Portuguese flag in an attempt to again escape licensure requirements of the East India Company and the South Sea Company. His attempt to escape these restrictions was matched by his apparent attempts to mislead Portlock and Dixon when he and his first mate gave conflicting accounts of their fur purchases for the 1786–1787 season.

Meares was rankled by what he considered a coerced contract with Nathaniel Portlock and George Dixon to leave the Pacific Northwest in 1787 and also by William Beresford's comments in 1789 in Dixon's *Voyage* on the abuse of alcohol on the *Nootka.* Meares defended his actions in his own published account in 1790. On the accompanying map, he removed the name Dixon's Straits, which Joseph Banks had bestowed in honor of its discoverer, and replaced it with Douglas Entrance in honor of his own partner.

The embittered exchange between Meares and Dixon played out in three well-publicized letters between December 1790 and February 1791, at which time their rivalry became bound up with other publications on the Nootka Sound Controversy and Britain's conflict with Spain. These documents are a reminder that the loss of lives in reckless pursuit of commercial gain was often overshadowed by publicly aired personal animosities and the drama of international events.

47. James Hanna's Conflict with Natives, 1785

INTRODUCTION

Violence often accompanied encounters between Natives and Euro-American maritime explorers and fur traders. Contemporary accounts of such conflicts between fur traders and Indians are a reminder of the human cost bound up with an activity motivated principally by the pursuit of profit in a remote area of the world. The authors of these accounts, however, may not have directly observed the event, and their interpretations of what happened may have been influenced by either misunderstanding or their interest in commercial success. While it may be difficult to find an accurate description of events in these documents, they can reveal some of the biases and presumptions of their authors. These different sources describe part of the encounter between James Hanna, the first fur trader to arrive on the Northwest Coast after James Cook, and Natives in Nootka Sound. Hanna's journal entries for this period have been lost, and historians have relied on the reports of contemporary newspapers and other travelers to the region.

Reports on the China Trade

Letter from China, London Chronicle, *September 21, 1786*

[Y]ou will be astonished when I tell you, that the whole out-fit, with the vessel, did not cost them 1000L. and, though she was not more than one month on the coast, the furs she collected were sold at Canton for upwards of 30,000L. . . . It is astonishing that this business hath not been taken up long before this directly from England, as there is a full description of it in the publication you sent me of Capt. Cook's last voyage: it is fully expected that when the astonishing value of this trade is known in England, that the [East India] Company will send out some of their China ships to trade for furs on that coast, and to try to open a trade with Japan for the disposal of them. . . . [I]t would procure them vast quantities of silver and gold, and the furs would sell for 300 percent more than they do in China.

London World, *October 6, 1788*

Soon after her [the *Sea Otter*'s] arrival, the natives . . . tempted probably by the diminutive size of the vessel . . . and the small number of her people, attempted to board her in open day; but were repulsed with considerable slaughter. This was the introduction to a firm and lasting friendship. Capt. Hanna cured such of the Indians as were wounded; an unreserved confidence took place—they traded fairly and peaceably—[and] a valuable cargo of Furs was procured.

William Beresford,* Voyage Round the World but More Particularly to the North-West Coast of America, *1789

Soon after his [James Hanna's] arrival, the natives attempted to board his vessel in open day, but were repulsed with considerable slaughter: this affair secured the friendship of the Indians, for they afterwards traded quietly and peaceably. Captain Hanna is said to have procured a valuable cargo of furs.

***Alexander Walker,* An Account of a Voyage Made from China to the N.W. Coast**

The Savages commenced the Attack in this manner. They surrounded the Schooner with their Canoes amounting to 16 or 20, from whence they assailed her with all kinds of Missile Weapons; but finding this Mode not likely to succeed, they attempted to board her with long knives. This effort seems not to have been very obstinate, nor was there much difficulty in repulsing them. It is remarkable that the Evening before the Engagement, the Inhabitants traded with Captain Hannah as usual, and so little was his suspicion of them, that they would probably next day have taken him by surprise, had not a party of Strangers who were also trading with him, given Information of the design. . . .

The Accuracy of this [the preceding interpretation of events] may be doubted. It seems from subsequent events and from Circumstances which have transpired, that Capn. Hannah's People were the original transgressors. He is said to have Punished some of the Savages by firing on them, for a petty theft they committed and they made an Attack by surprise as usual in revenge. . . .

As an instance of the singular disposition of these People, the Battle was not long over, when the Savages who had been guilty of so recent an Act of perfidy, came again along side, and offered their Skins to Sale as if nothing had happened. The effects of this Engagement may in some Measure account for the extraordinary dread they expressed at our fire arms, but the Secrecy and cunning with which they concealed the transaction from us, forms a strong Feature in their character.

Sources: *London Chronicle,* September 21, 1786; *London World,* October 6, 1788, quoted in W. Kaye Lamb and Tomás Bartroli, "James Hanna and John Henry Cox: The First Maritime Fur Trader and His Sponsor," *BC Studies,* no. 84 (Winter 1989–1990), 14; William Beresford, *A Voyage Round the World,* 2nd ed. (London: Geo. Goulding, 1789), xvii–xviii; Alexander Walker, *An Account of a Voyage to the North West Coast of American in 1785 & 1786,* edited by Robin Fisher and J. M. Bumsted (Seattle: University of Washington Press, 1982), 199–200.

ANALYSIS

While all of these documents relate to the tragic event at Nootka Sound in August 1785, there are few details. Instead, the accounts focus primarily on the commercial outcome of the exchange and the character of the local Natives. In the *London Chronicle* article, the author expressed astonishment at the enormous profit realized by the sale of Hanna's furs in Canton but made no mention of the conflict and loss of life. This was the first news of British activity in the Pacific Northwest since the 1784 publication of Cook's *Voyages* and confirmed James King's prediction in that year that a maritime fur trade could be immensely profitable with little cost. In the *London World,* the summary of Hanna's experience suggested that Natives made an unprovoked attack on the *Sea Otter* based on their assumptions about the "diminutive" size of the ship and the small number of men on board. Because of their naïveté, they were "repulsed with considerable slaughter," an outcome that demonstrated British superiority. Following Captain Hanna's beneficent ministering to the wounded, the paper reported, the Natives were won over to an "unreserved confidence," and a "valuable cargo of furs" was obtained.

In 1789–1790 George Dixon, Nathaniel Portlock, and John Meares took up the same refrain in their published travel accounts, highlighting that Hanna obtained his furs not in spite of a deadly encounter with Natives but as a direct result of having demonstrated his superior strength. The show of force was necessary to achieve the desired commercial outcome.

Walker wrote his account during a visit to Nootka Sound within weeks of Hanna's departure. He did not interview Hanna, and his original manuscript eventually was lost, so he reconstructed his narrative from his notes and rewrote his draft, probably between 1822 and 1832. Perhaps with hindsight, Walker was able to suggest that "Capn. Hannah's People were the original transgressors," but he also emphasized the "perfidy" and "cunning" of the Natives.

Other newspaper reports and travel accounts recounted the deadly clash between Hanna and Natives at Nootka Sound, including reports by Maquinna, a principal chief, and Spanish commandant Esteban Martínez. All of these documents should be interpreted with an appreciation of the biases of their authors.

48. John Meares's Map, 1790

INTRODUCTION

Some fur traders plying the Northwest Coast after 1785 included maps in their published accounts of their voyages. No map caused as much reaction as this one by John Meares, one of three maps and 10 plans that Meares included in Voyages . . . from China to the North West Coast of America, *published in London in November 1790. The book was a handsome deluxe edition with high-quality paper and a long subscription list, which suggested that Meares enjoyed solid financial backing. It was reprinted in London the next year, followed by a French edition in*

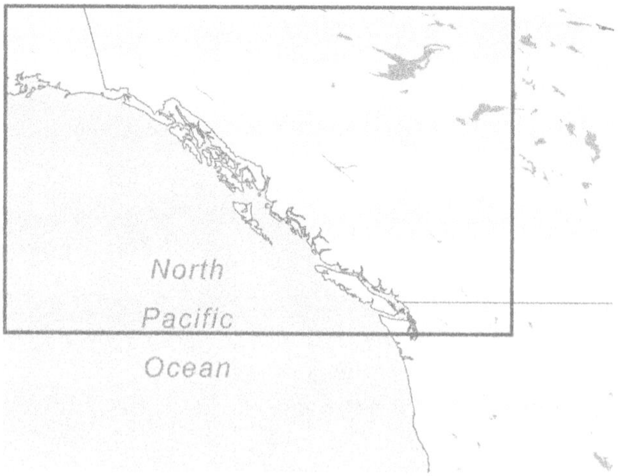

1794 and an Italian edition in 1796. By late 1790, the interested British public was familiar with many accounts about the idea of a Northwest Passage across the North American continent. The title of Meares's map—"A Chart of the Interior Part of North America Demonstrating the Very Great Probability of an Inland Navigation from Hudson's Bay to the West Coast"—may have intrigued readers, including those who had dismissed the Northwest Passage based on reports from James Cook's third voyage.

Meares was a figure of considerable interest in 1790. He had sailed to the Pacific on commercial ventures twice, in 1786 and 1788, and he was a central player in the Nootka Sound Controversy upon his return to England in April 1790. Following publication of his book in late 1790, he waged a public and vindictive exchange of comments with fur trader George Dixon. Meares was something of an authority on matters related to the Northwest Coast, and his published account and maps probably attracted attention.

ANALYSIS

For centuries, explorers searched for a navigable water passage through or above the North American continent—usually called the Northwest Passage or the River of the West. In the 16th century, the principal motivation for discovering the passage was to enable transit from Europe across the Atlantic and around or through the North American continent to Asian markets. In the late 18th century, another motivation was to discover a navigable river route from Hudson Bay to the Pacific, allowing more direct communication between British commercial interests in North America and markets in Canton and Macao. By the time reports of Cook's voyages were published in 1784, the probability of a Northwest Passage seemed remote. By 1790, however, new geographical information, much of it provided by fur-trading mariners, had rekindled interest in Bartholomew de Fonte's alleged discovery of a strait into the interior at latitude ~55° north and Juan de Fuca's reported claim that there was an entrance at 47–48°. And as several geographers noted, bad weather had frequently prevented Cook from making good observations along the coast between latitudes ~50° and 54°.

In London, a principal proponent of the Northwest Passage idea was Alexander Dalrymple, the well-respected hydrographer to the East India Company. He accumulated the available information on maritime explorations of the Northwest Coast and western regions of North America, including material from Hudson's Bay Company records. By 1789 he believed that a Northwest Passage was likely, either in the far north entering the Pacific near Cook's River or between the Strait of Juan de Fuca at ~47° and somewhere to the northeast beyond Haida Gwaii (Queen Charlotte Islands). Meares and Dalrymple were of common mind about the possibilities of a transcontinental passage, and Meares incorporated Dalrymple's ideas and his own observations to produce this highly conceptualized map.

In the far north, Meares used dotted lines to indicate a possible connection between an interior and stylized Slave Lake with either Cook's River or Prince William Sound. Farther south, he depicted Haida Gwaii (named Queen Charlotte Islands by George Dixon in 1787 and first depicted on his published map of 1789) and the Princess Royal Islands (named by Charles Duncan in June 1787 and first published on Aaron Arrowsmith's map in 1790). Between latitudes ~47° and ~56°, the relationship of these islands and mainland North

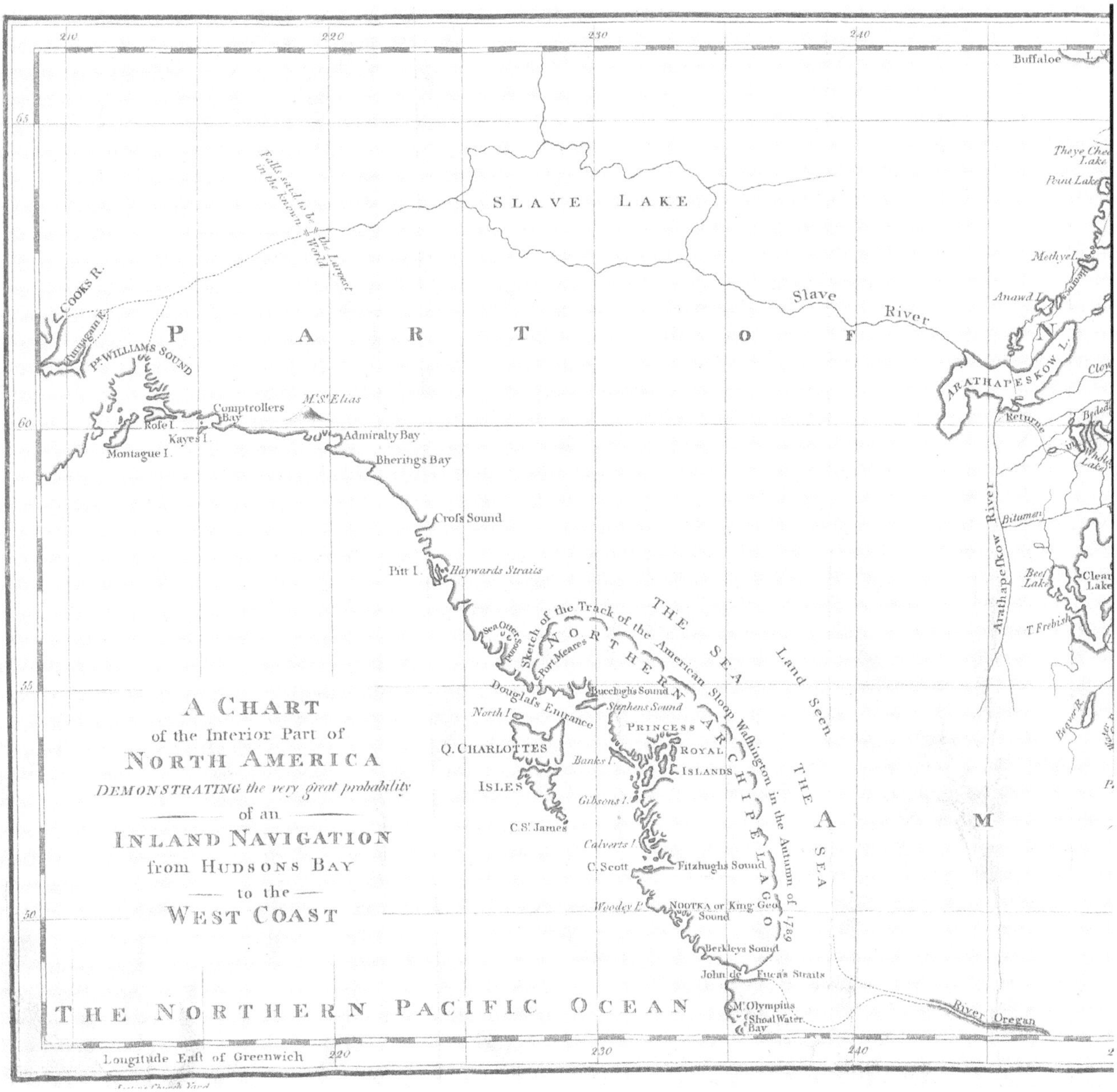

"A Chart of the Interior Part of North America Demonstrating the Very Great Probability of an Inland Navigation from Hudson's Bay to the West Coast." (Private collection of James Walker)

America was as yet unexplored, and it would not be until mid-1792 that British and Spanish explorations determined that Vancouver Island was in fact an island.

Beyond the "Northern Archipelago," Meares noted the presence of "The Sea" and, farther to the east, "Land Seen." In the absence of clear delineations of where archipelagoes stopped and main land began, readers of this map might have surmised that a ship could sail farther to the east toward Hudson Bay than previously thought. In his *Voyages,* Meares cited his own observations to support these conjectures. On July 13, 1788, he had sent a longboat with 13 men to reconnoiter the Strait of Juan de Fuca. Meares reported that "though the boat had not succeeded in the principal object of our expedition, yet it did not return without being able to communicate some knowledge of the straits of de Fuca up the strait and at that distance from the sea it was about 15 leagues. She had sailed near 30 leagues broad with a clear horizon stretching to the East for 15 leagues more. Such an extraordinary circumstance filled us with strange conjectures as to the extremity of this strait which we concluded at all events could not be at any great distance from Hudson's Bay" (Meares 1790, 179).

At the lower edge of the map, Meares noted and named Shoalwater Bay (present-day Willapa Bay in Washington state) at 46°43′ and Mt. Olympus, and above that location Juan de Fuca Strait, first sighted by fur trader Charles William Barkley in 1787. Using creative geography, Meares made the River Oregon (the River of the West) empty into either "The Sea" or Juan de Fuca's Straits. Perhaps the most audacious element of Meares's map was the legend "Sketch of the Track of the American Sloop Washington in the Autumn of 1789," which referred to Meares's belief that John Kendrick in the *Lady Washington* had entered the Strait of Juan de Fuca in 1789 and sailed in a great arc, exiting at Douglas Entrance (present-day Dixon's Entrance) at 54–55°.

Reviews of Meares's nearly 600-page book were mixed. Within two months, Dixon and Meares became antagonists and exchanged bitter comments in a series of public letters. On December 1, 1790, Dixon challenged Meares to produce the source for his statement in *Voyages* that "the track of this vessel [*Lady Washington*] is . . . of great moment as it now completely ascertains that Nootka Sound and the parts adjacent are islands and comprehended within the Great Northern Archipelago. The sea also which is seen to the East is of great extent and it is from this stationary point and the most Westerly parts of Hudson's Bay that we form an estimate of the distance between them" (Meares 1790, lvi). Dixon continued: "I cannot think of anything it [the map] resembles so much as the mould of a good old housewife's butter pat" (Howay 1969, 40). On January 1, 1791, Meares defended himself, writing that he had obtained his information from a close associate of John Kendrick, a claim that Kendrick himself never made.

49. Aaron Arrowsmith's Map of the Northwest Coast, 1790

INTRODUCTION

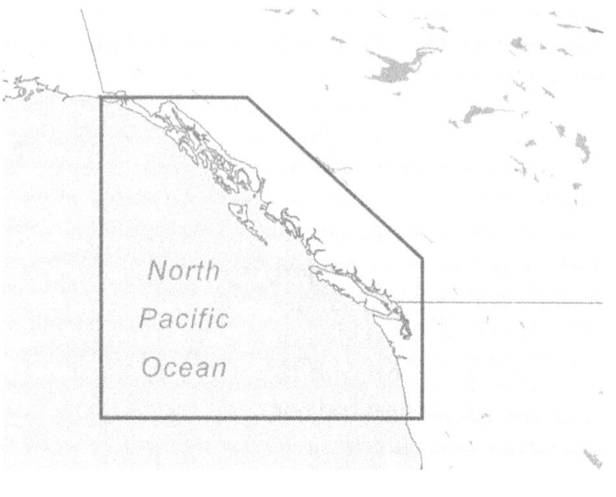

The principal motive for fur-trading expeditions to the Pacific Northwest was profit, but their success depended on determining the location of Natives who could provide the furs. Profit and mapping were tightly linked in the enterprise, and many early European fur traders in the Pacific Northwest published privately funded accounts that included maps of the region. Some British officials incorporated these accounts into their own work, but it is unclear how much of the geographical information reached the public.

More widespread dissemination of geographical information occurred through the works of commercial cartographers, such as London-based Aaron Arrowsmith, a mapmaker, engraver, and publisher whose firm produced influential maps of

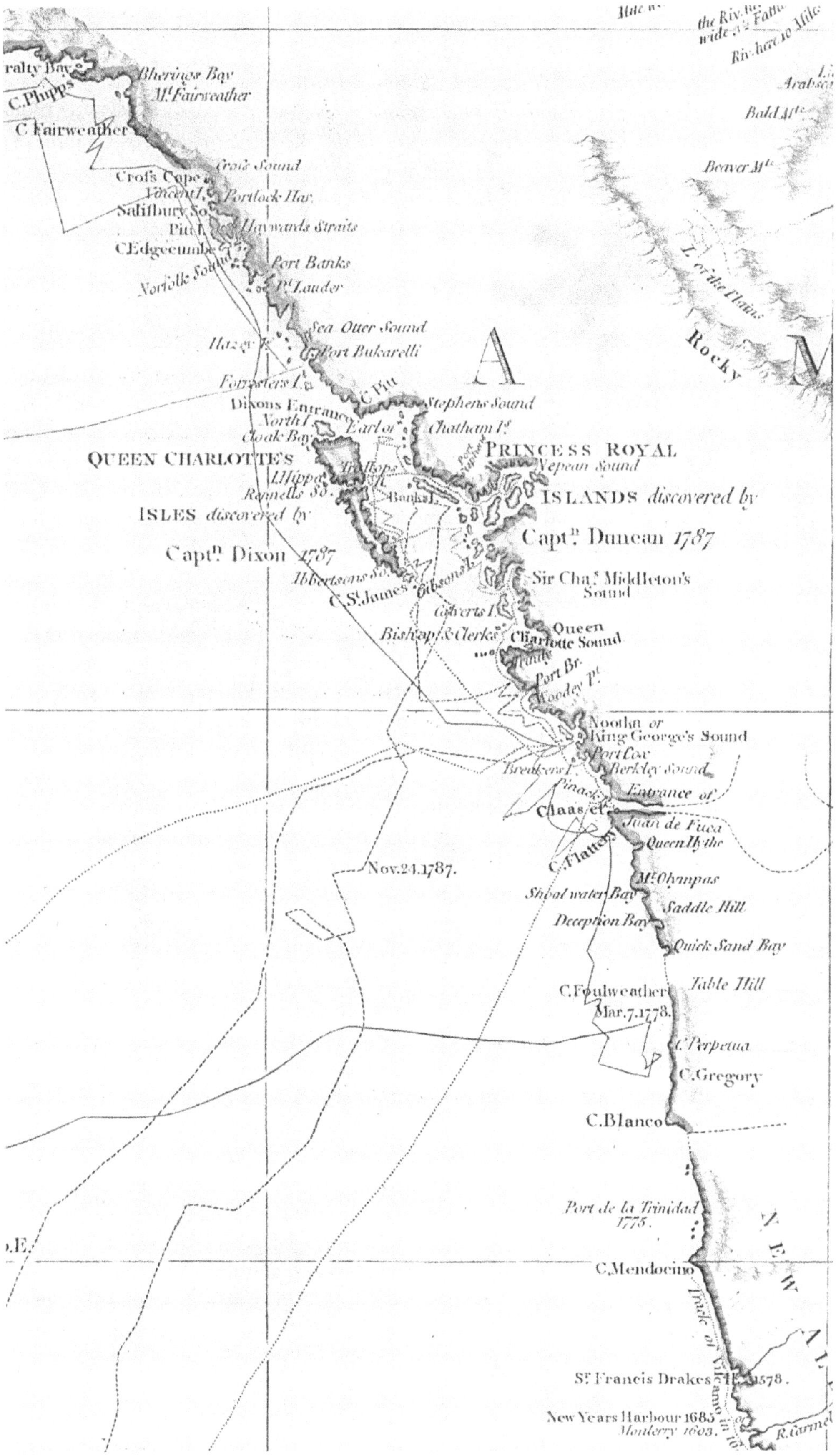

Detail of "Chart of the World on Mercator's Projection Exhibiting All the New Discoveries to the Present Time," 1790. (National Library of Scotland)

the world into the 19th century. This impressive map, published in April 1790, was Arrowsmith's first, and it established his international reputation as a source of the most recent geographical information from maritime discoveries in the Pacific Northwest. Arrowsmith's maps, one of his biographers noted, "were the most sought after and highly prized on three continents" (Martin and Martin 1999, 113). He was named hydrographer to the Prince of Wales in 1810 and hydrographer to the king in 1820, prestigious titles that most assuredly enhanced his reputation and his sales.

ANALYSIS

The map is titled "Chart of the World on Mercator's Projection, Exhibiting All the New Discoveries to the Present Time, with the Tracks of the Most Distinguished Navigators since the Year 1790, Carefully Collected from the Best Charts, Maps, Voyages, Etc Extant and Regulated from the Accurate Astronomical Observations Made in Three Voyages Perform'd under the Command of Captn. James Cook, in the Years 1768, 69, 70, 71, 72, 73, 74, 75, 76, 77, 78, 79 and 80," an extensive title that established the authority for Arrowsmith's information. The British public held Captain Cook in high esteem, and by 1790 most readers would have had some knowledge of British commercial pursuits in the Pacific Northwest. In addition, Arrowsmith intended the title to attract a market eager for the most recent cartographic information. The map, which consisted of eight copperplate engraved sheets, measuring 51 by 78 inches when joined, was published seven or eight times until 1808, always with revised or new information. On the full map (not shown here), Arrowsmith incorporated some 30 tracks of explorers from 1492 to 1787. In the Pacific Northwest he highlighted the results of Cook's voyages, but he also prominently noted the recent findings of several British fur traders, including George Dixon, Nathaniel Portlock, Charles Duncan, John Meares, James Colnett, and William Douglas. Dixon, on the *Queen Charlotte,* had traded on the Northwest Coast in 1786–1787 and had published his account with a chart of the Northwest Coast in 1789. On his map, Arrowsmith credited Dixon with the discovery of Haida Gwaii (Queen Charlotte Isles) in 1787. Arrowsmith also incorporated several of Dixon's place-names that remain today, including Port Banks, Dixon's Entrance, North Island (also called Langara Island), and Cloak Bay (named for the large number of fur garments obtained there). Farther north, he included names given by Dixon's sailing partner Nathaniel Portlock, on the *King George,* including Portlock Harbor.

A legend on the mainland opposite Haida Gwaii (Queen Charlotte's Islands on the map) noted the Princess Royal Island discovered by Captain Duncan in 1787. Nearby, Arrowsmith tracked Duncan's route in the *Princess Royal* and that of his sailing partner James Colnett on the *Prince of Wales.* Duncan skirted the eastern side of Haida Gwaii and named several places that Arrowsmith used, including Bishops and Clerks Island, Sir Charles Middleton's Sound (present-day Fitzhugh Sound), and two names that are still used, Napean Sound and Calvert Island. Arrowsmith also used place-name information he obtained from John Meares on his trading expedition to the coast in 1788 and 1789 (for example, Port Cox in Clayoquot Sound on Vancouver Island) and from William Douglas's expedition in 1788, Sea Otter Sound, which is still in use. Arrowsmith amended this information on subsequent editions of the map. For the 1799 edition, for example, he incorporated the most recent discoveries of George Vancouver and eliminated references to Duncan's less complete descriptions.

Arrowsmith published his map at a time when news of the dispute between Britain and Spain at Nootka Sound was a well-known political item. The dispute was over the Spanish seizure of British vessels and the imprisonment of British sailors over an alleged violation of Spanish claims of sovereignty in the Pacific Northwest. Pamphlets, newspapers, and maps such as Arrowsmith's brought the events of what was known as the Nootka Sound Controversy into public consciousness. Several times during the heated exchange of letters between John Meares and George Dixon in 1790–1791, Arrowsmith's map was referenced as an authority of the geography of the Pacific Northwest. Alexander Dalrymple also made use of the map, and the foreign secretary, the Duke of Leeds, used the map in 1790 as part of his case for the territorial sovereignty of Great Britain in the Pacific Northwest.

Arrowsmith's map should be studied within this broader context. In 1790, Spanish claims in the Pacific Northwest threatened British commercial and imperial interests. Arrowsmith may only have intended to produce a map whose commercial success depended on the inclusion of "all the new discoveries to the present time . . . carefully collected from the best charts, maps, voyages, etc extant," but his venture earned him a reputation as well as profit. While this map was an accredited source, it also became a tool of political purpose, an authoritative construct of a British identity in the Pacific Northwest. It was hardly possible to read the place-names and legends on the map and know that explorers other than the British had been there.

Further Reading

Beresford, William. *A Voyage Round the World; but More Particularly to the North-west Coast of America.* London: Geo. Goulding, 1789, https://archive.org/stream/cihm_50493#page/n425/mode/2up.

Clayton, Daniel W. *Islands of Truth: The Imperial Fashioning of Vancouver Island.* Vancouver: University of British Columbia Press, 2000.

Cook, Warren L. *Flood Tide of Empire: Spain and the Pacific Northwest, 1543–1819.* New Haven, CT: Yale University Press, 1973.

Harlow, Vincent T. *The Founding of the Second British Empire, 1763–1793.* 2 vols. London: Longmans, Green, 1964.

Howay, F. W., ed. *The Journal of Captain James Colnett aboard the* Argonaut. Toronto: Champlain Society, 1940.

Howay, Frederic W. "A List of Trading Vessels in Maritime Fur Trade, 1785–1794." *Transactions of the Royal Society of Canada,* 3rd series, 24(2) (1930): 111–149.

Howay, Frederic W., ed. *The Dixon-Meares Controversy.* Bibliotheca Australiana Extra Series. 1929; reprint, New York: Da Capo, 1969.

Howay, Frederic W., ed. *Voyages of the Columbia to the Northwest Coast, 1787–1790 and 1790–1793.* Portland: Oregon Historical Society Press, 1990.

Martin, James C., and Robert Sidney Martin. *Maps of Texas and the Southwest, 1513–1900.* Austin: Texas State Historical Association, 1999.

Meares, John. *Voyages Made in the Years 1788 and 1789 from China to the N.W. Coast of America.* London, 1790.

Nokes, J. Richard. *Almost a Hero: The Voyages of John Meares, R.N., to China, Hawaii and the Northwest Coast.* Pullman: Washington State University Press, 1998.

United States Senate. Doc. 335, 32d. Congress, 1st Session, August 11, 1852.

Wagner, Henry R. *The Cartography of the Northwest Coast of America to the Year 1800.* Amsterdam: N. Israel, 1968.

Woollen, William Watson. *The Inside Passage to Alaska, 1792–1920.* Cleveland, OH: Arthur H. Clark, 1924.

Expedition of Jean-François de Galaup, Comte de La Pérouse

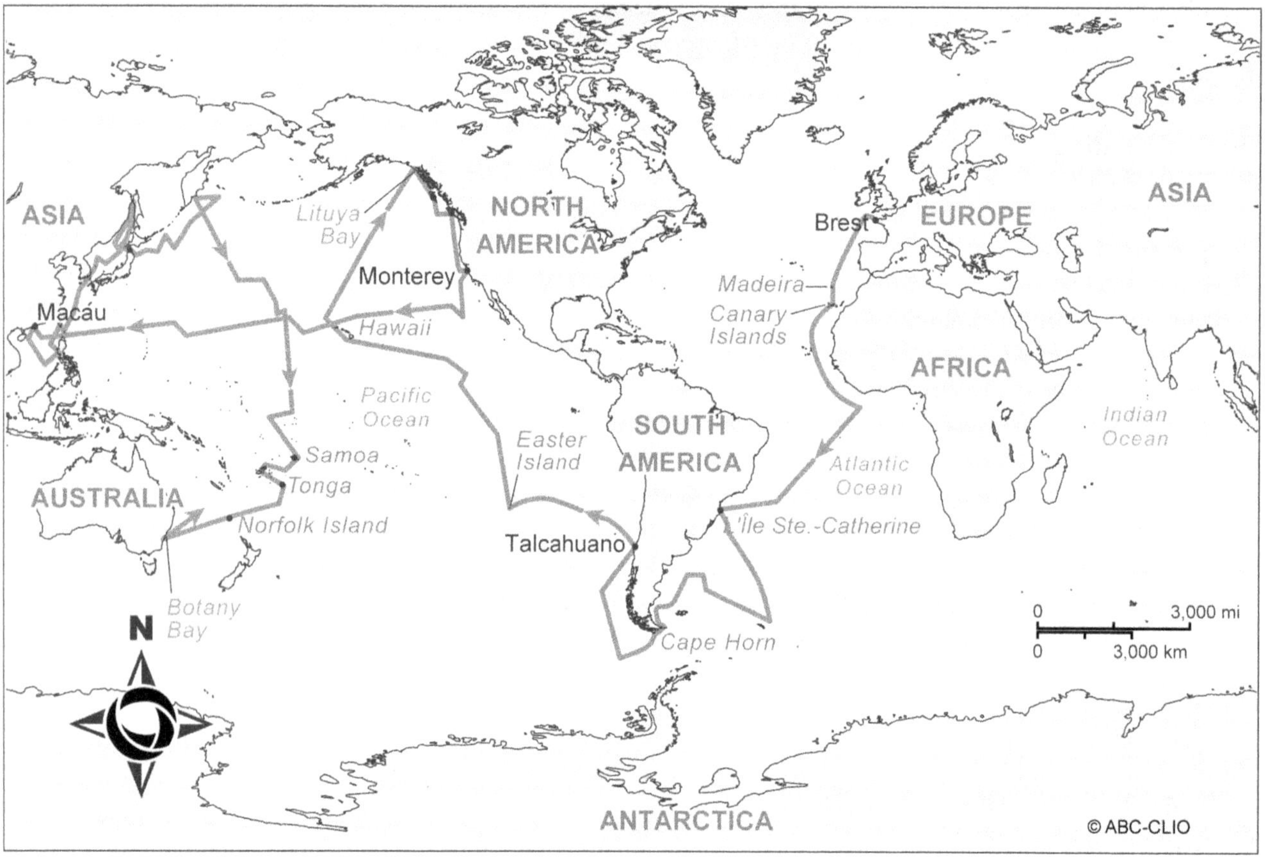

CHAPTER 6

Jean-François de Galaup, Comte de La Pérouse, Following Cook, 1785–1788

Jean-François de Galaup, Comte de La Pérouse, was the most successful French mariner to explore the Pacific Ocean in the 18th century. He was born Jean-François de Galaup in 1741 in a small town in southern France on the River Tarn, nearly 1,000 kilometers from the French naval port of Brest. The eldest of 10 children, Jean-François grew up in a bourgeois household, the son of a father who longed for his family to achieve noble status. Although his son's choice of a naval career at age 15 in 1756 did not sit well, he nevertheless supported his son. To enhance Jean-François's advancement in the navy, his father bestowed him with an amended family name that suggested nobility, La Pérouse, which he borrowed from a farmstead owned by the family. As Jean-François gained experience and reputation, the ploy appeared to work, for by the 1780s he was frequently referred to as Comte La Pérouse, even though he did not hold that title.

As a garde marine, La Pérouse lived, studied, and drilled at Brest, his port of attachment for his entire career. He studied mathematics, astronomy, hydrography, navigation, and seamanship, which included the practice and mastery of machines, shipboard devices, and weapons. His first service was during the Seven Years' War on the supply ship *Célèbre,* which sailed to Louisbourg on Cape Breton Island in northern Nova Scotia in May 1757. During the war, while sailing on the *Formidable* in 1759, La Pérouse saw military action and was captured by the British, only to be freed in a prisoner exchange in 1760. After the war ended he received a commission to ensign in 1764, and in 1767 he received his first command, the *Adour,* a slow storeship that he captained for two months on coastal duty. His most important break, though, came in 1768, when he served on the *Turquoise,* commanded by Charles d'Arsac de Ternay, that completed extensive survey work among islands off the French Atlantic Coast. Ternay became a friend and advocate for La Pérouse, whom another superior at the time described as an intelligent officer with a bright future.

By 1772 La Pérouse was in the Indian Ocean on the island of Mauritius, a key waypoint for ships sailing between the Atlantic and the East Indies. France had established an outpost on the island in 1715 at Port Louis, where Ternay was the commander. La Pérouse was dispatched on supply and survey missions in the Indian Ocean to bolster economic security at the French base, and he conducted a 10-month mission to French outposts in India in

In this painting by Nicolas-André Monsiau, La Pérouse (left) and Marquis de Castries (right) flank Louis XVI as he points out the planned expedition's route in the Pacific. The meeting took place in 1785, but Monsiau did not complete the painting until 1817. (Imagno/Getty Images)

1773–1774. His accomplishments had been important, but in peacetime there were few opportunities for advancement, so La Pérouse remained in Mauritius, even buying 150 acres of land with a fellow officer.

It was there that La Pérouse met and fell in love with Louise Eléonore Broudou, the 19-year-old daughter of Abraham Broudou, manager of the French naval storehouse and hospital in Port Louis. Although La Pérouse knew that his father would disapprove of his relationship with Eléonore and that his family already had selected a bride for him from an aristocratic family, he persisted in his romance. His assignments took him away from Mauritius and from France, giving him little chance to play the suitor's role, but his romantic attachment to Eléonore endured. Finally, in 1783 nearly 10 years after they first met, Jean-François and Eléonore married. He was 42 years old, and she was 28. His naval career limited the periods when they could live together, and in the end they had only 5 years of marriage before La Pérouse died in 1788.

The American Revolutionary War provided the opportunity for La Pérouse to advance. Louis XVI (1774–1792) had become king of France at the age of 20, and his chief ministers recognized the chance to contest British sea power through an alliance with the American rebels. By the time the French agreed to the Treaty of Alliance with the Americans in 1778, La Pérouse was a lieutenant and had met and worked with Charles-Pierre Claret de Fleurieu, director of ports and arsenals. The two met at Versailles when plans plotted out by Ternay and La Pérouse to attack British shipping in India had attracted Fleurieu's attention. Five years older than La Pérouse, Fleurieu came from a similar background, and their meeting sparked a recognition of their mutual interests in technology and navigation. Even though Fleurieu had risen higher and faster in naval ranks than La Pérouse, the two had an enduring relationship of cooperation and mutual admiration.

Inspired by James Cook's exploits, French mariner and navigator Jean-François de Galaup, Comte de La Pérouse (1741–1788), led a four-year exploration of the Pacific (1785–1788). He explored the Northwest Coast in 1786, before charting Asian Pacific and South regions. La Pérouse and his expedition were lost at sea in the southern Pacific in 1788. (The Print Collector/Print Collector/Getty Images)

Before the Ternay–La Pérouse plan to attack the British could materialize, the Ministry of the Navy followed the implications of the 1778 Franco-American alliance and changed its primary targets to British ships in the Atlantic to better aid the American rebels. As a result, La Pérouse sailed in the West Indies and along the Atlantic Coast of America for two years, harassing British ships. Interspersed were brief return voyages to Brest that allowed visits to Eléonore, who had left Mauritius for Nantes, France, in 1776.

In May 1778, La Pérouse took command of the frigate *Amazone* and raided the British in the North Atlantic, capturing 12 ships. His success as a raider brought him command of a new ship in 1780, the *Astrée.* But French losses in the West Indies by 1782 triggered a reassignment, and La Pérouse became captain of the 74-gun frigate *Sceptre,* which along with the *Astrée,* captained by Fleuriot de Langle, was charged with striking Hudson's Bay Company installations in Hudson Bay. The stunning and bloodless capture and destruction of company forts in August 1782 dealt an important blow to British commerce. For that campaign and other attacks on British shipping, the new U.S. Congress made La Pérouse a member of the Society of Cincinnati in May 1783. His success also brought a reward from Louis XVI—a pension of 800 livres, a little over one-fourth of his captain's pay.

The war over, La Pérouse returned to France, where he awaited reassignment and in 1784 began serving as Fleurieu's assistant in Paris. La Pérouse waited for an opportunity, and it came to him through Fleurieu and the first suggestions that France might launch a voyage of discovery in the

DID YOU KNOW?

In 1785 Jean-François La Pérouse used subterfuge, sending French spies posing as Spanish agents to London to get valuable, even unique, maritime instruments for use on his expedition to the Northwest Coast.

Pacific, one modeled on the great third voyage of James Cook. Fleurieu and the Ministry of the Navy were interested in the Northwest Coast, partly because Cook's survey was incomplete and partly because the region had become the center of a profitable fur trade with China. By February 1785 the minister of the navy, Charles Marquis de Castries, informed La Pérouse that he had forwarded his name to the king as commander of the expedition. Fast-paced planning between February and July involved La Pérouse, his naval superiors, and direct discussions with Louis XVI. By the end of July instructions had been finalized, the ships refitted for duty, and the officers and crews selected. On August 1, the expedition set sail from Brest for the Pacific.

The 44-year-old La Pérouse commanded the *Astrolabe,* while his friend Fleuriot de Langle captained the *Boussole.* The ships, two similar corvettes, were refitted merchant ships, each displacing 450 tons and carrying 115 crew. Ten scientists accompanied the expedition. Also included on the ships were 20 pigs, 200 hens, and 30 sheep plus timbers, iron, and naval stores to make repairs. The journey took the *Astrolabe* and *Boussole* to Tenerife, the Falkland Islands, and around Cape Horn to Concepción, Chile, where La Pérouse altered his instructions, which had directed him to investigate southern sea islands not visited by Cook. La Pérouse reasoned that there was not time to conduct such a survey, so he sailed to Hawaii and then directly to the Northwest Coast.

By July 1786 the expedition had sighted Mt. Fairweather at 58° north on the Alaskan coast, a landmark Cook had marked on his map. Coasting south, La Pérouse encountered Lituya Bay, where he remained for nearly a month surveying the unusual opening in the coastline, partly because he thought it might lead to an extensive interior waterway. That hope proved wrong, but this visit constituted the expedition's only lengthy hydrographic investigation of the coast and the only extended contact with Native people. His interaction with Tlingit at Lituya Bay left La Pérouse with mixed feelings. While he marveled at the artisanship and skill of the Tlingit at sea, he found their culture repulsive and their personal habits repelling. Far more eventful was his personal loss, when 21 members of a survey crew died in boat accidents on July 13. The tragedy stalled the expedition until the end of July, when the ships left Lituya Bay with bereavement and a personal sense of failure that La Pérouse never shook during the rest of the voyage.

La Pérouse coasted south to Monterey on the California coast and then across the Pacific to the Mariana Islands and Macao, near the mouth of the Pearl River in China. There he tried unsuccessfully to sell furs that the crew had acquired at Lituya Bay before sailing north to Kamchatka on the Siberian coast. He sent Jean-Baptiste Barthélemy de Lesseps overland from Kamchatka to St. Petersburg, Russia, with his journals, objects from the expedition, and maps—an extended trek that took more than a year to complete.

From the Siberian coast, La Pérouse sailed south to Samoa, where a second tragedy struck the expedition. On December 11, 1787, on a routine shore landing at Tutulia to acquire freshwater, La Pérouse's close friend and fellow explorer Fleuriot de Langle and 11 others died in armed conflict with Samoans. La Pérouse bemoaned this horrible incident, but he restrained himself from retaliating. "I had the opportunity to destroy, send to the bottom of the sea and break up a hundred canoes in which there were more than five hundred people," he wrote in his journal, "but I was afraid to attack the wrong victims and the call of my conscience saved their lives" (Dunmore 1995, 404). Nonetheless, in his description and notation about the episode he laid most of the blame on de Langle, noting that he nearly had ordered him to remain on ship rather than go ashore. The loss of his friend pained La Pérouse greatly, but he determinedly and professionally continued the expedition according to his instructions.

In March, La Pérouse repaired to Botany Bay in Australia, where he sent more documents back to France and planned out the remainder of his survey of South Pacific islands. The expedition left Botany Bay in mid-March and then vanished, ending abruptly and in mystery. The *Astrolabe* and *Boussole* should have returned to Brest by spring 1789, but no word had reached Paris six months later, and Claret de Fleurieu worried about the expedition's safety. He began to organize a search mission, and the French government alerted English, Spanish, and American officials to have their navies report any information they might gather on La Pérouse and his ships.

Eléonore moved to Paris in the winter of 1791, when the government authorized a rescue mission. By April, though, the French had received news of ship wreckage in the Admiralty Islands that likely was the *Astrolabe* and *Boussole*. The French Revolution made political decisions difficult, and the king nearly fled the country in June. After appeals for aid, Eléonore persuaded the French National Assembly to appropriate funds for her personal support while she continued to hope for good news from the rescue mission. That expedition, commanded by Antoine Bruni d'Entrescasteaux, did not search the Admiralty Islands until January 1793, where they found no physical evidence of La Pérouse, his crew, or his ships. Searchers tried but failed to locate the ships' wreckage during the 19th century. Finally in the 1960s, using advanced underwater technologies and diving gear, the remains were located near the Santa Cruz Islands, west of the Solomon Islands in the South Pacific.

> **DID YOU KNOW?**
>
> For decades after the *Astolabe* and *Boussole* vanished in the South Pacific in 1788, French mariners searched for La Pérouse's lost ships, but it was not until 2008 that divers found the ships' remains near Vanikoro in New Caledonia, northeast of Australia.

During the expedition La Pérouse had sent journals, documents, and charts to Paris in batches, but it was not until army general Baron Louis Marie de Milet de Mureau worked on them that publication became a possibility. He completed his work during the worst of the violent revolutionary period, and the account of the great expedition was published in three volumes in 1797.

50. The King's Instructions, 1785

INTRODUCTION

The background for the French exploratory expedition to the Pacific Ocean in 1785 began in the wake of a very successful British exploration in the late 1770s under James Cook. Jean-François de Galaup de La Pérouse, along with many other French mariners, was an admirer of Cook and could imagine no greater experience than sailing in Cook's wake and extending his own knowledge of the hydrography of the Pacific Ocean. French officials, including minister of the navy Charles Eugène de la Croix, Marquis de Castries, and Louis XVI, saw great promise for an official expedition that would confirm French prominence in the world of science and open new opportunities for commerce and even colonization. The Marquis de Castries probably began discussions with the king in 1784, but the first suggested plan came from William Bolts, a Dutch adviser to international merchants who had earlier tried to interest Austria and Russia in financing a Pacific expedition to counterbalance British commercial interests. In January 1785 Bolts sent a letter to Castries in which he argued that France could establish a foothold on the Northwest Coast. On the basis of Bolts's plan, Charles-Pierre Claret de Fleurieu wrote to Louis XVI, who embraced the idea. Fleurieu, Castries, and the king participated in planning the expedition, which was mostly completed by April. The first part of the final five-part document of instructions, dated July 1785, concerns activities that were planned to take place in May 1787, after the expedition rounded Cape Horn, sailed into the Pacific, surveyed Australian waters, and proceeded to Hawaii.

Private Instructions from the King to the Sieur de la La Pérouse, Captain in the Navy, Commanding the Frigates *Boussole* and *Astrolabe*, June 26, 1785

Part I
Plan of the Voyage

He will make without delay for the north-west coast of America; and for this purpose he will, proceed to the latitude of 30° north. . . .

He will be particularly careful to examine those parts, which were not seen by captain Cook, and of which no idea can be formed from the accounts of the Russian and Spanish navigators. He will make a strict search in the parts not yet known, to see whether there be not some river, or some narrow gulph, forming a communication, by means of the interior lakes, with some parts of Hudson's bay.

He will pursue his search as far as Behring's Bay, and Mount Saint Elias; and will visit the harbours of Bucarelli and los Remedios, discovered by the Spaniards in 1775.

Part II
Subjects Relating to Politics and Commerce

He will likewise enquire in what latitude furs may begin to be procured; what quantity the Americans can furnish; what commodities are best adapted for the trade; what conveniences may be found for making a settlement on this coast, if this new trade should offer the French merchant sufficient advantages, to induce him to engage in it, with the View of exporting the furs to China, which is said to be a ready market for them.

Part III
Operations Respecting Astronomy, Geography, Navigation, Physics, and the Different Branches of Natural History

The most important object for the safety of navigation is, to determine with accuracy the latitudes and longitudes of the places at which he may touch, or in sight of which he may sail.—With this view he will recommend to the astronomer employed on board each frigate to attend with the greatest exactness to the time-keepers, and to neglect no favourable opportunity of ascertaining on shore, whether they have kept the regularity of their going at sea. . . .

The sieur de la Pérouse will cause accurate charts of all the coasts and islands he shall visit to be drawn; and if, they be already known, he will examine the accuracy of the description and charts given by other navigators.

For this purpose, as he sails along coasts, and in sight of islands, he will cause their bearings to be accurately taken, by means of the reflecting circle, or the azimuth compass; and he will remember, that the bearings, which may be employed with most security for the construction of charts, are those, in which the situation of one cape, or any other remarkable object, can be rectified by those of another.

Part IV
Of the Conduct to Be Observed Towards the Natives of the Countries Where the Two Frigates May Land

At his arrival in each country, he will endeavour to conciliate the friendship of the principal chiefs. . . .

He will endeavour to learn what wares and commodities of Europe they prize the most highly. . . .

He will never send a long-boat, or any boat whatever, ashore, without it's being equipped with it's guns, and provided with musquets, cutlasses, pole-axes, and ammunition, in sufficient quantity, and commanded by an officer, whom he will order never to lose sight of the boat committed to his charge, and always to leave a party to guard it.

Part V
Precautions to Be Taken for Preserving the Health of the Crews

The sieur de la Pérouse is not ignorant, that one of the precautions most efficaciously contributing to preserve the health of seamen, is a constant attention to extreme cleanliness in the ships, and in their persons. For this purpose he will have recourse to every known method, such as ventilators, fumigations, perfumes, to renew and purify the air in the hold and between decks. Every day, when it can be done, he will order the hammocks and clothing of the crews to be exposed to the open air; and that the sailors and others, who compose them, may not be negligent of personal cleanliness, he will divide them into companies, the inspection and superintendance of which shall be shared among the officers of each frigate.

Source: J. F. G. de La Pérouse, *A Voyage Round the World, Performed in the Years 1785, 1786, 1787, and 1788; by the Boussole and Astrolabe, under the Command of J. F. G. de la Pérouse,* 3 vols., translated from the French, 3rd ed. (London: A. Hamilton for G. G. and J. Robinson, J. Edwards and T. Payne, 1807), 1:14, 23, 29, 35, 42, 45–46, 50–51, 54, 56.

ANALYSIS

The decision to launch a major scientific expedition to the Pacific Ocean in 1785 followed earlier French expeditions that represented continued interest in science and discovery. In the 1740s, Charles-Marie de La Condamine had conducted experiments in present-day Ecuador that proved Isaac Newton's theory of Earth's shape, thereby scientifically establishing an accurate measurement of latitude. Two decades later in 1766–1769, Louis-Antoine de Bougainville commanded a French circumnavigation that took him through the South Pacific, but the proximate reasons for the expedition were based on the French desire to duplicate and exceed Cook's exploration of the Pacific and to pursue the quest for the Northwest Passage. Pursuit of the fabled waterway through North America to Hudson Bay and the potential for France to participate in the Pacific fur trade with China were central to the first and second parts of the instructions to La Pérouse. The instructions, however, cautioned him not to pursue any interior waterway, because this would delay him from completing other parts of his mission.

The third part of the instructions, which cover navigation and astronomy, held sway, because it was the planners' purpose to describe and chart unknown places and increase knowledge. To achieve this goal, La Pérouse needed able scientists. He enlisted Joseph Lepaute Dagelet, an astronomer and mathematician; Robert de Lamanon, a geologist; Joseph Hugues La Martinière, a botanist; and Louis Dufresne, a naturalist. To acquire the best instruments, Paul Mérault de Monneron, a graduate engineer and captain in the corps of engineers, undertook a spy mission to London in March 1785, where he disguised himself as a Spanish trader and tried to purchase sophisticated instruments and naval stores recommended by a veteran of Cook's third voyage. Joseph Banks, head of the Royal Society in London, learned of the ruse, but rather than expose Monneron he quickly helped him acquire two "dipping needle" compasses that Cook had used for the French voyage. For Banks, aiding the French effort in expanding Cook's explorations trumped nationalistic pride.

It is clear that King Louis played a significant role in planning the expedition, but historians differ on how much his wishes determined the course and shape of the voyage. Whatever his desires were, he endorsed Claret de Fleurieu's original plan in 1785. Jean-Nicolas Buache de la Neuville, the royal geographer, also reviewed the plan, as did the minister of the French Navy, the Marquis de Castries. The king had ruled out making the expedition both commercial and scientific because he recognized the political hazard in conducting commerce with royal vessels. Nonetheless, he embraced the potential for French commerce in the Pacific and agreed with Fleurieu that France needed to contest British sea power in the region.

The king's personal instructions emphasized the safety of crew and officers and also directed La Pérouse on how he should engage with Native people, especially in positioning the French for potential commercial development. This part of the instructions drew on the experiences of earlier French expeditions, but it also revealed the uncertainty among the French elite about the character of non-Europeans. Enlightenment philosophers had labeled Natives in American and Pacific lands "noble savages," arguing that seemingly unsophisticated peoples were naturally moral and uncorrupted, unlike so-called civilized populations. Experienced mariners who had encountered Natives had more critical viewpoints, and La Pérouse in particular often interjected harshly critical comments on Native behavior in his journal, from the ships' first encounter with Pacific peoples at Easter Island to their extended stay among the Tlingit on the Northwest Coast.

The threat of scurvy, addressed in the fifth part of the king's instructions, was a feared plague on any long deepwater voyage. The specter of debilitating outbreaks of scurvy, which could cripple sailors and jeopardize voyages, made captains, crew, and sponsors eager for remedies. All seafaring nations addressed the problem, and most recognized that fresh food, clean water, and clean living conditions were the best preventatives.

51. Sighting the Northwest Coast, 1786

Following the route laid out in his instructions, La Pérouse sailed from Brest in July 1785, first to Madeira, then to Tenerife around Cape Horn, stopping at Concepción, Chile, in March 1786.

He knew that following the planned route in the Pacific would jeopardize his survey of the Northwest Coast and that sailing north could get him to Alaskan waters by summer. The king and the minister of the navy had given La Pérouse latitude to decide his course, so he abandoned the instructions and sailed to Easter Island, making Hawaii by late May 1786. He bypassed planned investigations of the South Pacific to survey the areas not investigated by Cook. By letter from Concepción in March, La Pérouse alerted Marquis de Castries that he had a "new plan of campaign" to carry out the spirit of his instructions while fundamentally changing them. "Since my departure from Europe," La Pérouse explained, "all my thoughts had been directed towards the routes followed by the old navigators" (Dunmore 1995, 473). This indicated how important he viewed his investigation of rumored locations in the North Pacific, especially pursuing the blank spots on Cook's rendition of the Northwest Coast. La Pérouse's journal records his first sighting of the coast on June 23, 1786.

La Pérouse's First Sighting of the Northwest Coast

Of this part of America, as far as Mount St. Elias, in the latitude of 60°, captain Cook had only a transient view, Nootka Sound excepted, where he made some stay: but from Mount St. Elias to the point of Alashka, and thence to Icy Cape, that celebrated navigator explored the coast with that courage and perseverance, of which all Europe knows him to have been capable. To explore that part of America, therefore, which lies between Mount St. Elias and Monterey Bay, must be a work of high importance to commerce and navigation: but it would require some years, and we will not deny, that, having only two or three months to bestow on it, on account of the season, and still more of the vast plan of our voyage, we must leave much of the minutiae to the navigators that may follow us. Many ages in all probability will elapse, before all the bays and harbours of this part of America are perfectly known. . . .

As we advanced to the north, and approached America, we observed sea-weeds of a species entirely new to us. A ball of the size of an orange terminated a stalk forty or fifty feet long, resembling an onion run to seed, but much superior in size. . . . As the mist cleared away, a long chain of mountains covered with snow burst at once upon our sight, which we might have discerned thirty leagues farther off, had the weather been clear. We distinguished in these the Mount St. Elias of Behring, with it's summit rising above the clouds.

The sight of land, after a long voyage, usually excites feelings of delight; but on us it had not this effect. The eye wandered with pain over masses of snow, covering a barren soil, unembellished by a single tree. The mountains appeared to be at a very little distance from the sea, which broke against the cliffs of a table-land three or four hundred yards high. This plain, black as if burned by fire, and totally destitute of verdure, formed a striking contrast with the whiteness of the snow. . . .

I have already mentioned a table-land three or four hundred yards high, serving as a base to vast mountains, a few leagues within it, and soon we perceived to the eastward a low point covered with trees, which appeared to join the table-land, and terminate at a short distance from a second chain of mountains which appeared farther to the east. We were all unanimously of opinion, that the table-land terminated at this low point covered with trees, and was an island separated from the mountains by an arm of the sea, the direction of which must be east and west like that of the coast, and in this supposed channel we expected to find a convenient shelter for our Vessels.

Towards this point I steered my course, keeping the lead constantly going, and never found less than forty-five fathoms of water, with a muddy bottom. . . . I had sent my barge, as early as ten in the morning, under the command of Boutin, to reconnoitre this channel or bay: Messrs. de Monti and de Vaujuas had been sent from the Astrolabe for the same purpose. . . .

[T]he sky appeared very lowering; every thing indicated, that the dead calm which obliged us to come to an anchor, would be succeeded by foul weather. At length, at nine in the evening, our boats returned, and the three officers unanimously reported, that there was neither channel nor river; and that the coast merely formed a pretty large semicircular bay to the north-east. . . . M. de Monti had landed with great difficulty; and, as he was the commanding-officer of this little division of boats, I gave the place the name of *Baie de Monti.*

Source: J. F. G. de La Pérouse, *A Voyage Round the World, Performed in the Years 1785, 1786, 1787, and 1788; by the Boussole and Astrolabe, under the Command of J. F. G. de la Pérouse,* 3 vols., translated from the French, 3rd ed. (London: A. Hamilton for G. G. and J. Robinson, J. Edwards and T. Payne, 1807), 2:61–66.

ANALYSIS

La Pérouse had focused his attention on James Cook's exploration of the Northwest Coast because of the enormity of the achievement and his own ambition to complete a survey of the region for France. The *Astrolabe*'s route, from north to south along the coast, reversed the tracks of Cook and Spanish explorers. In addition, La Pérouse's instructions warned him off resurveying areas already documented by Cook, a reference to Prince William Sound and Cook's Inlet in the Gulf of Alaska. La Pérouse's mention of the coast from Mt. St. Elias to Monterey reflects two ambitions. First, he hoped to end speculation about a Northwest Passage through North America, which Cook had nearly concluded did not exist. La Pérouse was also skeptical, but he nonetheless pursued the question. Second, he saw an opportunity for France to establish a port on the Northwest Coast that would not conflict with existing Russian, English, and Spanish claims.

Before encountering the Mt. St. Elias landscape, La Pérouse had made landfall at Easter Island in early April 1786 and at Hawaii in late May. The verdant landscapes of both island groups and his engagement with Native people there had not prepared him and his crew for the fog, mist, turbulent sea conditions, and foreboding appearance of the Northwest Coast. The conditions, even in June, made close navigation along the coast difficult. In May 1778, poor conditions had prevented Cook from putting into shore for 10 days as he sailed north from Sitka to Prince William Sound. La Pérouse faced similar conditions in 1786, but he was determined to fill in the missing details on Cook's charts.

The seaweed that La Pérouse encountered alerted him to the likelihood of land. The strange kelp, which reminded him of an overgrown onion, was probably *Nereocystis luetkeana,* a species of seaweed that flourishes in Alaskan waters. In recent times the kelp has been harvested commercially for use as a fertilizer, but Tlingit in the 18th century would have harvested it for quite different uses: for fishing lines fashioned out of the plant's "hair," a tough, silky, threadlike growth, and for food, harvested by slicing the hollow kelp tube.

La Pérouse's instructions guided his survey work, which included a specific directive: "He will investigate with the greatest care whether, in parts that are not yet known, there might not exist some river, or some narrow gulf, that communicate with some part of Hudson Bay by way of the interior lakes" (Dunmore 1994, cxxx). He had great doubts about claims made by earlier navigators, especially the purported voyage of Juan de Fuca and the rumored explorations of Bartholomew de Fonte, whom La Pérouse refers to as Fuentes. Although both mariners' exploits turned out to be fabrications, La Pérouse could not afford to overlook their claims of discovery. After he left the Lituya Bay area, for example, he recorded in his journal in August 1786 that he imagined that some Spanish sailor had penetrated the coast "and that this single real event might have been the basis for the ridiculous tale of Fuentes" (Dunmore 1994, 148). When they reached present-day Yakutat Bay, which he named Monti Bay, he sent three boats to survey the opening in the headland. The report indicated no interior river, hence no entrance to a Northwest Passage, but the officers' surveys also did not disclose the extensive inlets and waterways that emptied into the bay.

52. Map of Port des Français, 1797

INTRODUCTION

The present-day name of Port des Français (Harbor of the French) is Lituya Bay, located at latitude ~58°38′ north on the southeast coast of Alaska. This map of the bay is from the atlas that accompanied the first English edition of A Voyage Round the World, in the Years 1785, 1786, 1787, and 1788, *by J. F. G. de La Pérouse (1798), published one year after the first French edition and in the same year as* Voyage of Discovery *by George Vancouver. Although not as detailed as Vancouver's charts, La Pérouse's atlas contains 69 beautifully engraved maps, plans, and views, including this unique plan of Port des Français. Remarkably, therefore, English readers in 1798 had access to two sets of cartographic material—one French and the other British—engraved to the highest standards. Both sets of journals and atlases were reissued in England and France, so images of the port were widely disseminated, and the well-publicized tragedy that occurred there meant that readers in England and France knew about the harbor. For years after, whalers in the area often referred to the harbor as Frenchman's Bay.*

The maps in this English edition were produced by a different engraver than the one employed in Paris. Both, however, skillfully used the relatively recent method of engraving hachure lines to depict relief and elevation of the many mountains in this part of southeastern Alaska. The lines were cut so that their lengths indicated slope direction, with progressively closer spacing and a denser and wider breadth indicating elevation. Even though this was a planar or vertical view of the region, readers could appreciate the mountainous topography.

ANALYSIS

An appreciation of this map requires a broader understanding of the events of July 1786, when La Pérouse explored this inlet. It was James Cook's third voyage that had motivated him and the sponsors of the French expedition to the Northwest Coast. La Pérouse admired Cook and frequently referred to his surveying, navigational, and mapping skills and his success in maintaining the health of his crew. In a letter to the minister of the navy, La Pérouse wrote that "I have scrupulously avoided changing the names which Captain Cook bestowed on the various

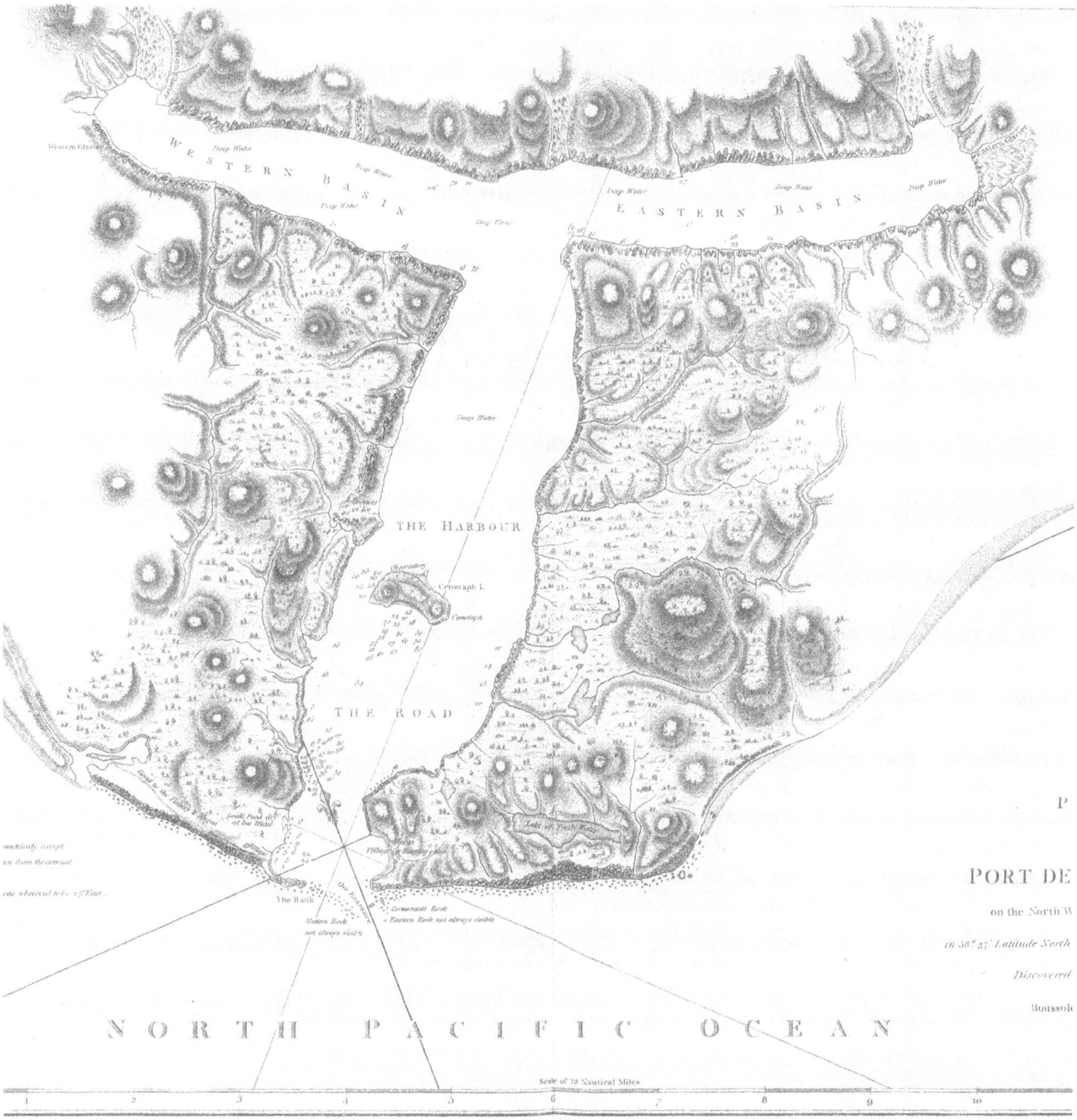

"Plan of Port des Français on the North West Coast of America." (Private Collection of James Walker).

headlands he discovered, but you will notice, My Lord, that we saw the coast of America from much nearer than that celebrated navigator, and so we were permitted to name ports, bays, islands and inlets he did not even suspect. . . . [F]rom Nootka to Monterai [Monterey] I sought only the points which Captain Cook had not been able to survey and which remained as dotted lines on his chart" (Dunmore 1995, 491). On July 2, 1786, La Pérouse came upon the entrance to a harbor that he recognized was one of those "dotted lines." With some difficulty but great anticipation, he had discovered an important harbor unknown to Cook. La Pérouse entered "this harbor [that] had never been seen by any navigator—it was 60 leagues N.W. of Puerto de los Remedios, which is the limit of the Spanish navigations 200 leagues from Nootka and at about the same distance from [Prince] Williams Sound; thus if the French government had any plans for a trading post along this part of the American coast, no other nation could raise the slightest objection" (Dunmore 1994, 103).

Over the next few days, La Pérouse assessed the Port des Français area and its resources: "I think there is no country where the sea otter is more common than in this part of America, and I would not be surprised if a trading post with a trading range of a mere 40 or 50 leagues would collect ten thousand skins in a year" (Dunmore 1994, 107). And later, "We had already been to the end of the bay which is perhaps the most extraordinary place on earth. . . . This is the channel by which we planned to enter into the heart of America; we thought that it might lead to a great river running between the mountains, which had its source in one of the great lakes in northern Canada. Such was our dream" (Dunmore 1994, 109–10). Survey crews, however, soon encountered impenetrable ice barriers that created a cul-de-sac, eliminating the bay as a Northwest Passage.

After nearly two weeks La Pérouse was ready to leave Port des Français, but tragedy struck on July 13 when 21 officers and men in two boats were swamped while conducting surveys near the mouth of the bay. All drowned in what was perhaps the greatest single loss of life from noncombatant or disease causes during 18th-century exploring voyages to the Pacific Northwest, aside from the loss of the expedition's men in 1788.

La Pérouse delayed his departure from the harbor until July 30. He named the island he had "purchased" Isle du Caenotaphe in honor of his men, a name that endures. The name of the bay today is Lituya Bay, a Tlingit-derived name bestowed by Russian visitors as early as 1802. Nearby Cape Beautemps, named by La Pérouse, is now known by the name originally given by Captain Cook, Cape Fairweather. Also nearby, Mt. La Perouse in Glacier Bay National Monument was named in 1874, a tribute to the French commander who at one time considered Port des Français to be a place of the greatest potential and the greatest loss in the Pacific Northwest.

53. Lituya Bay Tragedy, 1786

INTRODUCTION

Coasting south from their first sighting of North America, the Boussole *and the* Astrolabe *found a promising inlet that might be the access to a lake, a river, or another interior waterway—a Northwest Passage. It had appeared on Cook's chart as an undefined place, and La Pérouse determined to explore and map it. But it turned out that the enticing bay had veiled dangers. "There was little wind and tide was ebbing so fast," La Pérouse wrote in his journal, "that we could not counter it. The Astrolabe was driven away fairly fast and I was compelled to drop anchor in order to avoid being carried by a current about whose direction I was unsure. . . . The fierce current, which our officers had not mentioned, had made me very doubtful about this harbor" (Dunmore 1994, 102). Nonetheless, de Langle thought the bay promising and persuaded La Pérouse to enter on the tide, which they did and anchored in just under 20 feet of water. Their anchorage was dicey because of the tidal currents and the danger of being swept onto the rocky entrance.*

La Pérouse ordered soundings and investigations to find a better anchorage. It was during additional surveys that tragedy struck. The description of events comes from a survivor, Charles-Marie Fantin de Boutin, a ship-of-the-line lieutenant who participated in the survey.

Charles-Marie Fantin de Boutin's Account of Tragedy at Lituya Bay

On 13 July at 5:50 a.m. I left the *Boussole* in the small boat; my orders were to follow Mr [Charles] Descures who commanded our Biscay boat and Mr [Edward La Borde] de Marchainville was to join us, in charge of the *Astrolabe*'s Biscay boat. . . . [T]o take soundings in the bay . . . [Descures] was expressly forbidden to expose to the slightest danger the boats under his command and to go near the pass if the seas were breaking or even if there was any swell. . . . I saw that the seas were breaking right across the pass and that it would be impossible to approach. At the time Mr Descures was ahead of me, oars raised, and seemed to be waiting for me, but when I had come within a musket shot he went on . . . [and] I was never able to catch him up. . . . I could see the *Astrolabe*'s boat a quarter of a league away inside the bay; Mr Descures then hailed me, saying with a laugh, 'I think that the best thing we can do is to go and have lunch, as the seas are breaking horribly in the pass.' I replied that this was quite evident and that I imagined our work would be restricted . . . but having glanced towards the E. he saw that we were being carried by the tide; I saw it too. . . . All at once I found myself among the biggest waves which almost swamped my boat. . . . I finally realized that if only I just made 50 *toises* [100 meters] to the E. I would find less dangerous waters; I made every effort to succeed, struggling to starboard between each wave, and at 7:25 I was out of danger. . . . I had seen our Biscay boat submerged among the breakers. . . . It was impossible for the shipwrecked men to get away from . . . such a fierce current as they found themselves in, and they must have been dragged away by the tide which flowed out to sea until 8:45. And anyhow, how could even the best swimmer resist for more than a few moments to the fierceness of those waves? . . .

After all these details, I feel I should explain the reasons for Mr Descures' actions. . . . [H]e thought he was keeping at a distance which was more than adequate for safety. . . . I cannot say how easy such a mistake was or why it was not possible to estimate the strength of the current. . . . Without wishing to list all the reasons that contributed to make us feel so fatally confident, I must say that on the day we entered the bay our boats took soundings in every direction for more than two hours without sensing any current; it is true that when our frigate were entering they were driven back by the ebb tide but . . . on 11 July our two commanding officers went themselves . . . to take soundings in the pass. . . . So one must conclude that on 13 July the violence of the current was due to special circumstances.

Source: John Dunmore, trans. and ed., *The Journal of Jean-François de Galaup de La Pérouse, 1785–1788,* Vol. 1 (London: Hakluyt Society, 1994), 116–120. Used by permission of the Hakluyt Society.

ANALYSIS

Sounding surveys were essential to charting bays, harbors, and naval obstructions. Taking soundings required sailors to use small boats and physically measure water depths, using weighted lines. The long sounding lines—the oldest navigational tool, in use since the sixth century—had graduated markings of varied lengths. Sounding weights carried the lines to the seafloor to accurately measure water depth. At Lituya Bay, the survey party used 700-foot sounding lines.

A Biscay boat was a small double-ended shallop, a slow sailing vessel used by Basques in the Bay of Biscay whale fishery. Charles Gabriel Morel d'Escures, a ship-of-the-line lieutenant and decorated officer, had directed the outfitting of the *Astrolabe* and was expert in ship craft. Éduoard Jean Joseph de Laborde Marchainville, a close associate of Fleuriot de Langle and ship-of-the-line lieutenant, was valued on the *Astrolabe* as an officer because he "had scientific training which La Pérouse lacked" (Dunmore 1985, 201). His brother also perished in the Lituya Bay disaster, a double tragedy, La Pérouse noted, that violated "a strict rule never to send the two brothers to carry out the same duties" (Dunmore 1994, 115).

Lituya Bay is one of the most unusual and dangerous sea enclosures in the world. The tide averages 15 feet and can create currents twice a day of more than 15 miles per hour. Slack-water conditions, when ships can pass safely, last for less than 30 minutes, and ebb tide conditions can produce dangerous waves. In addition, earthquakes and glaciers calving ice at the interior edges of the bay can generate huge waves. The highest sea wave ever recorded—1,740 feet—roared through Lituya Bay on July 9, 1958, scouring a glacier and denuding thousands of tall firs on the promontory. In July 1786, Boutin and his fellow sailors unknowingly ventured into the most dangerous spot on the Northwest Coast.

On the morning of July 13, the survey party worked its way from the *Astrolabe* and *Boussole* west toward the bay's opening. La Pérouse had anchored near an island in the middle of the bay, where he had erected an observation camp. Trade with some Tlingit had been ongoing from the ships' arrival. While La Pérouse had little good to record about his interactions with Tlingit chiefs, he did enter into a contract with one who offered to sell the island to the French for "several ells of red cloth, some axes, adzes, iron bars and nails" (Dunmore 1994, 111). When the disaster struck, however, La Pérouse embraced Tlingit aid in finding a surviving sailor. "The gestures of these rough men [Tlingit] indicated that they had seen the loss of the two boats and that any help would have been in vain. We filled their arms with presents and tried to get them to understand that all our riches would belong to the one could save a single man" (Dunmore 1994, 115).

Poor weather forced La Pérouse to remain at Lituya Bay until July 30, when the *Astrolabe* and the *Boussole* left to continue the expedition. On leaving, La Pérouse buried a message at a hastily erected monument on the island, which he named Isle du Caenotaphe (present-day Cenotaph Island): "At the entrance to this port, twenty-one sailors perished. Whoever you may be, add your tears to ours" (Dunmore 1994, 121).

54. The Tlingit Narrative of the Lituya Tragedy, 1886

INTRODUCTION

The disaster in Lituya Bay, where 21 sailors lost their lives, changed the course of the La Pérouse expedition and created great sorrow among the captains and crews of the Boussole *and the* Astrolabe. *The event also affected those Tlingit who witnessed the capsizing and drowning, and the story became part of Tlingit oral tradition. A century after the tragedy, anthropologist George Emmons reported that "Cowee, the principal chief of the Auk qwan of the Tlingit people, living at Sinta-ka-heenee, on Gastineaux Channel, told me the story of the first meeting of his ancestors with the white man, in Lituya Bay, where two boats of the strangers were upset and many of them were drowned" (1911, 296). When Emmons recorded the story in 1886, he quizzed Tlingit elders about their knowledge of La Pérouse and other Europeans who had visited their world generations before and discovered that no one knew the explorers' identities or the names of specific expeditions. Nonetheless, comparisons of the Tlingit story fit the European descriptions of the disaster.*

The story first describes the Tlingit canoe journey from Icy Straits in present-day Glacier Bay, southeast of and more than 50 miles in distance to Lituya Bay. The conclusion links the Tlingit accident with the French disaster.

Native Account of the Meeting between La Pérouse and the Tlingit

One spring a large party of Thluke-nah-hut-tees from the great village of Kook-noo-ow on Icy Straits, started north, under the leadership of three chiefs—Chart-ah-sixh, Lth-kah-teech, and Yan-yoosh-tick.

In entering Lituya, four canoes were swallowed by the waves and Chart-ah-sixh was drowned. The survivors made camp and mourned for their lost companions. While these ceremonies were being enacted, two ships came into the bay. The people did not know what they were, but believed them to be great black birds with far reaching white wings, and, as their bird creator, Yehlh, often assumed the form of a raven, they thought that in this guise he had returned to earth, so in their fright they fled to the forest and hid. Finding after a while that no harm came to them, they crept to the shore and, gathering leaves of the skunk cabbage, they rolled them into rude telescopes and looked through them, for to see Yehlh with the naked eye was to be turned to stone.

As the sails came in and the sailors climbed the rigging and ran out on the yards, in their imagination they saw but the great birds folding their wings and flocks of small black messengers rising from their bodies and flying about. These latter they believed to be crows, and again in fear they sought the shelter of the woods.

One family of warriors, bolder than the rest, put on their heavy coats of hide, the wooden collar and fighting head-dress, and, armed with the copper knife, spear, and bow, launched a war canoe. But scarcely had they cleared the beach when a cloud of smoke rose from the strange apparition followed by a voice of thunder, which so demoralized them that the canoe was overturned and the occupants scrambled to the shore as best they could.

Now one nearly blind old warrior gathered the people together, and said, that his life was far behind him and for the common good he would see if Yehlh would turn his children to stone, so he told his slaves to prepare his canoe, and, putting on a robe of the sea otter, he embarked and paddled seaward. But as he approached the ships the slaves lost heart and would turn back, and all deserted him save two, who finally placed him alongside. He climbed on board, but being hardly able to distinguish objects, the many black forms moving about still appeared as crows, and the cooked rice that they set before him to eat looked like worms, and he feared to touch it. He exchanged his coat of fur for a tin pan and with presents of food he returned to the shore. When he landed the people crowded about surprised to see him alive, and they touched him and smelled of him to see if it were really he, but they could not be persuaded to eat the strange food that he had brought to them.

After much thought the old man was convinced that it was not Yehlh that he had gone to and that the black figures must be people, so the natives, profiting by his experience, visited the ships and exchanged their furs for many strange articles.

It was at this time that two boats were lost at the mouth of the bay and many of the white men were drowned.

Source: G. T. Emmons, "Native Account of the Meeting between La Perouse and the Tlingit," *American Anthropologist* 13(2) (April–June 1911): 297–298.

ANALYSIS

Cowee's story only includes the notation that white explorers lost their lives at the mouth of Lituya Bay, but there is more about European explorers than is first evident. The story centers on the encounter between the Tlingit and the unusual geography of Lituya Bay, a place that lived frighteningly in Tlingit geography. The name comes from the Tlingit name Ltu.áa, meaning "the lake within the point," a designation that describes the enclosed geography of the bay, which fronts three glaciers and meets the Pacific Ocean through a narrow opening. The name also reflects the enchanted character of all places in Tlingit culture, where place-names indicate both physical and social realities. In Lituya, the Tlingit identify the home of "the man of Lituya," a monster who protects his deep watery environs by shaking the seas whenever people approach and annoy him. The power of the monster in Tlingit culture is represented in effigy by a pipestem that portrays a frog forcing water into the mouth of a bear, a graphic reference to the power of the convulsive tides and overpowering currents in Lituya Bay.

Chart-ah-sixh and his companions had been on a long trade journey to Yakutat Bay when they suffered the wrath of the monster in Lituya Bay. Their object was the acquisition of copper, which they fashioned into implements, weapons, and adornments. Copper also provided wealth, and the Tlingit traded it with Haida and other Native groups south of their homeland for canoes and other goods. The appearance of the *Boussole* and the *Astrolabe* after the Tlingit had lost men to the Man of Lituya posed another possibility: that they were in greater peril at the mercy of Yehlh, their creator. The French had nearly wrecked as they entered Lituya Bay, and their working the sails on the ships and anchoring suggested something ominous to the Tlingit, who saw their options as limited. The courageous among them prepared for combat. They dressed in elk-hide armor, wooden helmets, and collars and armed themselves with copper daggers and spears. The fiery blast from the ship's cannon dissuaded the war party, however, and they retreated to safety.

La Pérouse recorded the encounter with Tlingit people much differently, noting that they made "friendly signs to us, laying out and waving white coats and various types of furs," a prelude to almost daily encounters that centered on trade (Dunmore 1994, 102). He learned from observation, however, that the Tlingit approached the Lituya Bay entrance with great caution and "never ventured except at slack water" and that they gestured in relief when they successfully entered the bay (Dunmore 1994, 125). He also heard their story of loss and saw the funereal cremation of the few Tlingit victims who had been found.

In the Tlingit story, the fearsome wrath of the Man of Lituya, the arrival of the French, and the courage of the Native travelers are connected to a significant geographical place. The story also mixes the survival of loss among the Tlingit with the discovery that La Pérouse, the ships, and the sailors were nothing more than visitors and that the visitors had desirable objects to trade.

55. Tlingit Trade Exchange, 1786

INTRODUCTION

The instructions that La Pérouse carried on the voyage included a specific reference to engagement with indigenous peoples. Similar to the instructions that English and Spanish governments gave leaders of their 18th-century Pacific expeditions, Marquis de Castries, the minister of the navy, cautioned La Pérouse to engage Native people in a friendly, accommodating manner. There was an expectation that congenial relations could easily be established and that the result would be agreements that would benefit both groups. Above all other cautions, the instructions required La Pérouse to "forbid [his crew] ever to use force to take from the inhabitants anything they might refuse to hand over voluntarily" (Dunmore 1994, cxlviii).

The point could not have been more finely drawn: the French government knew well the potential for commercial gain from trading North American furs to Asian buyers. The instructions charged La Pérouse to "ascertain the latitude at which one can begin to buy furs, the quantities the Americans can provide, which goods and items would be most appropriate to buy furs with" (Dunmore 1994, cxxxiv). To that end, La Pérouse understood the larger challenge, as the instructions pointedly put it: "He will make every honest endeavour to establish links with the country's natives" (Dunmore 1994, cxlvi).

Trading with Tlingit at Lituya Bay

During our forced stay at the entrance of the bay, we had been continually surrounded with the canoes of the savages, who offered us fish, skins of otters, and other animals, and different little articles of their dress, in exchange for our iron. To our great surprise they appeared well accustomed to traffic, and bargained with as much skill as any tradesman of Europe. Of all our articles of trade, they appeared to have no great desire for any thing but iron: they accepted indeed a few beads; but these served rather to conclude a bargain, than to form the basis of it. We at length prevailed on them to take pewter pots and plates: yet these had only a transient success, iron prevailing over every thing. They were not unacquainted with this metal. Every one had a dagger of it suspended from the neck. . . . This weapon had a sheath of tanned leather, and appeared to be their most valued moveable. As we examined these daggers very attentively, they informed us by signs, that they made use of them only against the bears and other wild beasts. Some were of copper, but they did not appear to give a preference to these. This metal is pretty common among them: they use it chiefly for collars, bracelets, and various other ornaments; and they also point their arrows with it.

It was a matter of great question with us whence they procured these two metals. There was no improbability in supposing this part of America to contain native copper, which the Indians might reduce into sheets or ingots: but native iron perhaps does not exist, or at least is so rare, that it has never been seen by the majority of mineralogists. It cannot be admitted, that these people are acquainted with the method of reducing iron ore to the state of metal. . . . We brought away many specimens of this iron, which is as soft and easy to cut as lead: perhaps mineralogists may be able to point out the country and mine that produce it.

Source: J. F. G. de La Pérouse, *A Voyage Round the World, Performed in the Years 1785, 1786, 1787, and 1788; by the Boussole and Astrolabe, under the Command of J. F. G. de la Pérouse,* 3 vols., translated from the French, 3rd ed. (London: A. Hamilton for G. G. and J. Robinson, J. Edwards and T. Payne, 1807), 2:78–80.

ANALYSIS

During the 26 days the expedition spent in Lituya Bay, La Pérouse interacted with Tlingit daily. He commented in his journal on their appearance and demeanor and specifically on their material culture, especially weapons, armor, and personal ornaments such as bracelets and necklaces. The coastal Tlingit hunted and gathered from a diverse environment, which included the pursuit of land mammals (bears, deer, and marmots) and sea mammals (seals, porpoises, sea lions, and otters). They hunted using bows and arrows, spears, slings, harpoons, daggers, and clubs. Fishing gear included snares, traps, dip nets, and ingeniously designed hooks, and they used stone, antlers, bone, shell, and copper for implements and hunting gear. La Pérouse took particular note of their weapons and reported that they hung daggers "around the neck." Etienne Marchand, who visited the Tlingit in 1790, described the men wearing leather holsters around their shoulders and reported that they were loathe to be without them. Made of iron, the daggers were often double-ended, with a short blade on one end and a longer blade at the other with a handle in the middle.

The Tlingit acquired worked iron and raw copper through trade with Eyak and Ahtna along the Copper River, north and west of their villages. Iron found its way into indigenous culture on the Northwest Coast by the mid-18th century, but there was more copper available, and many tribes used it for personal decoration and implements. The Tlingit worked the raw copper into effective personal weapons by sawing, hammering, and heating the raw metal and shaping it into a trapezoid shape. The stone "saws" were very hard flakes split from basalt stones and honed with an edge, which they used to shape copper implements.

The Tlingit had encountered Europeans before the French—Russians in 1741 and English in 1778—but their facility in trade, especially the hard bargaining described by La Pérouse, came from intertribal trade on the Northwest Coast. Their trading expertise, which so surprised La Pérouse, came from their role as middlemen for coastal and interior peoples. With interior Athapaskan peoples, the Tlingit traded seaweed, eulachon oil, and shellfish for furs, elk hides, and clothing, and they acquired canoes from the Haida, far to the south. As early as 1774, Spanish explorers reported copper trade items among the Haida that might well have come from the Tlingit.

In general, however, La Pérouse found little to praise about the Tlingit. He particularly found the women's adornment of a labret, or wooden disk, in the lower lip to be unbecoming. He called them "the most repulsive on earth," but elsewhere he characterized Tlingit skill in fashioning implements and clothing as comparable to French artisans (Dunmore 1994, 137). He concluded that Tlingit had "made much greater progress in their crafts than in morals," and he left the region considering them to be far inferior to Europeans, even though they drove hard bargains in trade (Dunmore 1994, 139).

56. La Pérouse's Map of the Northwest Coast, 1797

INTRODUCTION

The "Chart of the North West Coast of America Explored by the Boussole *and* Astrolabe *in 1786" illustrates some of the remarkable elements of scientific maritime mapping methodology used in the late 18th century and the iterative nature of mapping a geographically complicated region such as the Northwest Coast. Most remarkable perhaps, the map was published in France during the tumultuous last decade of the century, when the country was riven by revolutionary politics and foreign wars.*

La Pérouse's published journal was accompanied by the Atlas du voyage de La Pérouse, *which contained 69 copperplate engraved charts, plans, views, and other representations. A general map*

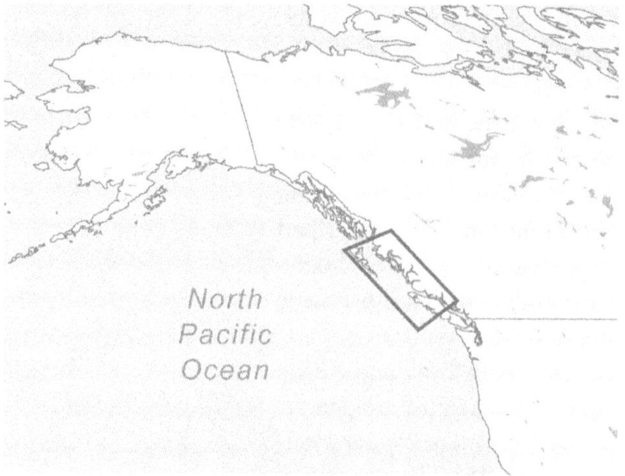

of part of the Northwest Coast was divided into three separate sections (this one is the second), all of which were drawn to a larger scale. Surely due to intentional design, this atlas contains many more detailed coastal maps than the atlas of James Cook, after whose voyage La Pérouse's expedition was patterned. The charts for the atlas were prepared by the royal geographer, Jean-Nicolas Buache de la Neuville. Buache was a member of the French Academy of Sciences, and his uncle was coauthor of the 1752 map that had brought public attention to the discoveries of Vitus Bering and Aleksei Chirikov.

This section of the map extends from latitude ~54° to ~50°, from Baie de Clonard (present-day Graham Island in Haida Gwaii) to just north of Nootka Sound on the northwest coast of Vancouver Island. On present-day maps this region covers the northern reaches of Haida Gwaii (Queen Charlotte Islands) group, along the western coast of these islands, and then across Queen Charlotte Sound and along part of western Vancouver Island. The map includes information from explorations conducted during August 13–26, 1786. Longitude was measured west of Paris.

ANALYSIS

In despair from the loss of 21 men at Port des Français, La Pérouse left that harbor on July 30 and headed south, intent on exploring areas on the coast that James Cook had not examined. There was lingering belief in the existence of a transcontinental water route, "some river or some narrow gulf," his instructions noted, that "communicate with some part of Hudson Bay by way of the interior lakes" (Dunmore 1994, cxxx).

In part because of the unanticipated delay at Port des Français, seasonal fog and winds obstructed the expedition's progress south. When poor visibility prevented charting, La Pérouse had to tack out to sea and wait for the fog to clear before returning to complete his task. The ship's jagged tracks near the top of this map represent La Pérouse's tacking maneuvers. Farther south, dotted lines and letters represent his use of triangulation surveying methods and show the exactitude of his work. The views of various mountains, drawn with hachure lines, convey impressions of elevations. Despite nearly 13 days of arduous charting nearly 275 miles of coastline, La Pérouse never knew that he was mapping island groups and not the North American mainland. Nevertheless, his map improved greatly on Cook's chart, which had represented the Northwest Coast with a straight line between these latitudes.

La Pérouse was the first navigator to chart this area with such detail, so French names predominate, although none survive on present-day maps. At the top of the map, La Pérouse named Baie de Clonard after one of his officers, and Cape Hector (present-day Cape St. James) honored the commandant at Brest. Charles-Pierre Claret de Fleurieu was a geographer and minister in Paris and a correspondent of La Pérouse, and all of the places named after him were likely located on the eastern side of Queen Charlotte Sound. Cape Buache, probably on the east side of present-day Moresby Island, was named after the royal geographer, Jean-Nicolas Buache de la Neuville. Iles de Sartine are the present-day Scotts Islands, which English fur trader James Strange named on August 1, 1786. Although Strange and La Pérouse were in the same vicinity at nearly the same time, they did not meet. Finally, Pte. Boisée is French for Woody Point (present-day Cape Cook) on the west coast of Vancouver Island. Cook had seen this point from a distance and applied the name Woody

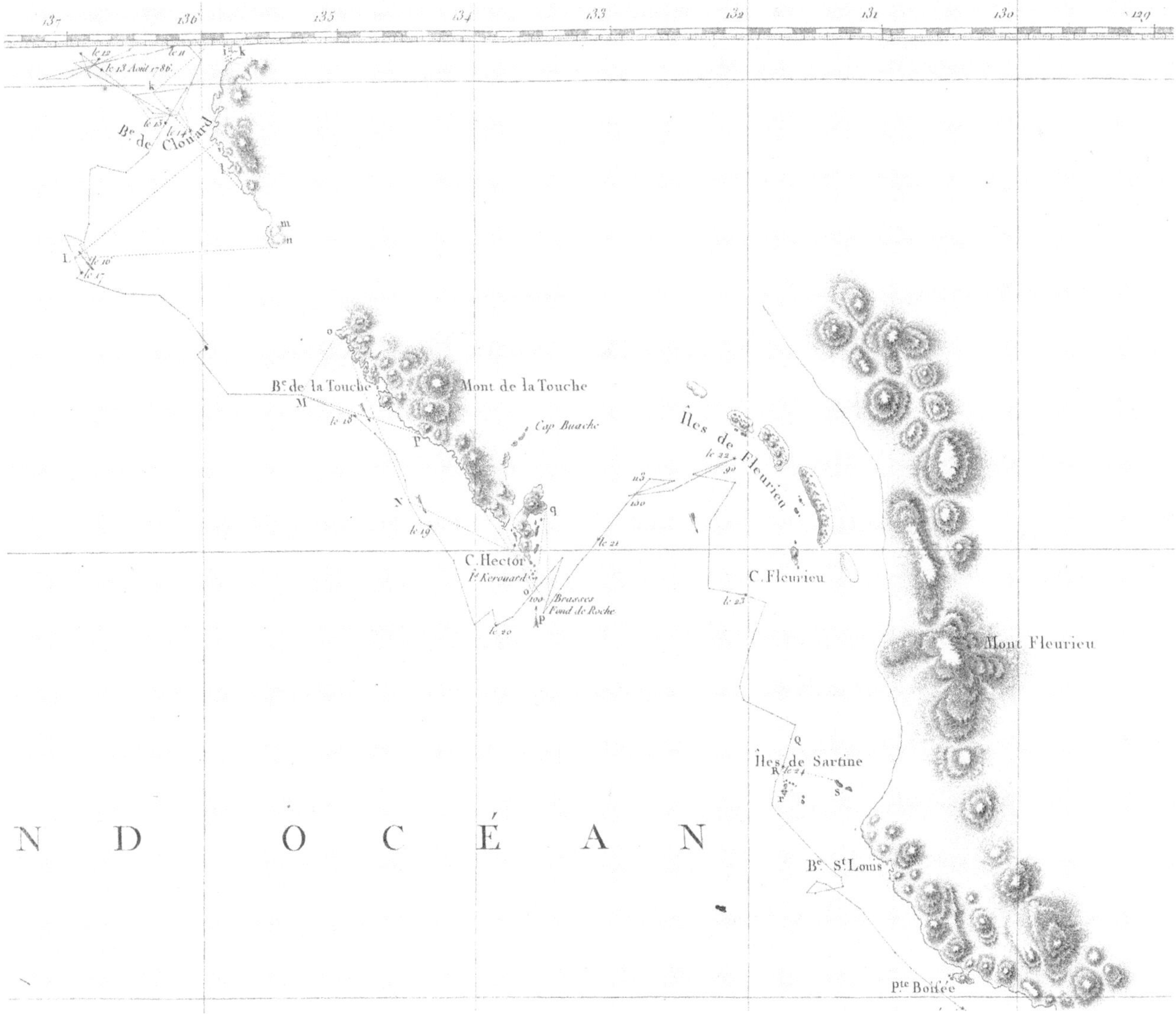

"Chart of the North West Coast of America Explored by the *Boussole* and *Astrolabe*." (Copyright © 2000 by Cartography Associates)

Point on his map, which La Pérouse carried on his chart. When published in 1797, La Pérouse's map was the product of highly skilled seamanship, the use of the latest scientific maritime navigational instruments, and artistic and technically expert engravers. It was created under the auspices of the royal geographer.

For the international community, proof of France's potential role in the geopolitics of the Pacific Northwest Coast depended on documentation of French exploration in the area. La Pérouse's maps and his account of the voyage had to be made public. It was a tortuous course to publication from the time the documents were sent to Paris, probably in September 1787 from Kamchatka, Siberia, to their appearance in 1797. In April 1791, the National Assembly had decreed "that the accounts and maps, sent by Mr de la Pérouse, of part of his voyage, as far as to Botany Bay, shall be printed and engraved at the expense of the nation; and that the expense shall be defrayed from the fund of two millions granted by the fourteenth article of the decree of August 3, 1790" (La Pérouse 1807, 3). But the politics of revolution intervened, including the arrest in August 1792 and the beheading the following January of King Louis XVI, who asked on the eve of his execution for the latest news about the search for La Pérouse. After nearly two years of delay, work began on preparing the notes, letters, journals, and maps for publication. George Vancouver had returned to England after his multiyear voyage of discovery, and it was probably not a coincidence that La Pérouse's account and maps were finally printed and distributed the year before publication of Vancouver's account.

57. Macao Trade Protocols, 1787

INTRODUCTION

Although the French expedition focused on hydrography, navigation, and natural science, the planned route took the Boussole *and the* Astrolabe *to Asia, principally for trade. Their first port was Macao, a Portuguese trading entrepôt that had been granted a license from China to purchase Chinese articles in exchange for European and other trade goods. In part, the expedition's route reflected the larger purpose of retracing portions of James Cook's third voyage, and part of that effort included selling furs in China that had been collected on the Northwest Coast. Once the furs were sold, La Pérouse was to sail north to Kamchatka before returning to the South Pacific to conduct more scientific investigations.*

As in other ports where La Pérouse had opportunity, he wrote extensive letters to his superiors in France, detailing events, describing conditions, and explaining decisions he made as commander of the expedition. This letter was one of many to the Marquis de Castries, minister of the French Navy, the most important French official to oversee naval operations and the one who carried primary responsibility for success of the expedition. He reported directly to Louis XVI.

La Pérouse to Marechal de Castries, Minister of Marine, January 26, 1787, Macao

I appended the list Mr du Fresne had drawn up, as well as the price he hoped to get, but all our hopes were dashed by events and the two hundred and six thousand nine hundred and nineteen livres, Mr du Fresne's asking price per his estimate enclosed herewith, was reduced to nine thousand five hundred piastres or a little less than fifty thousand *livres,* and the deal was concluded by Mr [Philippe] Vieillart, French Consul, as you will see from the enclosed letter. I wished, I wanted even, that the King's man should be aware of this affair, as my feelings and those of the two staffs would have suffered from the slightest lack of authenticity. I have had the honour of notifying you of all the officers' refusal to share in this, even when these otter skins were thought to be worth four times what they sold for. They even made a point of keeping for themselves not even enough material to make a button for their uniform, and I make bold enough to give my word of honour that no trader has ever been more faithfully dealt with in a matter of business in which he was directly involved; Mr du Fresne himself, whom I had appointed supercargo for the two ships, and to whom I had granted a commission of three per cent on the sale price, refused to accept it and wanted even the value of his endeavours to be distributed. . . .

I believe that I have the honour of pointing out at this stage, My Lord, that although the expenditure incurred by the two frigates is modest we do use fairly large sums, but these are merely advance drawings on the pay or the keep of both frigates who, in accordance with your decrees, are entitled to request payment once two months are due, and so our expenses are in no way out of the ordinary and consist only of the campaign's normal outgoings which can be paid out equally well during the voyage or on our return. Moreover our accounts are quite in order; we took two thousand piastres on each frigate when we sailed from France and Mr Vieillart had two thousand piasters loaned to us in the King's name when we reached China. . . .

Although the memoir on otter skins, which I had the honour of sending you, concerns only America and the ease with which one can obtain this article of trade on that coast, the circumstances surrounding the sale we concluded in China will prove, My Lord, that all the speculations carried out in Europe would end in ruin. Nine hundred and ninety-eight furs, three hundred of them whole and of the best quality, fetched a mere fifty thousand *livres,* and at the price reported in Cook's third voyage this same quantity would have produced at least a hundred thousand, but an English company from Bengal made six voyages to Cook's River and [Prince] Williams Sound this year, and although these expeditions resulted in very few furs, the mere rumour of these voyages brought about an extraordinary drop in prices. I will find out in Kamchatka whether Russian trade in northern China has suffered as much as the Canton trade.

Source: John Dunmore, trans. and ed., *The Journal of Jean-François de Galaup de La Pérouse, 1785–1788,* Vol. 2 (London: Hakluyt Society, 1994), 501–502. Used by permission of the Hakluyt Society.

ANALYSIS

During his stay at Port des Français in July 1786, La Pérouse had collected and traded for pelagic and land furs from the Tlingit. He intended to follow the model set by Cook's voyage in 1778—trading the furs in China and reaping a high profit. By the late 18th century the fur trade had become part of a global exchange, with furs harvested from across North America, including the Northwest Coast, finding their way to lucrative Asian markets. The maritime fur trade from North America to China, however, faced limitations. China condoned trading with foreigners only in Canton and Macao, ports on the South China Sea, and in Kyakhata (present-day Altanbulag, Siberia), a Russian trading city on the border with China in Mongolia. The system required coordination in port with Chinese traders and representatives of foreign governments. Portugal had established Macao in 1587, but China controlled the trade by 1685 and oversaw non-Chinese exchanges during the fur trade era. In both Macao and Canton, foreign government authorities, such as French consul Vieillard, managed trade relationships. Vieillard, however, had little interest in the expedition and the furs, and he negotiated a poor exchange rate. With few options, La Pérouse left Louis Dufresne, the supercargo, with instructions to get the best return possible.

Commanders of official voyages, such as this scientific expedition to the Pacific, assigned officers as managers of basic shipboard operations. On merchant vessels, a supercargo had responsibility for the valued cargo and for trading in port, and he often represented the owner of the cargo in the negotiations for sale. Although the expedition was scientific and not commercial, La Pérouse needed a supercargo to handle the collected furs from the Northwest Coast, and he contracted with Dufresne, the principal naturalist on the voyage. Dufresne was a poor sailor and miserable at sea. He had pleaded to leave the voyage in Chile, but La Pérouse had him stay aboard and become supercargo to justify his wage. It is also likely that Dufresne had initially joined the expedition because of his knowledge of animal furs and the fur trade. He fulfilled his duties, and La Pérouse allowed him to leave the expedition in Macao. Dufresne headed back to France, carrying documents and "four very fine otter furs, and two lengths of material made by the Indians of Port des François" as presents for Queen Marie Antoinette, plus "eight otter skins" for King Louis XVI (Dunmore 1995, 500).

58. Uncertain Remedies for Scurvy, 1788

INTRODUCTION

La Pérouse wrote letters to government officials at intervals, usually when he was in port and had access to ships that could take mail and cargo back to France. Writing to the director of ports and arsenals, Claret de Fleurieu, La Pérouse communicated his sadness over the death of his longtime friend Paul Antoine Fleuriot de Langle and other sailors in the South Pacific. It had been the second tragedy on the long voyage and one that begged for explanation. In his letter La Pérouse barely describes the attack on Tutuila, Samoa, that killed 12 of his men, but he does offer an extended discussion of scurvy, because it was the problem of scurvy that set up the conditions that led to the loss of life.

La Pérouse to Claret de Fleurieu, Botany Bay, February 7, 1788

Minor symptoms of scurvy disappeared when we used the fresh foodstuffs I obtained at the Navigators Islands [Samoa]. I swear to you, my dear friend, that sea air is not the cause of this sickness, but rather the foul air in the tween-decks, when it is not renewed, or even more the poor quality of the food. . . . It follows from this that the decomposition of the humours, the blood &c. and equally extracts of cochlearoa [a medicinal herb] and all remedies contained in flasks are nonsense. . . . And I consider malt, spruce-beer, wine, sauerkraut &c. to be antiscorbutics only because these substances, liquid or solid, suffer very little deterioration, and finally are a suitable food for man; they are however insufficient to cure scurvy, but I believe they must slow its advance. . . .

My theory on scurvy therefore comes down to these aphorisms, which are not from Hippocrates: any type of food suitable for man and able to make up for daily losses; external air introduced as often as possible in the tween-decks and the hold; the damp caused by fog regularly countered by fumigations and even with braseros [fire pit heater]; finally, cleanliness and frequent inspections of the sailors' clothes.

I have no confidence in Captain Cook's observation on the corruption of water in barrels. I believe that what is in good condition when loaded on board, after passing through the two or three stages of decomposition known by every sailor, which cause it to smell for a few days, thereafter keeps in excellent condition, and as light as distilled water, because all the heterogeneous matter has precipitated and remains as a sediment at the bottom of the barrels, and at the moment I am writing to you . . . I am drinking water from Port des Français . . . which is excellent. It is however this erroneous opinion, which I have never shared, that caused our misfortunes at the island of Mahouna [Tutuila]: how can one go against a captain when he assures you that all his crew will suffer from scurvy within a fortnight if he does not have any fresh water. . . .

These precautions [safely storing longboats] were needed against the Indians of New Holland who, although very weak and not numerous, are, like all savages, very ill-natured and would set fire to our boats if they had means of doing so. They threw spears at us one minute after receiving our presents and signs of friendship. I am a hundred times more angry against the philosophers who so praise them as against the savages themselves. [Jean-Robert] Lamanon, whom they murdered, was telling me the day before he died, that these men were worth more than us. A rigid follower of the King's instructions, I have always behaved towards them with the utmost moderation; but I would not undertake another campaign of this kind without asking for different orders.

Source: John Dunmore, trans. and ed., *The Journal of Jean-François de Galaup de La Pérouse, 1785–1788,* Vol. 2 (London: Hakluyt Society, 1994), 538–539. Used by permission of the Hakluyt Society.

ANALYSIS

The threat of scurvy plagued all long deepwater voyages before the scientific discovery of its cause and prevention. The symptoms rendered sailors increasingly weak, first in muscle strength, then with loss of weight, and finally because of internal bleeding and neuropathy. Scurvy attacked soldiers in land campaigns as well, but the disease seemed endemic on long sea voyages. Britain, Spain, and France all recognized the threat that scurvy posed to their fleets, but no solution to the problem had been found by the mid-18th century.

In the 17th century, British mariner John Hawkins collected information on scurvy deaths, noting that 10,000 sailors had died of the disease. In 1753 Dr. James Lind published *A Treatise of the Scurvy,* identifying the lack of fresh vegetables as a potential cause and arguing that citrus fruits were most effective in preventing and treating scurvy. La Pérouse had a copy of Lind's book on the voyage. Nonetheless, many doctors disagreed with him, and a debate about the cause of the condition continued well into the 19th century. In the third edition of his book, for example, Lind seems to discount citrus as a preventative: "I do not mean to say that lemon juice and wine are the only remedy for the scurvy; this disease, like many others, may be cured by medicines of very different, and opposite qualities to each other, and to that of lemons" (Bartholomew 2002, 11). Although it was not until 1932 that researchers identified a lack of vitamin C as the cause of scurvy, mariners had long observed that some vegetables, certain herbs, and citrus fruits seemed to be antiscorbutics. La Pérouse was not alone in casting about for scurvy's cause, and he also was among many who suggested wildly different treatments and cures.

It was the search for freshwater as an antiscorbutic that led to the loss of life at Tutulia. The *Astrolabe*'s captain, Paul Fleuriot de Langle, was convinced that freshwater prevented scurvy, and he insisted on going ashore despite disagreement from La Pérouse, who nearly ordered him to remain on ship. "Mr de Langle was a man of such excellent judgment and of such ability," La Pérouse wrote in his log, "that these considerations above everything else finally caused my own will to bow to his, and I promised him that we would stay on short tacks during the night and would send the next day our two longboats and two boats ashore armed as he considered appropriate and that everything would be placed under his orders" (Dunmore 1995, 397). It was a fateful decision.

On December 11, 1787, Langle took a small party in longboats with water barrels into the bay and toward shore. A low tide made landing difficult, and Langle found himself and a few men a distance from his boats and vulnerable to attack. Some Tutuilas took advantage, perhaps attracted by French arms, and slew Langle and 11 others. La Pérouse, an enlightened mariner who had read and digested the works of French philosophers, disagreed with the image of the "noble savage" that Lamanon had embraced. In his letter, La Pérouse turned his rage on the philosophers' ideas in his remorse for a lost colleague.

Further Reading

Bartholomew, Michael. "James Lind and Scurvy: A Reevaluation." *Journal of Maritime Research* (January 2002): 1–14.

Cruikshank, Julie. *Do Glaciers Listen? Local Knowledge, Colonial Encounters, and Social Imagination.* Vancouver: University of British Columbia Press, 2005.

Dunmore, John. *Pacific Explorer: The Life of Jean-Francois de La Pérouse 1741–1788.* Annapolis, MD: Naval Institute Press, 1985.

Dunmore, John, ed. and trans. *The Journal of Jean-François de Galaup de la Pérouse, 1785–1788.* 2 vols. London: Hakluyt Society, 1994, 1995.

Emmons, G. T. "Native Account of the Meeting between La Pérouse and the Tlingit." *American Anthropologist* (April–June 1911): 294–298.

La Pérouse, J. F. G. de. *A Voyage Round the World, Performed in the Years 1785, 1786, 1787, and 1788, by the* Boussole *and* Astrolabe, Vol. 1. 3rd ed. London, 1807.

Expedition of Alejandro Malaspina

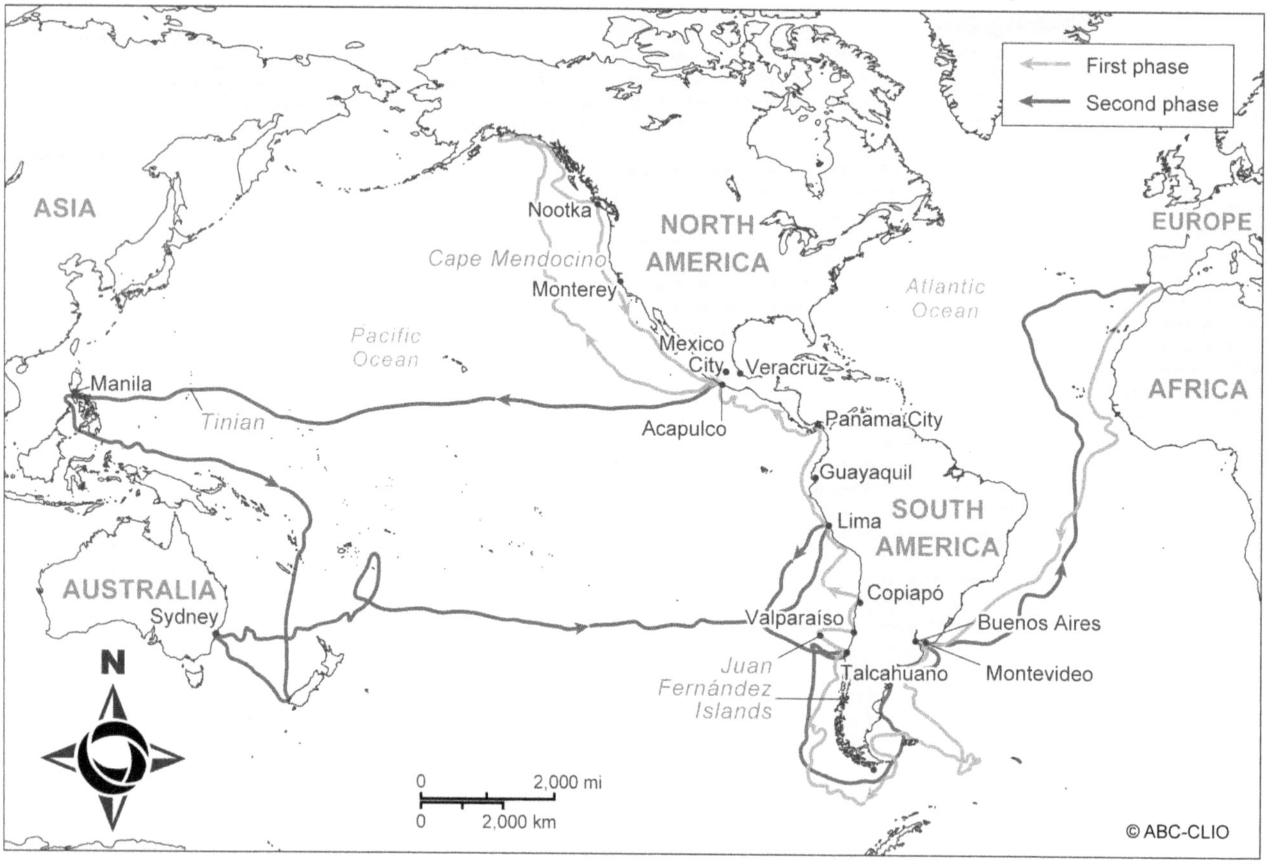

CHAPTER 7

Alejandro Malaspina, Spanish Enlightenment on the Northwest Coast, 1788–1791

Alejandro Malaspina is among several highly significant 18th-century Spanish explorers who expanded his nation's information about the North Pacific. He advocated Enlightenment science, politics, and economics viewpoints and purposes that he applied in his explorations of the Pacific Ocean, first in a circumnavigation of the world in 1786–1788 and second in a grand voyage of discovery in 1789–1794. Malaspina's contributions to Spanish cartography justified the royal investment in his expeditions, but his political missteps and eventual incarceration kept most of his discoveries from broad dissemination until the late 19th century. Nonetheless, he and two of his officers—Dionisio Alcalá Galiano and Cayetano Valdés y Flores—played significant roles in expanding cartographic knowledge about the Northwest Coast from Monterey to the Gulf of Alaska.

Malaspina was born into an aristocratic family in Mulazzo, Italy, about 75 miles east of Genoa and 60 miles southwest of Parma, near the seacoast. The family heritage stretched back to the Middle Ages in northwestern Italy, where Malaspina aristocrats held land and prominent regional political positions. By the 18th century the family also had close ties to Bourbon Spanish in the Duchy of Sicily, where young Alejandro moved with his family in 1762. It was Malaspina's good fortune to receive benefits from both Spanish and Italian authorities. He studied at Clementine College in Rome from 1765 to 1773 and became a member of the Order of St. John in Malta in 1773–1774. At Clementine, Malaspina embraced science, empiricism, and the Enlightenment's emphasis on reason, and he learned navigation and general seamanship in Malta. In 1774 at age 20, he gained admission to midshipman school at Cádiz, Spain's premier naval base, and rose rapidly in rank. He served on ships in the Mediterranean and on voyages to the Philippines, rising to the rank of frigate captain in 1782. By the time he was 30, he had secured command of the Royal Philippines Company frigate *Astrea,* which he sailed around the world in 1786–1788.

Malaspina became interested in exploring the North Pacific. He thoroughly studied the 1770s voyages of James Cook and the 1780s explorations of Jean-François de La Pérouse. In league with José de Bustamante y Guerra, Malaspina proposed a multiyear scientific exploration of the Pacific patterned on the expeditions of the two great explorers, including intensive surveys on the Northwest Coast. Carlos III approved and financed the voyage.

Juan Ravenet painted this watercolor of Malaspina and Bustamante conducting a gravity experiment at Port Egmont in the Falkland Islands in December 1789, using a Licot pendulum to measure the earth's circumference at specific latitudes. (DeAgostini/Getty Images)

Bustamante held the same rank as Malaspina but was younger and from a middle-class family in northern Spain, and it was Malaspina whom Spanish officials recognized as the expedition's leader.

Malaspina oversaw the construction of two expedition ships, the *Descubierta* and the *Atrevida*—nearly identical at 109 feet long, with a 29-foot beam, and displacing more than 300 tons. The ships were launched in April 1789. The mission was focused on scientific discovery, so each vessel carried fewer armaments than usual to free up space to store specimens, laboratories for experiments, and technical gear. Nonetheless, the expedition was military and had to be prepared for defense, so each vessel carried 14 six-pounders and 2 four-pounders, plus 8 six-pounders in the holds as ballast. Each ship carried 102 crew, including 14 officers. Finding men willing to sign on for a multiyear voyage proved difficult, and the desertion and sickness rate meant that Malaspina had to replace seamen throughout the expedition. Only about 40 percent of the original crew completed the voyage, with many new crew members recruited in ports in the Western Hemisphere.

This voyage of discovery was perhaps the most scientifically prepared of any European maritime expedition in the 18th century. Malaspina and Bustamante had full authority to select their officers, which they did on the basis of their sea experience and their chartmaking and hydrography capabilities, plus their knowledge of astronomy, mathematics, and natural sciences. While King Carlos III preferred that the captains choose Spanish-born officers over men born in Mexico or South America, they ended up with a few officers who were not born in Spain, including naturalists Antonio Pineda, born in Guatemala, and Alférez Viana, born in Uruguay. Malaspina and Bustamante also selected Bohemian botanist Tadeo Haenke, who worked closely with Luis Neé, a French-born botanist at the Real Jardín Botánico in Madrid. Haenke hustled from Vienna to join the expedition and then missed the sailing from Cádiz. In haste and determined, he made it to the east coast of South America but again was just days late and failed to intercept the ships. Finally, after a grueling tramp over the Andes he boarded the *Descubierta* in Valparaiso, Chile.

One of Malaspina's most important selections was Felipe Bauzá y Cañas, the expedition's chief of charts and maps. With considerable expertise in geography and hydrography, Bauzá caught Malaspina's attention because he was a respected teacher of cartographic drawing at the naval school at Cádiz and had experience in harbor installations and fortifications. Bauzá oversaw astronomical observations, coastline sea soundings, and chartmaking on the voyage and took charge of the expedition's maps and natural history reports when the expedition returned.

After considerable preparation and the acquisition of scientific instruments and materials, the expedition left Cádiz on June 30, 1789. Malaspina included antiscorbutic citrus fruit in their foodstuffs, and he required that the ships be cleaned regularly and that the air between decks be tested using a eudiometer, which measured the relative flammability of gases. At ports of call, Malaspina reprovisioned the ships with fresh foods, replacement livestock, and available grains. "The method of preparing dried beef on all our Pacific coasts and in Río de la Plata is well known," he recorded in his journal. "The animal is killed and divided into several slabs; the meat is then left in brine for twenty-four hours and then exposed to the Sun until it is completely dry. The meat tastes very good and not too salty" (David et al. 2001–2004, 2:241). While in port, he also found it necessary to replace crew members because of illness or desertion.

The track of the expedition took the *Descubierta* and *Atrevida* to the Canary Islands, to the Cape Verde Islands, and then to Montevideo on September 20, 1789. After two months Malaspina sailed around Cape Horn to Chile, arriving in early February 1790, and by December the expedition had visited ports in Peru and Panama. They arrived in Acapulco, Mexico, on March 27, 1791. Malaspina expected to sail to Hawaii, return to Mexico to pick up crew, head to the Northwest Coast, and then make his way to Kamchatka in Siberia. Orders from Spain changed his plans, however, sending him directly to the North Pacific Coast to investigate a potential Northwest Passage. Royal geographer Jean-Nicolas Buache de la Neuville had delivered a paper in Paris in 1790 that endorsed Lorenzo Ferrer Maldonado's purported exploration of the Northwest Passage in 1588. Malaspina had serious doubts about Maldonado's geography and, by implication, the geographical ideas of Buache and the Delisle family of cartographers, mapmakers who often had included a Northwest Passage on their maps.

In April 1791 Malaspina and Bustamante prepared to sail north to the Gulf of Alaska, leaving Acapulco in early May with a new artist on board, Tomás de Suría. On June 23 the two captains came in sight of the Northwest Coast at near 56° north, where they recognized landforms from maps they

DID YOU KNOW?

Tomás Suria, artist with Alejandro Malaspina, described his shipboard sleeping accommodation: "From my breast to the deck, which was my roof, the distance is only three inches. . . . I am forced to make for myself a roll of cloth to cover my head, although this suffocates me, but this is a lesser evil than being attacked by thousands of cockroaches."

carried by James Cook and Juan Francisco de la Bodega y Quadra. Rather than make landfall, Malaspina continued north, as he noted in his journal, "with the desire to reach the vicinity of latitude 60° as soon as possible, towards which our investigations must be directed" (David et al. 2001–2004, 2:99). On June 27 they struck the coast at Port Mulgrave, a cove in present-day Yakutat Bay named by George Dixon in 1787 and potentially the location of Maldonado's fabled strait. For nine days the captains remained at Mulgrave, where they interacted with Tlingit with mostly good relations except for a near conflict over a stolen pair of seaman's pants. They also investigated the inlet and determined that it could not be Maldonado's strait. From Port Mulgrave, Malaspina sailed north to Prince William Sound, where inclement weather forced them south to the Mt. St. Elias region of the coast and finally to Nootka Sound on present-day Vancouver Island. They anchored at Santa Cruz de Nuca, the Spanish garrison location, on August 11, 1791.

At Nootka Sound, Malaspina and his officers had their most extensive contact with Native people on the Northwest Coast. They arrived in the middle of the Nootka Sound Controversy between Spain and Britain, and Malaspina, as the king's representative, had to evaluate the condition of the Spanish garrison and their relationships with the Mowachaht villages in the area. Although Malaspina could not resolve the controversy on the ground—that was left for Vancouver and Bodega to accomplish in 1792—he made amiable contact with chiefs Maquinna and Tlupananú, which effectively improved the Spanish situation at Nootka.

José Cardero, a servant on the expedition and amateur artist who had developed considerable skill, drew this likeness of a Nootka woman in 1792, showing details of her adornment. (Peter Horree/Alamy Stock Photo)

At key locations on the Northwest Coast and at nearly all other ports on the expeditionary route, Malaspina undertook extensive scientific and hydrographic observations. Under Felipe Bauzá's guidance, an observatory tent at each location became the base of activity, which included careful lunar observations, triangulated measurements of landforms, computation of chronometer accuracy, and preliminary chartmaking. Surveys by Tadeo Haenke and Luis Neé and artistic renderings of people and landscapes by Tomás Suría and José Cardero made up the bulk of materials and visual representations of areas visited by the expedition. The charts and maps provided important information for the Spanish Royal Navy, although many of them remained unpublished until the 19th century. Nonetheless, Malaspina and his officers fulfilled the signal and specific purposes of his voyage, especially the survey of the Northwest Coast, the determination of the existence or nonexistence of Maldonado's strait, and a scientific exploration of important regions of the Pacific Ocean.

Malaspina coasted south to Monterey in California and reached the port at Acapulco on October 18, 1791. Orders from Madrid required a new Spanish effort to resolve the Nootka Sound Controversy, a task that Bodega took on. In addition, Spain's viceroy in Mexico, Juan Revillagigedo, sent two ships—the *Sutil* and the *Mexicana*—north to explore the Strait of Juan de Fuca as a possible opening to the

Northwest Passage. As Malaspina prepared to sail to Manila and the western Pacific Ocean, Dionisio Galiano left the expedition to captain the *Sutil*. In late December the *Descubierta* and the *Atrevida* sailed for Manila, arriving in March 1792 and reaching New Zealand by March 1793. From there Malaspina's track took him east to Peru, where the two ships sailed separately around Cape Horn to Montevideo before sailing together to Cádiz. They arrived in their home port in September 1794.

> **DID YOU KNOW?**
>
> Malaspina's maps and reports went unpublished for 80 years after his great voyage, because his foolhardy criticism of the Spanish government landed him in prison for treason in 1795, and his logs and records were withheld from public view.

Malaspina's life after the great expedition turned a much different corner than he expected. Rather than basking in the heroic plaudits that came on his return, he offered serious and critical suggestions about how Spain should conduct its foreign policy and manage its colonies. Caught in the crosscurrents of royal bureaucracy and court politics, Malaspina angered Manuel Godoy, Spain's first minister, who punished him for his insubordination. Godoy carefully orchestrated Malaspina's trial in November, which resulted in court orders that sent him to the Castillo de San Antón prison at La Coruña, where he remained until 1803. Napoleon Bonaparte interceded on Malaspina's behalf in 1803 and secured his release from prison but only on condition that he leave Spain permanently. At 48 years old, Malaspina left for a quiet life in Parma, the region of his birth, where he died in 1810.

The first published account of Malaspina's expedition—*Memorias sobre las observaciones astronomicas,* by José Espinosa y Tello—appeared in 1809. Ensign Francisco Viana privately published his own journal of the expedition in Montevideo in 1849, and José Bustamante's journal appeared in 1868. It was not until 1885 that Malaspina's diary appeared, edited by Pedro de Novo y Colson, and was recognized as an official account of the voyage.

59. Malaspina's Plan for a Circumnavigation, 1788

INTRODUCTION

The plan for a Spanish maritime expedition to the Pacific Ocean in 1788 originated with two young officers, 34-year-old Alejandro Malaspina and 29-year-old José Bustamante. Their proposal caught the attention of Antonio Valdés, minister of the Spanish Royal Navy, and King Carlos III (1759–1788), who approved the voyage on October 14, 1788, just two months before his death. Carlos III had embraced Enlightenment thought and pushed Spain to modernize government, society, and science. His son, Carlos IV, also supported the expedition and retained Valdés, but Carlos IV lacked administrative ability and was a much less effective king than his father.

Bustamante had seen action in sea battles and had commanded transport ships across the Atlantic to Mexico, becoming a frigate captain in 1784 at age 25. Malaspina, a frigate captain by 1782, had participated in the battle at Gibraltar, and in 1784 he had undertaken his first voyage to the Pacific on the Asunción, *which took him to Manila. Considered a top Spanish navigator, Malaspina circumnavigated the world in 1786–1788 as captain of the supply ship* Astrea. *That voyage took him from Cádiz to Manila as part of a monopolistic commercial enterprise, the*

Real Compañia de Filipinas, which hoped to establish sugarcane and spice industries in the Philippines. The experience taught Malaspina the importance of colonial trade for Spain, the role of colonial administration, the challenges of long sea voyages, and the crucial part that detailed and accurate navigation charts and harbor maps played in efficient and safe seafaring.

Alejandro Malaspina to Antonio Valdés, Madrid, September 10, 1788

For the past twenty years two nations, the English and the French, in noble competition, have undertaken voyages of this sort in which navigation, geography and humanity itself have made very rapid progress. The history of human society has thus been founded on much wider research; natural history has been enriched with almost endless discoveries; and possibly the most exciting victory has been the preservation of health in the course of long sea voyages through different climates while facing the most challenging labours and dangers.

The proposed voyage would aim to accomplish these objectives, and this part, which can be termed the scientific aspect, would certainly be undertaken following earnestly in the wake of Cook and La Pérouse.

But a voyage undertaken by Spanish navigators must necessarily involve two other objectives. One is the making of hydrographic charts covering the most remote regions of America and the compilation of sailing directions capable of providing safe guidance to inexperienced merchant mariners. The other is the investigation of the political status of America both in relation to Spain and to other European nations.

The commercial conditions in each province or kingdom, with individual consideration of their natural products and manufactures; their preparedness and capacity to resist an enemy invasion and conversely their ability to provide forces to attack the same enemy; the condition of the ports most suited for maintaining reciprocal trade; and finally the important activities of shipbuilding and supply of naval equipment, are aspects whose careful and secret investigation should be of some interest to the State. Our report would be adjusted to different political axioms concerning national prosperity whose acceptance or rejection must be subjected to the adjudication of reputable assessors called upon to examine them. . . .

His Majesty's Navy will be able to provide all the personnel for this commission, except for the two botanists or naturalists and the two topographical draughtsmen, for which posts it should be possible to find volunteers without undue difficulty in Madrid. As for the class of ships . . . the three principal requirements of safety, convenience and economy can easily be combined. The complement required for each of the ships can be kept down to about one hundred men. The actual detail of each vessel, such as the rigging, inboard arrangements, number of boats, their features and their equipment, and, finally, the quantity and quality of victuals, are too much to go into at present, and in any case cannot be precisely determined until such time as His Majesty shall decide on the scope of the proposed expedition.

To carry out the plan outlined below might take approximately three and a half years from 1 July 1789, the date by which the ships could depart if His Majesty were to give his approval now, either as proposed or with modifications, allowing eight months to those who will have to implement it. . . .

The two corvettes will sail from Cádiz on the 1 July 1789 and will proceed to Montevideo, where a fresh rating of the chronometers, such astronomical observations as are possible, and all kinds of natural history studies, will be carried out. . . . From there the Malvinas will be surveyed. . . . [A] course will be set to round Cabo de Hornos. . . . Finally we will drop anchors at Chiloé, which might be towards the end of 1789. The whole of the year 1790 will be employed on the western coasts of America from Chiloé to San Blas. . . .

The first three months of 1791 will be taken up with a survey of the Islas Sandwich. After following the coast of California, the voyage will continue to the north between Asia and America as far as the snows will allow. After calling at Kamchatka (if the Government thinks fit), we shall continue to Canton to sell sea-otter furs for the benefit of the crews.

Source: Andrew David et al., eds., *The Malaspina Expedition, 1789–1794,* Vol. 1 (London: Hakluyt Society, 2001), 312–313. Used by permission of the Hakluyt Society.

ANALYSIS

Malaspina's projected expeditionary circumnavigation of the world came directly from his own dedication to Enlightenment-inspired scientific and political investigations. In the introduction to his account of the voyage, he wrote that "the compilation of a hydrographic atlas for long-range navigation by Spanish ships, whether for the exchange of goods between the colonies and the mother country, or for more extensive trade with countries outside Europe, was a good enough reason in itself to send to the Pacific Ocean the ships and men needed for the job" (David et al. 2001–2004, 1:lxxxi). The relative urgency for a Spanish exploratory expedition, as Malaspina argued, came directly from the accomplishments of James Cook, Jean-François de Galaup La Pérouse, and other navigators who had furthered their nation's imperial designs and knowledge of the world's oceans.

Malaspina focused on politics as much as science. He aimed to further Spanish colonial interests, even to reform the administration of the colonies, while he embraced larger ambitions for the benefit of humanity. "Would that I were able, in the exercise of so great a duty," he wrote, "to apply the utmost energy in philanthropy; the greatest assiduity in the study of nature, the highest impartiality in examining the nature and rights of man, and bring them all to bear on social questions!" (David et al. 2001–2004, 1:lxxxii). There was more to Malaspina's ambitions than revealed in the plan. He had compiled the lengthy *Axiomas de políticos sobre la América* that expressed his commitment to Enlightenment thought, especially to the ideas of Scottish philosopher Adam Smith's *Inquiry into the Nature and Causes of the Wealth of Nations* (1776), which argued for free trade and development of individual labor as a source of national wealth. Malaspina saw and recorded his experiences on his great voyage through a biased viewpoint—a personal meld of Enlightenment ideas that he hoped would lead toward a reform of Spanish national policies.

Malaspina had remarkable latitude in planning the expedition, including the choice of key officers. He selected men who had served with him or had special experience that he valued: Dionisio Acalá Galiano, a student of Spanish geographer Vicente Tofiño; Cayetano Valdés, the nephew of the minister of the navy; Francisco Viana, who had sailed with Malaspina in the *Astrea;* and Felipa Bauzá, an accomplished cartographer. Malaspina also secured the right to choose the best available scientists, including non-Spanish specialists such as French naturalist Luis Neé and Bohemian botanist Tadeo Haenke. Naturalist Antonio Pineda, aboard the *Descubierta,* had charge of scientific investigations.

The plan laid out a route that covered investigations of South America on Atlantic and Pacific coastlines, stops at Acapulco and San Blas on the coast of Mexico, a near circumnavigation of Australia, and a return around the Cape of Good Hope home to Cádiz. But for many reasons, including international politics, Spain altered the route and instructed

Malaspina and Bustamante to proceed directly from San Blas to the Northwest Coast. The change made the planned circumnavigation unworkable, so after their surveys on the coast Malaspina and Bustamante paused at Acapulco and headed back to Spain by way of Cape Horn. They left Galiano and Valdés to conduct a second season of surveys in the Northwest in 1792.

60. In Search of the Strait of Anian, 1789

INTRODUCTION

Once he received royal approval of the voyage, Alejandro Malaspina left few things ignored in his preparations. The first priority was the choice of ships. Rather than sail in refurbished vessels, he had two identical corvettes built at La Carraca shipyard at Cádiz: the Descubierta *(Discovery), launched on April 8, 1789, and the* Atrevida *(Audacious), launched on April 28, 1789. Each three-masted ship was 109 feet long with a 29-foot beam and displaced more than 300 tons. Following the expedition's focus on science, the ships included special compartments for the storage of specimens, a 14-foot-deep hold, and a laboratory. For protection, the ships also carried 14 six-pound cannon on deck, an additional 8 six-pounders in the hold, and 19 marines and gunners. Malaspina chose the officers, scientists, and specialists for the expedition and directed key officers to aid him in researching earlier navigators' experiences in the Pacific along with new astronomical technologies and equipment. Reflecting the expedition's avowed political mission, he instructed officers to research social, economic, and governmental conditions in Spain's colonies in the Western Hemisphere.*

Alejandro Malaspina to Antonio Valdés, Carraca, June 9, 1789

Among the valuable documents obtained by Don José Espinosa from the Archivo de Indias, worthy of first place must be the account of the voyage of Ferrer Maldonado concerning the passage or communication between the Pacific and Atlantic Oceans. This journal, summarized by the Spanish author of *Historia política de los establecimientos ultramarinos de las naciones europeas,* bears all the hallmarks of authenticity; and to the extent that I have studied the Introduction to the third voyage of Captain Cook and the voyage of La Pérouse on the coasts of California, both navigators seeking to discover this passage, it would appear that it has so far eluded all foreign searches and that there is scope for a new exploration. But it would be rash of me, following behind navigators with such a reputation for daring, to vaunt myself in the eyes of Europe as undertaking this important business with greater success, so long as the public is unaware either that there is a specific order from His Majesty that we should try with equal determination to follow those who have preceded us, or that there is some certainty not taken into account so far concerning this voyage and its appearance of authenticity, which together with the explorations already carried out, leave room for a new search.

Of course the journal or narrative in question require that the lower regions of America from 60° to 65° be closely explored, since Ferrer himself declares that the mouth of the Strait of Anian is very difficult to find; and it is also necessary that the ice proves less of an obstacle than usual, as it made a passage beyond the 71st parallel impossible in the years 1778 and

'79, when to follow in the wake of Ferrer Maldonado it is necessary to reach 75°. But I can confidently assert that if this passage is to be sought anywhere it must be done from the Pacific rather than from the Atlantic or Labrador region.

Without Your Excellency's express permission I have decided not to communicate a copy of this memorandum or of Ferrer Maldonado's sailing directions to the Academies of Paris or London. But it would seem to me appropriate, so as not to offend the navigators who have preceded us and not to commit us to excessive effort, that at this time of undertaking this exploration (if it should please His Majesty) the foundations on which it rests should be made public. In case of overwintering in the Bahía Bucareli it would be necessary to have the right equipment, and to take decked launches from San Blas.

Source: Andrew David et al., eds., *The Malaspina Expedition, 1789–1794,* Vol. 1 (London: Hakluyt Society, 2001), 318–319. Used by permission of the Hakluyt Society.

ANALYSIS

The search for the Northwest Passage, often called the Strait of Anian, had long been an object of explorations of the North Pacific. Perhaps the prime motivation for Malaspina's expedition was to emulate the achievements of James Cook and Jean-François La Pérouse, who had explored the Pacific a few years earlier and had greatly expanded scientific and navigational knowledge of the region. While the North Pacific had not been the prime focus of Malaspina's planned circumnavigation, earlier surveys of the Northwest Coast by Spanish explorers Juan Pérez in 1774 and Bruno de Hezeta in 1775, plus unresolved questions about the reported sighting of the Northwest Passage by Lorenzo Ferrer Maldonado in 1588, prompted renewed Spanish interest in the region.

In January 1789, three of Malaspina's officers researched records, maps, and other materials in the Real Archivo de Indias in Madrid in preparation for the voyage. José Espinosa y Tello uncovered a seemingly authentic report of Maldonado's discovery of an opening in the Northwest Coast that confirmed a passage through the continent. The document that Espinosa brought to Malaspina was Maldonado's memoir, which he had offered to the king in 1609. Espinosa also went to Paris, where he acquired journals written by La Pérouse and scouted navigational instruments. Espinosa was among several officers who had significant training in survey methods, including the use of navigation instruments and chartmaking. Their training came at the hands of Vicente Tofiño, the premier mapmaker in Madrid who had directed coastal surveys in Spain, Portugal, and the North African coast. Tofiño, self-taught in mathematics and astronomy, was an instructor at the Spanish naval academy and rose to Rear Admiral by the late 1780s. In addition to Espinosa, he also trained two other Malaspina officers, Dionisio Galiano and Felipe Bauzá.

Malaspina was a keen student of cartography and navigational science. He combed Cook's *A Voyage to the Pacific Ocean* (1784), noting the captain's recommendations for setting aside shipboard space to store scientific instruments and collections. Malaspina read thoroughly the work of Alexander Dalrymple, the Scottish hydrographer who had produced hundreds of maps for the East India Company, and also studied a Dalrymple pamphlet published in 1789 that focused on charts produced by mariners who had sailed along the Northwest Coast. From the pamphlet, Malaspina determined the areas that Spain had not yet explored and the potential for a Northwest Passage.

Malaspina's need for the best instruments and technical publications available prompted him to press Naval Minister Antonio Valdés to purchase chronometers, quadrants, books, and equipment from British instrument makers and publishers. Valdés, in turn, sent £1,000 to the Marques del Campo, the Spanish ambassador to London, to purchase the instruments, and Dalrymple and Alexander Aubert, an astronomer and a fellow of the Royal Society, helped find and secure the latest devices. As a result, the Spanish acquired several important natural history books on Western Hemisphere flora and fauna. They also had four John Arnold chronometers, numbers 10, 61, 71, and 72; an Arnold pocket chronometer number 105; a specially designed astronomical quadrant made by Jesse Ramsden; and eight sextants.

61. Malaspina and Bustamante at Acapulco, 1791

INTRODUCTION

Nearly two years after leaving Cádiz, Malaspina reached Acapulco on the Atlantic Coast of Mexico in March 1791. The Descubierta *and* Atrevida *had stopped at Montevideo, rounded Cape Horn, and visited Concepción, Chile; Callao, Peru; and Panama. A week out from Panama, Malaspina worried that they were behind schedule and split the expedition in mid-January 1791, sending Bustamante and the* Atrevida *on to Acapulco and then to San Blas, the Spanish naval base established in 1768 on the Mexican coast at latitude 21° north. Malaspina surveyed the Nicaraguan and Mexican coastlines and arrived at Acapulco on March 27.*

Bustamante remained in San Blas for two weeks, where he repaired boats before returning to Acapulco in April. In Acapulco, Malaspina assessed the expedition's critical needs and traveled to Mexico City for discussions with Viceroy Juan Revillagigedo. Malaspina returned to Acapulco on April 19. Two officers, José Espinosa and Ciriaco Cevallos, had arrived from Spain and traveled cross-country from Veracruz and Mexico City to San Blas, bringing with them new instruments, equipment, and news of potential war with Britain over the conflict at Nootka Sound. They accompanied Bustamante to Acapulco, arriving on April 20. Malaspina directed hasty preparations for departure following new orders from Antonio Valdés in Madrid to proceed directly to the Northwest Coast.

Journal of Alejandro Malaspina, Acapulco, April 20–21, 1791

I had the pleasure of seeing the *Atrevida* appear, already just a short distance from the port. Our separation had lasted almost four months and since the season was now well advanced for the forthcoming cruise, this timely reunion would contribute greatly to its success. These were powerful reasons to prevent us from mutually succumbing to a friendly complacency. . . . not a moment was lost in making the necessary arrangements and preparations in both ships.

Capitán de Navío Bodega y Quadra, with much zeal and eagerness, had provided him [Bustamante] with a number of charts giving all the information about our [Spain's] most recent discoveries and also those which he hoped would be carried out in the course of this year in accordance with the instructions. . . .

On the following day, the 21st, Tenientes de Navío Don José Espinosa and Don Ciriaco Cevallos handed over to me the books and instruments they had been entrusted to bring . . . two Arnold pocket chronometers numbers 344 and 351, which had been sent from England . . . together with a pendulum, also constructed in London. This was to be used, as I was informed in an official letter from his Excellency the Ministro de Marina, to repeat as far as was possible the experiments with gravity at different parallels of the Earth, not only to arrive at an agreement on measurements based on consistent observations which would eventually be checked in Europe precisely on the 45th parallel, but also in order to continue the investigations as to the true shape of the Earth. . . .

[April 25] The ships' complements were now short of five marines. . . . Nor was the *Descubierta*'s crew complete and on board the *Atrevida* three Filipino sailors, recently admitted to the service and trusting no doubt to their familiarity with the area around Acapulco, though it was known to be the least favourable for desertion, had dared to desert once more. In view of these developments I requested replacements from the troops at the castle, if there were any who would volunteer. . . .

At this point I also thought it would be opportune to examine all the losses from the two crews since our departure from Cádiz. . . .

The only thing left to do before we made our departure was to draw up the plan for the coming cruise and to inform the commanding officer of the *Atrevida* of the instructions and the corresponding places of rendezvous. The plan had to take into account not just the orders and instructions I had just received regarding the verification of the passage of one Ferrer Maldonado from the Atlantic to the Pacific Ocean and then from there to the Atlantic, but also the considerably advanced season, [and] any work previous to our own so as to combine possible hydrographic advantages.

Source: Andrew David et al., eds., *The Malaspina Expedition, 1789–1794,* Vol. 2 (London: Hakluyt Society, 2003), 65–66, 71, 74–75. Used by permission of the Hakluyt Society.

ANALYSIS

Sealed orders from Antonio Valdés dated December 1790 arrived in San Blas by courier in early April 1791, directing the captains to sail promptly to the Northwest Coast rather than survey the Hawaiian Islands. The new orders were partly the result of a scientific paper delivered in Paris by Jean-Nicolas Buache de la Neuville in November 1790, which strongly argued for Maldonado's discovery of the Northwest Passage at 60° north. The nephew of the great cartographer Phillipe Buache, Jean-Nicolas Buache was the geographer to the French crown and the operator of a respected French cartography business. His paper appeared to legitimize the document that Espinosa had uncovered in the Spanish archives in 1789 during preparations for the voyage. Valdés urged Malaspina to personally inspect the claims.

The new orders also focused Malaspina's attention on the expedition's original scientific ambitions, which included nautical surveys and chartmaking, astronomical observations, and naturalist investigations. Bustamante took advantage of his brief stay in San Blas to query Juan Francisco de la Bodega y Quadra, commander of San Blas and an experienced sailor on the Northwest Coast. Bodega wrote a specific sailing guide advising the best approach to the area described by Maldonado in 1588, a course that would take Bustamante and Malaspina to the coast near Mt. St. Elias.

The captains used the instruments that Espinosa and Cevallos had added to the expedition's equipment to test calculations of Earth's shape, which had been established by French scientists in present-day Ecuador in 1737. Malaspina used a pendulum onshore at Acapulco on April 21, 1791, to count oscillations in 60-swing increments to 300 oscillations, while an astronomical clock kept time in a boat not far from shore. As Malaspina described it, the officer at the pendulum "was to call out so loud so that at the same time the officer in charge of the astronomical clock could indicate the second or fraction of a second to which it corresponded. . . . A final comparison of the total number of oscillations with the mean time elapsed according to the astronomical clock would determine the gain or loss of the pendulum compared to the astronomical clock over the period of the observation" (David et al. 2001–2004, 2:66). The experiment tested the pendulum time imparted by gravity in comparison to the chronometer's mechanical time.

In some ways, Malaspina's greatest challenge in Acapulco was finding replacement crew members, for he had lost nearly 70 percent to a range of causes, from desertion and disciplinary actions to accident and disease. He calculated his losses: of the *Descubierta* deserters, there were 2 warrant officers, 28 seamen, 8 marines, and 2 marine gunners, and of the *Atrevida* deserters, there were 5 warrant officers, 39 seamen, 4 marines, 3 seamen gunners, and 2 servants. When added to losses from death, illness, and discipline—21 on the *Descubierta* and 29 on the *Atrevida*—the total losses were 143 at Acapulco. Many crew had been replaced in other ports, but Malaspina still had to fill in with available men in Mexico and ended up with at least 20 percent non-Spanish.

62. Strait of Anian Investigations, 1791

INTRODUCTION

On his return to San Blas from surveying the Northwest Coast in the summer of 1791, Malaspina wrote a report for Antonio Valdés that specifically addressed the minister's reason for sending the expedition directly north from Acapulco. No seafaring nation could ignore the chance of discovering a water passage through North America, even if reports such as Maldonado's had not been confirmed. Early maps had included potential passageways between the Atlantic and Pacific Oceans, but they were not based on observation or use. They were, in effect, hopeful inclusions that suggested a shortcut through the continent.

Sixteenth-century cartographers Gemma Frisius, Gerard Mercator, and others depicted versions of a navigable water passage on world maps, and for generations English mariners searched the Atlantic Coast of North America for what was called the Northwest Passage. In 1745, the English Parliament offered a £20,000 reward to the mariner who found the passage. Maldonado's claim (1588), however, like that of Juan de Fuca (1592) and Bartholomew de Fonte (1640), fell into cartographic historian Henry Wagner's category of apocrypha. "Maldonado's [claim] never received any serious consideration," Wagner concluded, "either when it was first written or when it unexpectedly came to light at the close of the eighteenth century. Its career was brief; nearly everyone who read it stamped it at once as fictitious" (1932, 183). Nonetheless, it played a role in Malaspina's expedition.

Alejandro Malaspina to Antonio Valdés, San Blas, October 12, 1791

In compliance with the orders of His Majesty, which Your Excellency designed to send to me, bearing the date of December 23rd of last year, the sloops of the Royal Navy, *Descubierta* and *Atrevida,* set sail from the port of Acapulco on May 1st with the intention of making explorations in the latitude of 60°, to see if the passage from the Pacific to the Atlantic, which the explorer Ferrer Maldonado gave notice of, really existed. Comparing the views of Mr. de Buache with the original record of Maldonado and with the results of the English and Spanish investigations (I felt) that the existence of the passage actually seemed very improbable, and consequently it was somewhat difficult to direct the operations of the sloops in such a way that not the least doubt should remain on this question and so that six months time and labor should not be spent for nothing. . . . I finally decided that, even if the original purpose of our voyage was not realized, I should not fail to make good use of the time in matters of equal importance [surveying the northwest coastline].

I shall pass over for the moment scholarly speculation concerning the authenticity of the account in question, the possibility of a passage, and the reflections on both present by M. Buache to the [French] Royal Academy of Sciences . . . but I shall say to Your Excellency for the necessary justification of my measures that it did not seem possible to admit the supposition of the French geographer that the latitude of 60°, fixed by Ferrer Maldonado, is exact and that the last explorations of Captain [George] Dixon and of our ships from San Blas greatly altered the aspect of the problem from that which Mr. de Bauche had visualized on examining only the voyages of Captain Cook. In view of this and in agreement with Don Josef Bustamante I formed the plan of examining with the utmost carefulness the coast which stretched from Cape Fairweather by Behring Bay, Mt. St. Elias and Kayes Island to Cape Hinchinbrook in Prince William's Sound. This gulf had been carefully explored in the previous year by Don Salvador Fidalgo, the Russians already settled in Cook's River could not be ignorant of nor hide the existence of a channel of communication, and thus navigation any further west appeared just as useless as exploration of the coasts mentioned seemed useful. . . .

Nevertheless, by the 28th of July we were able to consider the orders of His Majesty completed, finding ourselves once more off Cape Fairweather with the assurance that there is not even a middle-sized river along the shores that leaves the least hope of any passage nor indeed do the closely joined mountain ridges of the interior admit of possibility . . . [list of places noted on Malaspina's map] and, finally, that the botanical excursions and lithological investigations of Don Tadeo Haenke and also the experiments with the simple pendulum in latitude 59°34′ will contribute to the advancement of the sciences, and advancement which His Majesty, their patron, eagerly desires.

Source: William Inglis Morse, ed., *Letters of Alejandro Malaspina* (Boston: McIver-Johnson, 1944), 44–45.

ANALYSIS

In all sciences, there is nothing so alluring as the proof of a desired outcome. In geography, the Northwest Passage through North America proves the point, for it had been rumored, fictitiously described, and avidly pursued for more than two centuries by the time of Malaspina's claims. Malaspina had given some credence to Maldonado's voyage, even though his embrace of Enlightenment philosophy and science should have warned him off. By the time Malaspina reached Acapulco, however, he had concluded that the idea of a Northwest Passage was absurd, and his survey confirmed his skepticism. Maldonado claimed that he had sailed in 1588 from the Atlantic near Labrador through a broad strait west-southwest to the Pacific and the Strait of Anian before returning to Spain. According to his account, which appeared in "A Relation of the Discovery of the Strait of Anian," the entrance cut through steep hillsides, could harbor up to 500 vessels, and had already been used by vessels from China. The paper delivered by Jean-Nicolas Buache de la Neuville, which had prompted the new sailing orders from Valdés, did not pass muster with José Bustamante, who labeled it false in his "Reflections," written in April 1791 before sailing north.

Although Malaspina avoided delving into scientific appraisals of Maldonado's voyage in his letter to Valdés, he wrote a lengthy critique in 1792. In his "Dissertation" on Maldonado, he wrote that "progress in geography, as in the other sciences, was dependent on reasoning and experiment" (David et al. 2001–2004, 2:468), and he harshly evaluated the account. In 12 specific and pointed criticisms, Malaspina wondered how anyone could take Maldonado's claims seriously. "The most serious geographical inconsistencies" (David et al. 2001–2004, 2:474), Malaspina pointed out, were misplacing Mt. Fairweather and Mt. St. Elias and describing the north coast as angling east to west, the opposite of its west-to-east contour. He wondered, "But why occupy so much time with this sort of inquiry? Or why should we believe Ferrer Maldonado's account could be true while no positive evidence is found of its origin. . . . [A]nd surely, even if the extremely costly searches intended to settle this hydrographic question were to cease at this moment, how great would be the confusion of a reader in the twenty-first century on reading . . . Captain Cook's third voyage, and afterwards the reflections of Captain [John] Meares" (David et al. 2001–2004, 2:477).

Malaspina approached the Northwest Passage question as a mariner, as a political thinker, and as a scientific analyst. He looked at the issue as a navigational and economic question. "Originally the question asked was whether there was a passage in the northern hemisphere *useful* for a communication between Europe and Asia," Malaspina offered; "this was soon replaced by another question, whether there was any communication between the Pacific and Atlantic oceans in the northern hemisphere" (David et al. 2001–2004, 2:478). He maintained, in the end, that it was probably "inconceivable that the clamour of geographers concerning such [discoveries] . . . will finally die down, and their insults in the name of science to modern navigation will cease" (David et al. 2001–2004, 2:478).

63. Port Mulgrave Map, 1802

INTRODUCTION

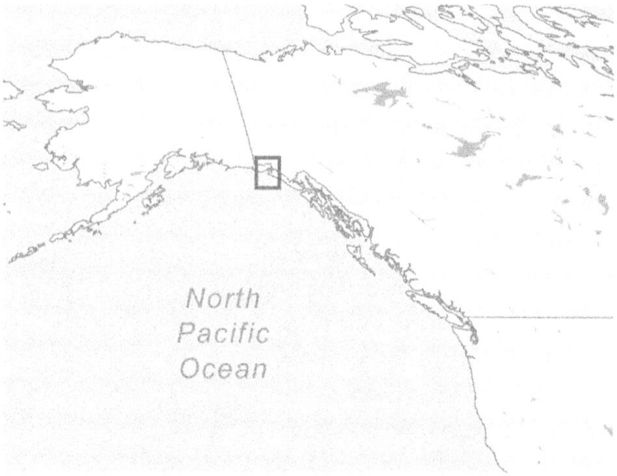

Officials of maritime nations exploring the Pacific Northwest during the 18th century expected cartographers to chart not only coastal surveys but also ports. All published accounts of Northwest Coast explorations included large-scale harbor maps, the results of the laborious and often dangerous work of gathering information both at sea and on land. During the last two decades of the 18th century, largely because of such maps, ports such as Port Mulgrave were well known to government and commercial mariners as practical harbors and geopolitically important locales. Documentation of these ports had to be precise, for they had to guide pilots to safety in time of distress. The accounts also identified the locations of Native villages, served as authoritative records of European possession rituals, and documented the search for a navigable water passage through North America.

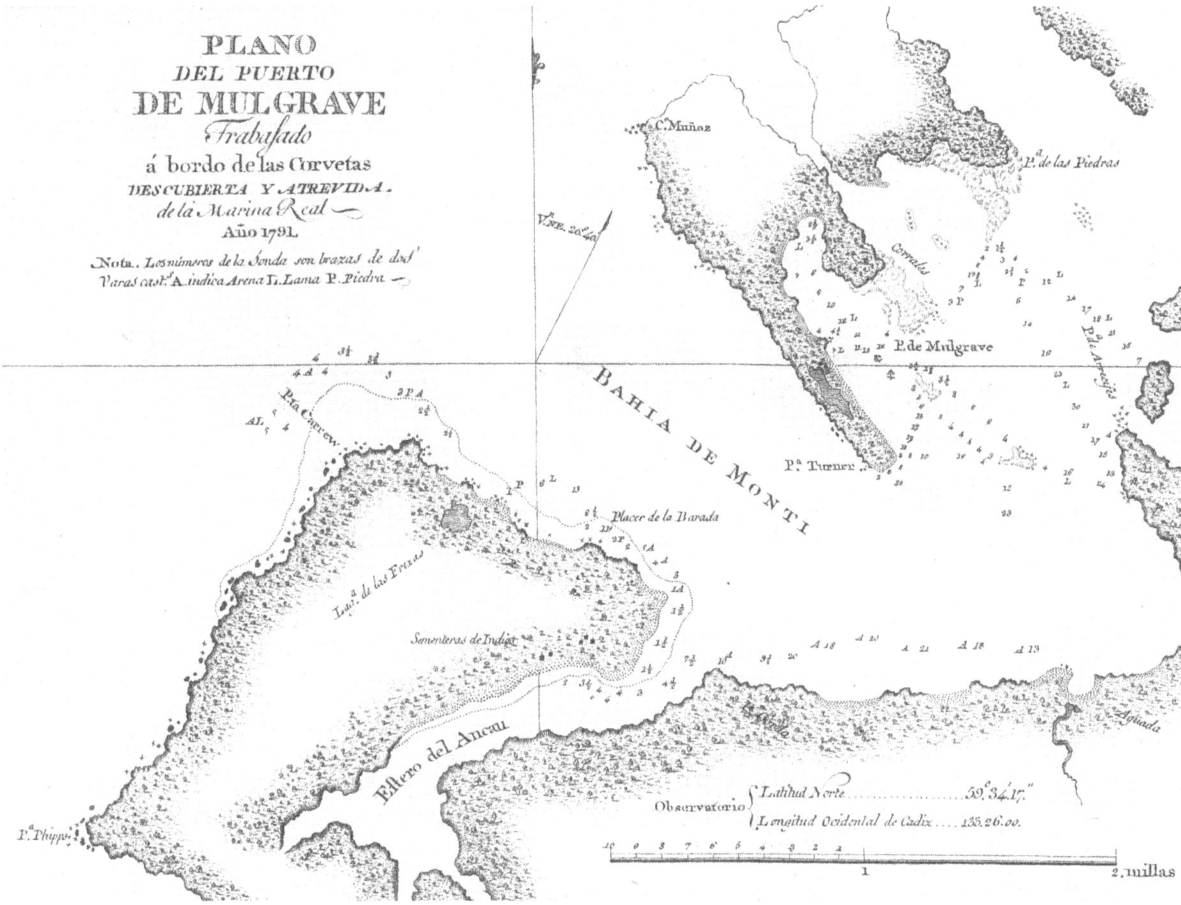

"Plan of Port Mulgrave." (Copyright © 2000 by Cartography Associates)

The full title of this map can be translated as "A Plan of Port Mulgrave Drawn on Board the Royal Navy Corvettes Descubierta *and* Atrevida *in the Year 1791." The soundings are noted in fathoms, and the bottoms encountered are identified by letters: A. for Sand, L. for Mud, P. for Rocks. It was one of five plans of ports that were included in the* Atlas Para El Vivage de por las Goletas Sutil y Mexicana, *which accompanied the official account of the voyage of Galiano and Valdés published in Madrid in 1802. Four of the charts were compiled during Malaspina's survey in 1791, but his records were confiscated because of his disgrace and arrest after the voyage, and his name (but not the names of his ships) was eliminated. Considering the vast amount of information that Malaspina compiled, the decision to publish these plans is an indication of how important the intelligence they contained was to Spanish officials.*

ANALYSIS

Malaspina's map of Port Mulgrave corresponds to part of the entrance of present-day Yakutat Bay in southeastern Alaska. Located at ~59°44′, the bay is in the traditional homeland of the Yakutat Tlingit people and is now part of the Wrangell–St. Elias National Park and Preserve in Alaska. By the time Malaspina arrived in June 1791, the bay was known to navigators from a map published in 1787 by English fur trader George Dixon, who made the earliest known European exploration of the bay. Malaspina had no commercial interest in the area, but he was still interested in the possibility of discovering a navigable route across the continent based on Maldonado's 1588 account of a Northwest Passage.

On June 27, Malaspina recorded the initial excitement of the crew as they entered Yakutat Bay: "By six in the morning, still enjoying a fair breeze, we had already cleared that point and had examined in great detail the entrance to Port Mulgrave. We then saw at the furthest point of Admiralty Bay [part of present-day Yakutat Bay], among the cordillera whose lower slopes are washed by the sea, an inlet whose mouth and course seemed to tally with the coast described by Ferrer Maldonado. At once imagination soon supplied a thousand reasons in support of hope. We considered everything that might agree with the description of the passage. As we rejected or attempted to justify what did not accord with the terrain described, each of us in the end, according to his own fantasy, either increased the length of the narrows to the north, or put down to distance and eyesight the absence of any features which could not yet be found to dispel any remaining doubts" (David et al. 2001–2004, 2:103–104).

Malaspina and his men spent 10 days surveying the bay, including 3 days exploring the promising inlet to the north and west. An impenetrable ice barrier in the inlet led to the name Desengaño (Disappointment or Disenchantment), which is retained today for a long arm of Yakutat Bay that reaches into Hubbard Glacier. One of the four other charts of ports in the 1802 atlas is Malaspina's "Plano del Puerto del Desengaño." During this time Malaspina recorded many interesting interactions with the Tlingit, but the principal object of his mission had been accomplished—to prove or disprove Maldonado's claim that there was a Northwest Passage. On July 2 Malaspina formally took possession of the region with minimal ceremony.

Although a renewed search for a Northwest Passage was the immediate reason for Malaspina's exploration, virtually all previous European voyages to the Pacific Northwest had been under orders to either begin or end their explorations as far north as approximately 60°, because that parallel was considered the most likely northern limit that a ship could reach before turning west toward the Aleutians. Juan Pérez reached to ~55° (1774), Juan Francisco de la Bodega y Quadra to ~58°30′ (1775), James Cook to ~61°30′ (1778), and Ignacio Arteaga y Bazán to ~60°13′ (1779). None of them encountered Yakutat Bay.

In 1786, Jean-François de Galaup La Pérouse reconnoitered the entrance to the bay and named it Baie de Monti, after one of his officers. The next year Dixon extended his search for fur-trading opportunities to the same latitudes and became the first European to enter Yakutat Bay. On his map of 1789, which Malaspina carried, Dixon named the bay after the Englishman Lord Mulgrave and named Point Carrew and Point Turner after crew members. Malaspina retained these names on his map, although he moved the place-name Mulgrave to the northeast within the bay and adopted La Pérouse's name Bahia de Monti for the entrance. The place-names Monti, Carrew, and Turner remain today. In 1788 a Russian vessel and the English fur traders James Colnett and William Douglas probably entered the bay, and Peter Puget of the Vancouver expedition explored the bay in 1794.

On the Malaspina map, the Estero del Ancau is the present-day Ankau River. A dotted line marks ships' tracks, and multiple soundings and notations of the nature of the sea bottom (sand, mud, rock), the presence of shoals (crosses), and the tree-lined shore add to the carefully plotted data accumulated by the surveying parties. As with most European-authored maps of the period, cartographers included little or no evidence of the presence of Native residents of the region.

Within a few years, from 1786 to 1794, Port Mulgrave was visited and charted by men from England, Russia, Spain, and France. This map is one indicator of the bay's potential commercial and geopolitical importance to Europeans.

64. With the Tlingit at Port Mulgrave, 1791

INTRODUCTION

Tomás de Suría was born in April 1761, likely in Madrid, and took art studies under Jerónimo Antonio Gil at the Royal Art Academy of San Fernando. At age 17, Suría went to Mexico with Gil and continued his studies at the Academy of San Carlos. In 1788 Suría married Maria Josefa Fernandez de Mendoza and by 1791 had two children. When Malaspina requested an artist for his expedition, word got to Gil, who offered Suría for the position from among his best students. Suría reached Acapulco on February 16, 1791, and by March he had been assigned to work under Antonio Pineda on the Descubierta. *Suría's journal is in two sections; only the first section has survived.*

Suría was a neophyte sailor and was not careful with names, but he was an excellent artist. Malaspina noted that Suría "recorded with the greatest accuracy everything which the skill of the artist could supply to make the narrative of the events more easily understood" (David et al. 2001–2004, 2:425). On returning to Acapulco, Suría spent eight months completing his work as an artist-engraver for Malaspina and in 1792 joined Galiano and Valdés on the Sutil *and* Mexicana *on their voyage to the Northwest Coast. Suría then returned to Mexico and worked as an engraver, replacing his mentor Gil as chief engraver at the mint and becoming involved in politics after Mexican independence. Suría died in 1835. In this document, he describes his first encounter with Tlingit at Yakutat Bay in June 1791.*

Journal of Suría's Voyage to the Northwest Coast, June 29, 1791

The 29th, the fog cleared up somewhat and the observatory, a field tent, was put ashore. The Indians continued to be very sociable and in return for clothes buttons, which they hang as pendants from their ears, gave us some rich, fresh salmon, at the rate of one for a button. This fish in the north is the most delicate thing which can be imagined. We could never satisfy ourselves with it notwithstanding that we ate an abundance of it. The chief, with his two sons, the eldest very ferocious and gigantic, asked the commander by signs to be allowed to come aboard. This the commander promised. He was entertained in the captain's cabin today. There was much to wonder at about these three men. The chief was an old, venerable and ferocious looking man with a very long gray beard, in a pyramidal form, his hair flaccid and loose on his shoulders. False hair over it in various locks, without any order or arrangement, made him look like a monster. A large lion skin for a cape was gathered in at the waist and left entirely bare his breast, arms, thighs, and endowments, very muscular and strong. All gave him a somewhat majestic air, which he manifested by speaking but little, measuredly, and with a sound which at times seemed to be the bellow of a bull. At other times it was softer and in speaking to his sons it was sweeter than in conversation. The elder of these was more than two yards tall, equally stout and muscular. He had his hair loose which, on account of its thickness, seemed like a horse's mane. It was very black like that of his beard. He was dressed in a black bear skin and very hairy, also in the form of a cloak which he fastened with some ornament, leaving bare at times all his nakedness, and passing to and fro over the quarterdeck, very proud and straight, his look full of ire, arrogant, and condescending. They both began to explain to us by signs that other ships had been there and that they had seen them. These we inferred were those of Dixon which were there in 1787. They also explained to us about some battle with very strange gestures and postures, which showed us they were very warlike. What we could draw from all their signs was that a short time before they had fought some other cacique who had killed the son of their chief. They showed us his helmet which was of a figure, and an extraordinary construction of wood, copper, and of straw cloth, and with a mask in front which appeared to be a wolf.

Source: Henry R. Wagner, "Journal of Tomás de Suria of His Voyage with Malaspina to the Northwest Coast of America," *Pacific Historical Review* 5(3) (September 1936): 249.

ANALYSIS

Malaspina knew from George Dixon's descriptions that only a few Tlingit inhabited Yakutat Bay—the village at Port Mulgrave had about 80 people—but he also knew that they had a reputation for sometimes being aggressive. The expedition picked an anchor spot, Port Mulgrave—a good location, Malaspina noted, because the Natives lived "nearby in numbers large enough to be studied without fear of disturbing their customs" (David et al. 2001–2004, 2:107). As the *Atrevida* and the *Descubierta* entered the bay, a group of Tlingit approached in canoes—two large vessels and one two-person kayak—singing songs and making welcoming gestures. They came aboard only when seamen got into the canoes as hostages, and then they traded furs for everything from clothing to metal items. Discovering that some Tlingit had made off with some iron, Malaspina banned them from the ships and set up a bartering area onshore.

Tomás de Suría went ashore with Felipe Bauzá, who established an observatory. The tent drew Tlingit in to continue trading, this time offering more salmon than furs in exchange for Spanish items; salmon traded at one fish for one iron nail. At the observatory, Bauzá established a baseline by using steel chain measurements and sightings of prominent land features to frame a triangle with the use of theodolites (devices that measured vertical and horizontal planes) to precisely measure distances. Dionisio Galiano, Juan José Vernacci, and Juan Concha—the expedition's best astronomers—established latitude and longitude using an astronomical clock and the Ramsden astronomical quadrant to compare daytime sun and nighttime moon sightings with chronometers that had been set in Panama and Acapulco.

Suría looked at the world he encountered through artistic eyes, remarking on the appearance, conditions, and reactions of both his cohorts and the Tlingit. On this day at Port Mulgrave, the action around the observatory tent buzzed. "The numbers of natives gathering around our tent increased at every moment," Malaspina wrote in his journal. "At one time they wanted to look through the telescopes and at another to come to the chests and examine their contents" (David et al. 2001–2004, 2:113). Suría took the opportunity to sketch several women and to make line drawings of Tlingit objects. He also drew a picture of a Tlingit chief, whom he described as "venerable and ferocious." The chief was so pleased with his portrait that he asked the artist to draw the mask of the man who had slain his son. In Suría's telling, an enemy had killed the chief's son, but Malaspina wrote that it was the chief's father who had been killed and that his ashes were placed near a pole at a ceremonial spot near the Spanish observatory tent.

The following day, the chief implored Malaspina to fire one of the *Descubierta*'s guns to frighten an enemy group that was approaching shore in canoes. Malaspina complied. While the Spanish witnessed what they thought was a near conflict avoided, the chief hailed the visitors and welcomed them ashore with warm embraces and singing—a ritualized welcoming of guests that included a mock battle.

65. Speculations on the Peopling of America, 1791

INTRODUCTION

On leaving Port Mulgrave, Malaspina sailed north to complete "the real object of the voyage"—to survey "the stretch of coast below the 60th parallel, just where the supposed passage of Ferrer Maldonado was to be found according to the speculations of M. de Buache" (David et al. 2001–2004, 138–139). It was also a part of the Northwest Coast that James Cook had not observed on his third voyage, and it was crucial that Malaspina investigate it thoroughly. As he noted in his journal in mid-July, "this coast had been drawn by Captain Cook and our own [Spanish] navigators from a great distance" (David et al. 2001–2004, 154). Even more concerning, Malaspina understood how important it was for Spain to compete against Britain on the Northwest Coast and that an accurate charting of the coastline would aid his country. Suría understood this reality as well. "The English had the advantage of us in these discoveries" of the hydrography of the Northwest Coast, he wrote. "Our commander, far from being disquieted by such demonstrations of the natural and inveterate spirit of heroism of our nation, gave as an example an invitation to a celebration in the cabin in view of the officers" (Wagner 1936, 264), by which he meant that Spain could not be satisfied with being in second place behind the British.

At the conclusion of the voyage, Malaspina sequestered Suría's journal and the journals kept by other members of the expedition, planning to include the information in his massive report to the king. As a result of Malaspina's political incarceration and the postponed publication of his expedition memoir, however, the publication of Suría's account was delayed. The first English translation appeared in 1936.

Journal of Suría's Voyage to the Northwest Coast, August 1791

Seeing the coast of this Puerto Bucareli so cut up into small islands, rocks, and farallons [pillar rocks], besides the innumerable islands of this archipelago which we surmised to exist, gives us occasion to think with some foundation that in some other time perhaps this coast extended for a long distance into the sea and that through some formidable earthquake it sank, leaving behind the many islands. This opinion is more confirmed if you make two reflections: first, if we compare America with the other parts of the world, Asia, Africa, and Europe, it will be noted that America from the Strait of Bering in 72° of latitude north to 60° south, where Cape Horn is located, covers a tremendous distance of thousands of leagues which occupy almost half the globe, considering the fact that this extent of latitude is very unequal to its longitude, as there are various isthmuses, for example, that of Panama, where the distance between the north and south seas is only thirty leagues. America occupies a very small extent of longitude compared to that of the two immensely wide seas which bathe it. In the South Sea, as a proof of its size, the direct course from Acapulco to Manila, that is from the coast of America to that of Asia, measures 3000 leagues. That this disproportion which I have shown is notable in this part of the world and is not seen in the others is manifest by what I have said. Possibly in ancient times America was wider and who knows if it did not connect with Asia, when the distance from one to the other was much less, and that in consequence the many islands which are close to both coasts were united to them. It is noticeable that some of them, as for example Guadalupe, which is in front of Old California, is fifty leagues distant from the coast, the Sandwich Islands are distant a thousand leagues, and those of Otaeti are the same, both settled by people. How could these people have come, lacking as they did a knowledge of navigation? On the other hand their language, customs, dress, and religion do not bear any resemblance to those of the uncivilized people of America. If their inhabitants had belonged to the cultivated nations of Europe who by some accident or shipwreck had reached these islands is it believable that there would not have remained some remnant at least of their primitive dialect? Is it not the fact that their language is original and different from all those known? If we reflect on their religion we encounter the same difficulty, as they are idolatrous like all the inhabitants of the northern regions of America, and if we enlarge somewhat on the matter, in view of the proofs just given, it can be considered that the havoc and ruin of this coast happened many centuries before the conquest. In view of the fact that in none of the writings of those times does any similar notice occur nor even by tradition of the Indians themselves, in my opinion, some continents became separated and lost when that terrible earthquake took place at the death of Jesus Christ, demonstrating even to a stubborn person the pain of seeing Our Maker suffer. Then would have been formed these islands, separating among them the inhabitants who before occupied the whole continent. For this reason little difference is noted in the customs of these islanders from that on this coast. Likewise, they are found to be heathen because when this event occurred all Americans were heathen, and many centuries were to pass before the Spaniards propagated among them the light of the evangel.

Source: Henry R. Wagner, "Journal of Tomás de Suria of His Voyage with Malaspina to the Northwest Coast of America," *Pacific Historical Review* 5(3) (September 1936): 270–271.

ANALYSIS

Suría wrote this extended speculation after describing the expansiveness and complexity of the Northwest Coast. As an artist with no previous experience in deepwater sailing, he took in the world that he saw with curiosity and a reliance on empirical knowledge. Assigned to Antonio Pineda, Suría drew detailed sketches of the flora and fauna the naturalist collected in Mexico, which added significantly to the scientific documentation of the expedition. Pineda remained in Mexico during the Northwest Coast voyage, so Suría worked closely on that journey with Felipe Bauzá, the expedition's primary cartographer. Suría also had close relationships with Joaquin Hurtado, the second pilot on the *Descubierta,* who tutored Suría in the fundamentals of navigation.

Expedition leaders enthusiastically embraced Enlightenment science, the powerful approach to knowledge that elevated rationality over superstition and physical evidence over imaginary stories. In his 1788 plan, Malaspina had emphasized the benefits of pursuing scientific discovery of the world as a defense of empire. He was following the lead of the Royal Botanical Garden, which had launched a systematic survey of natural history in New Spain based at the Botanical Garden of Mexico in Mexico City. Suría had contributed botanical drawings to that effort before joining the expedition in 1791.

Suría no doubt discussed his observations with other crew members, and he may have been influenced by Malaspina's philosophical opinions and interest in geology, especially metal ores. His speculative description of the geological history of the Northwest Coast also likely originated in his discussions with Pineda, who had investigated volcanic geology in Nicaragua before the expedition arrived in Acapulco. In 1785 James Hutton in Edinburgh had proposed a new theory of geologic history, later called Uniformitarianism, that explained landforms as the result of gradual and unending change in relationships between sea and land. Hutton challenged the biblical explanation of Earth's form, and it is clear from Suría's discussion that he held both scriptural and scientific ideas to be true. There is no evidence that he or his colleagues had read Hutton, but religiously based explanations of geological forms were becoming increasingly unsatisfactory in the late 18th century. Suría's speculations about earthquakes and human migration anticipated 19th- and 20th-century scientific discoveries and theories—by Charles Lyell in geology, Charles Darwin in biology, and Alfred Wegener on plate tectonics. Suría's reference to the origins of Native peoples in the Americas, however, was likely the result of reading José de Acosta's 1590 work, *History, Natural and Moral, of the Indies,* which proposed that North American Indians had migrated from Asia.

Nonetheless, Suría's emphasis on divine causation for earthquakes indicates that while he acknowledged the power of empirical science, he did not abandon his belief in religious ideology. In another section of his journal, for example, he agreed with Tadeo Haenke, the naturalist on the expedition, that landforms on Earth "cannot be the mere child of accidents which can often be seen in Nature but that of the Divine Providence" (Wagner 1936, 266).

66. A Plan of the Friendly Harbor, 1802

INTRODUCTION

This chart, "A Plan of the Friendly Harbor Situated on the West Part of the Entrance to Nootka in the Year 1791," shows a harbor on the northwest coast of present-day Vancouver Island. The view is from the northeast looking into the harbor. Friendly Harbor (or Cove), as it was known, is present-day Yuquot, a small and sparsely populated settlement on the southeast shore of Nootka Island and an epicenter for trading activity and geopolitics during the last 15 years of the 18th century. It is the ancestral home to the Nuu-chah-nulth (Mowachaht, Muchalaht), principally seafaring people who lived there from 4,300 years ago and who govern it today. Nootka Sound, the entrance to several broad inlets extending into the interior of Vancouver Island, sheltered more than 20 Native settlements identified by European visitors in the late 18th century. Friendly Harbor, at the entrance to the sound, was the most familiar of these villages, and it is represented in many published accounts of exploring expeditions and commercial enterprises. Friendly Harbor was also the site of the only non-Russian European settlement north of San Francisco on the Northwest Coast, the Spanish port of Santa Cruz de Nuca, from 1789 to 1795.

Members of the Malaspina expedition prepared this chart during their visit in August 1791. It was published in Madrid in 1802 in the atlas that accompanied Relación del viage de por las goletas Sutil y Mexicana, *the official report of the expedition of Alcalá Galiano and Cayetano Valdés. This and other maps of the region were drawn at a critical moment in the history of maritime exploration of the Pacific Northwest, when Spanish, British, American, and Native parties were simultaneously endeavoring to settle questions of sovereignty and trade disputes.*

ANALYSIS

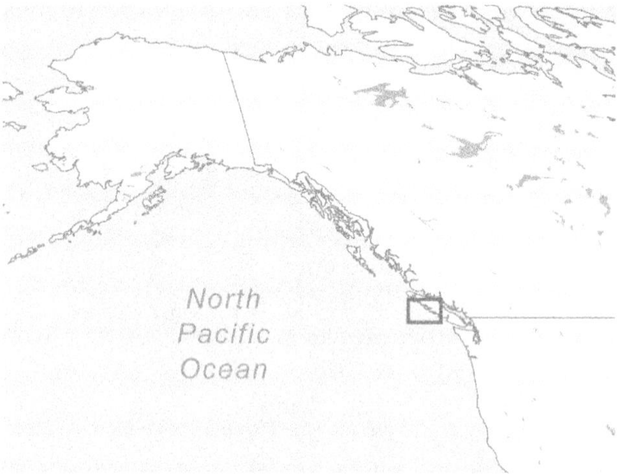

Although Esteban Martínez may have sighted the entrance to Nootka Sound during the 1774 expedition of Juan Peréz, Captain James Cook has been credited as the first European to enter the sound, in late March 1778. Cook originally named it King George's Sound but changed the name to Nootka based on his apparent misunderstanding that it was the Native name. Cook met with Maquinna, an important and influential chief who met every visiting captain through the rest of the century. The captain's account and atlas were published in 1784, including the first chart of King George's Sound (Nootka Sound). In 1786 English trader James Strange renamed the harbor Friendly Cove, reflecting its favorable location and profitable fur-trading activity. The name persisted on many published maps and charts, including those of the English trader John Meares in 1790.

In May 1789 Esteban José Martínez, using Cook's maps as a guide, was the first Spaniard to enter Nootka Sound. Fearful of suspected Russian intentions to occupy the area, Martínez built a small garrison in the cove and a protective battery atop an adjacent island. He took aggressive actions against British commercial interests, which created the Nootka Sound Controversy and drew international attention. The fortification was abandoned shortly thereafter but was quickly rebuilt by a detachment of Catalonian soldiers in 1790 under the command of Francisco de Eliza y Reventa.

When Malaspina visited Nootka Sound in August 1791, Spanish territorial claims in the Pacific Northwest were at a peak. The settlement at Nootka was fairly extensive, as reflected

A view of Friendly Harbor from the northeast. (Copyright © 2000 Cartography Associates)

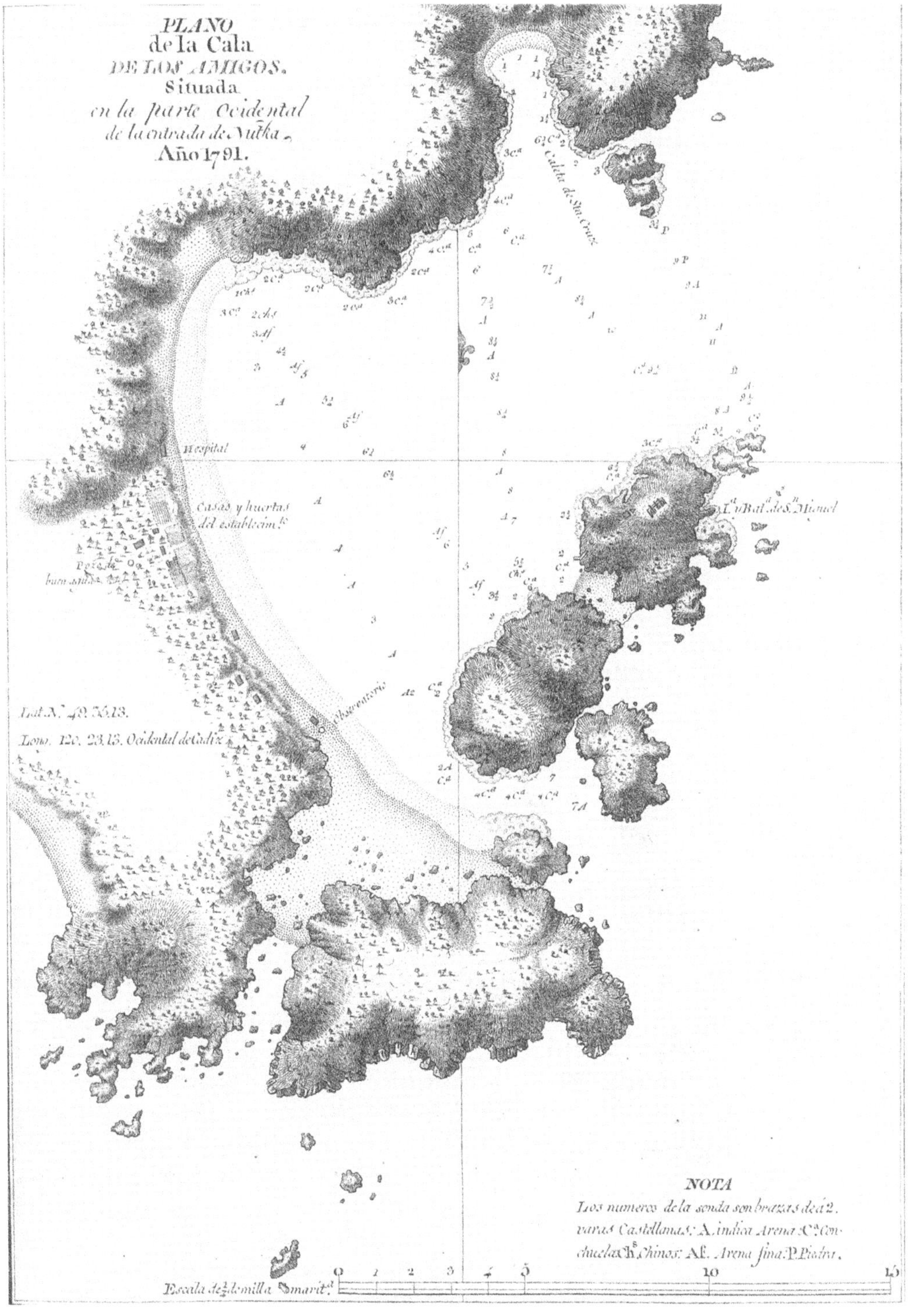

“A Plan of the Friendly Harbor.” (Copyright © 2000 by Cartography Associates)

by this chart and view of Friendly Harbor. Longitude was given west of Cádiz, and careful renderings of numerous offshore islands, a scale of miles, many soundings in fathoms, and records of the type of sea-bottom sediment (A for sand, C for shells, P for rocks) testify to Malaspina's meticulous documentation. This copperplate engraving is particularly notable for its artistic quality, especially in the use of shading and hachure lines to suggest topographical elevation of hills or mountainous areas. Place-names in the harbor document the Spanish presence. The battery on the island of San Miguel (present-day Observatory Island) protected the entrance of the harbor as well as a small *caleta* (cove), Santa Cruz de Nuca. Noted in the interior are the *casas y huertas* (houses and orchards or gardens) of the settlement, with a nearby hospital, observatory, and *pozo de buena agua* (well of good water).

The panoramic view by artist Manual José Antonio Cardero rendered a heavily fortified island and an oversized Spanish flag overlooking a ship, either Malaspina's *Atrevida* or Eliza's *Concepción*. In the background, Native canoes surround rather typical European-style buildings. In the foreground, Cardero re-created the day after Malaspina arrived at the harbor, when a large war canoe approached the Spanish ships. The canoe was propelled by 10 oarsmen on either side (Cardero illustrated 7) and carried Tlupananú, a chief of a nearby village, to welcome the Spanish.

In 1795, Spanish and British forces abandoned Nootka Sound and Friendly Harbor under terms of the 1790 Nootka Convention. Today in the community of Yuquot, a Canadian National Historic Site, there are few commemorative remnants of a late 18th-century European presence.

67. Malaspina at Nootka Sound, 1791

INTRODUCTION

The Atrevida *and the* Descubierta *arrived at Nootka on August 12, 1791, and anchored two days later in the harbor named by two British mariners, explorer James Cook and fur trader James Strange. Called Friendly Harbor, the name reflected their welcome by the Nuu-chah-nulth (Mowachaht, Muchalaht), who lived there. Spain had established a foothold at Nootka, which they named Santa Cruz de Nuca, and it was there in 1789 that their aggressive treatment of British mariners had precipitated a conflict between the two nations that nearly led to war. Malaspina knew about the controversy and wrote an official report on conditions at Nootka in 1791, which became part of the ongoing British and Spanish negotiations that eventually settled conflicting claims on the Northwest Coast without war. Nonetheless, the Spanish had anchored the frigate* Concepción *in the harbor and maintained an armed battery, and a small garrison onshore stood ready to protect Spain's interests at Nootka.*

Even though the expedition was not a colonizing venture, establishing friendly and reliable relations with Native people was a priority for Malaspina. He hoped to avoid the tense relations they had experienced at Port Mulgrave, but his officers approached Native groups in a defensive posture. The Mowachaht at Nootka had poor relations with the Spanish at the time of Malaspina's visit because of clashes between Indians and Esteban Martínez in 1789. In short, both the Spanish and the Mowachaht had good reasons to be cautious.

Journal of Alejandro Malaspina, Nootka, August 19–27, 1791

Our peaceful relationship with the natives was now more solidly established, although at the cost of various presents which were requested equally by the chiefs and their subjects, as well as a continual contribution of biscuit. The canoes no longer fled at the sight of our boats and we were daily surrounded by fishermen, with many fish of excellent quality, while some of the natives spent the night beside the observatory. There were few chiefs who had not visited us. We had received the same attention from chief Macuina [Maquinna], although there was considerable mistrust in his face, but he did not allow his three wives, who accompanied him, to come on board. Nor did I overlook on this occasion that he had his own personal interest with the sale of a slave girl that he had brought with him. The officers from the frigate *Concepción* usually bought them either for two muskets or for one or two sheets of copper. Dealings of this kind, too closely linked to questions of religion, morality and policy for discussion in a few lines, were at the time quite common in the establishment. As many as twenty-two children of both sexes had already been transported to San Blas or were soon to be taken there, entrusted for their future education and upbringing to one or another person of His Majesty's ships always chosen from among those who were of good conduct and married, with families in the Department.

The presents given to Macuina and to those who accompanied, or had preceded him, brought increasing numbers of natives to gather on the ships in the following days. Among them the chief Tlupananú always stood out. . . .

In return for the constant friendship towards the Spaniards . . . Tlupananú requested, as well as other presents, a large sail made of light canvas for his canoe. However, our desire to see the manoeuvres of that canoe was such that it restrained the impulses of gratitude on this occasion. . . .

Very early on the morning of the 23rd he finally came to visit us with the canoe we wished to see. It was paddled by some thirty men, whose chants, manoeuvres and skill surprised us in their first turns around the corvettes. . . . Don Tomás Suria drew from life all their actions. . . .

On the morning of the 27th we had another visit from Macuina, to whom we had sent various emissaries with invitations, hinting that we were anxious to ratify our friendship with him. . . . So many motives and even more the hope of a fine present, had indeed had their effect on Macuina's sentiments, as his face now showed clearly the desired change of heart. He took several cups of tea on board the *Atrevida,* a custom now well-established among his relations and subordinate chiefs. . . . Two canoe sails, four panes of window-glass, a sheet of copper, several *varas* [= 33 inches] of blue cloth and a few pieces of hardware, were the presents given him on board the *Descubierta.* On that occasion he confirmed the previous cession of the land for our present national establishment, assuring us that there would be lasting peace between us.

Source: Andrew David et al., eds., *The Malaspina Expedition, 1789–1794,* Vol. 2 (London: Hakluyt Society, 2001), 179–180, 186. Used by permission of the Hakluyt Society.

ANALYSIS

The first encounters between Mowachaht people and the expedition at Nootka were wary. Even so, Malaspina wondered why so few people were in evidence. He soon found out that Spanish-Mowachaht friction, especially between Chief Maquinna and the Spanish garrison, had disrupted relationships and political power among local chiefs. Malaspina did not recognize or understand the divisions of interest and power among Native chiefs at Nootka. The chiefs made alliances with competing European visitors—the British and the Spanish—to enhance their power. So it was that Maquinna's rival, Tlupananú, took advantage of Malaspina's arrival to display his importance by bringing an impressive war canoe into the harbor near the *Descubierta.* Because of Tlupananú's good relations with Spanish officers, especially Pedro Alberni, commanding officer of the Spanish garrison, Malaspina made an early and easy connection with him. Tlupananú took advantage, warning Malaspina that Maquinna was "crafty and overbearing and he looks on you with hatred and abhorrence" (Wagner 1936, 274).

Malaspina's later recorded success in dealings with Maquinna were self-congratulatory. Good relations with the chief, Malaspina realized, were important to maintaining Spain's small beachhead garrison at Nootka. In 1789 Esteban Martínez had killed a Mowachaht chief, Callicum, as a consequence of his seizure of a British vessel at Nootka. Callicum had established a relationship with the British, who guaranteed him protection against other Europeans. When Martínez had taken the British ship, Callicum had remonstrated against Martínez, who shot him dead in front of his family. Fearing for his own life, Maquinna fled and from then on limited his contact with the Spanish. By 1791, Malaspina hoped that he could entice Maquinna to reestablish favorable relations. Malaspina made some headway, but good relations came only after the chief met with Juan Francisco de la Bodega y Quadra and George Vancouver in 1792.

Spain had established the garrison at Nootka to defend their claims to control of the seas in the region, not as a colonization effort. In their relationships with Natives on the Northwest Coast, however, they often tried to "save" children by taking them to Mexico, Chile, or even Spain to Christianize and educate them in European ways. Spain had oppressed and enslaved Native peoples for more than two centuries in South America, so Malaspina might well have seen the paternalistic acquisition of Indian children in the late 18th century as an enlightened and progressive act, believing that Christianization offered them the greatest benefit.

68. Malaspina and Political Tragedy, 1794–1803

INTRODUCTION

Alejandro Malaspina and José Bustamante returned to Cádiz in September 1794, and Malaspina confidently expected to spend his time preparing his journals and papers from the voyage for publication. In December, King Carlos IV and Queen Maria Luisa received the two commanders and their principal officers at the Royal Seat of San Lorenzo El Escorial, 30 miles northwest of Madrid. Malaspina received praise for his accomplishment from the monarchs, but he desired a new opportunity to serve Spain. He perceived that Manuel Godoy, Spain's first secretary (prime minister) and a close associate of the king and queen, had blocked him from further service, and the two men became locked in an adversarial relationship.

Even though the king promoted Malaspina to the rank of brigadier in March 1795, his dispute with Godoy continued. The cause was politics or, more specifically, Malaspina's political opinions and his criticisms of state policies, which infuriated Godoy. At first Malaspina appeared to ignore or deny his poor treatment, as this correspondence reveals, but increasingly it became apparent that the state did not welcome his ideas or service. In November 1795 and without warning, Godoy had Malaspina arrested for state treason, summarily tried, and clapped into Castillo de San Antón at La Cornuña, where he remained imprisoned until 1803. Among other consequences of state imprisonment, Godoy confiscated Malaspina's journals and other written records of the voyage, materials that would not be published until 1885.

Alejandro Malaspina Letters, January–February 1796

Alejandro Malaspina to Paolo Greppi, December 19, 1794

I flattered myself that I would be allowed to be useful to this country at a time as tempestuous as the present. One single day would have been sufficient to expose my system. I have seen everything. I have been everywhere. I had hoped that no matter the chaos of the present system it would be realized that there is but a small step from the wrong route to the right one, from absurdity to sane philosophy. Everything seemed favourable. I felt attached to the most virtuous and wise in the country; people were obviously ready to listen to me and I was certain in my heart of my total dedication to the welfare of all, without selfishness or prejudice; but access to the Sultan is so difficult; everything around him so immersed in confusion and inaction. . . . So here I am condemned to the miserable craft of a writer while I burn to be active and to let my actions fit my principles.

Antonio Valdés to Manuel Godoy, January 19, 1795

The attached papers, numbered 1 and 2 [drafts of a treaty with France] were given to me in Madrid by D. Alejandro Malaspina following the interview he had with you. I had retained them in the belief that you would have no time to read them. However, I am forwarding them now in case there might be in them some element that could agree with your own thoughts, particularly as, for a Minister of State, nothing is to be ignored in matters of such importance. . . . I shall act as you may decide.

Manuel Godoy to Antonio Valdés, January 25, 1795

I have just read Malaspina's papers . . . [and] his handwriting is as bad as the substance, but the lack of principles and moderation in his ideas is such that I shall guard from showing any of this to the king and queen, so as not to sully the reputation of an officer whose career deserves recognition. I cannot help but telling you, for the sake of our service to God and to Their Majesties that, acting as if on your own accord, you should advise Malaspina to burn these drafts, if he has any, and keep silent for ever on this matter.

Alejandro Malaspina to Paolo Greppi, February 17, 1795

At this very moment it is touch and go whether I may be destined to matters of the greatest import concerning every aspect of the kingdom's prosperity, or be sent back to my old crafts of seamanship which I do not dislike in any case.

Source: Andrew David et al., eds., *The Malaspina Expedition, 1789–1794,* Vol. 3 (London: Hakluyt Society, 2004), 325–326. Used by permission of the Hakluyt Society.

ANALYSIS

Malaspina's correspondents included Paolo Greppi and Antonio Valdés, two men who shared confidences with him and honored his contributions to science, naval intelligence, and Spain. Greppi was the Spanish representative of a financial house in Milan and also Malaspina's personal financial agent and close friend. Valdés was Spain's minister of the navy, an honored officer, head of the Order of Malta in Spain, and promoter of Malaspina's service in the Spanish Royal Navy. In the December 19, 1794, letter, Malaspina calls Manuel Godoy "the Sultan," referring to Godoy's reputation as a powerful minister and his closeness to the king and queen. Malaspina evidently believed the false rumors that Godoy was having an adulterous relationship with Queen Maria Luisa, a charge made by Godoy's political enemies. More important, Malaspina considered Godoy to be a corrupt official who should be replaced.

Malaspina had angered Godoy by impugning the minister's policies and the broader administration of Spain's empire, especially the political management of the nation's South American colonies. His statement that he would "let my actions fit my principles" refers to his embrace of the liberal political viewpoint and axioms of John Locke, Jean-Jacques Rousseau, and other Enlightenment philosophers who denied the divine right of kings and advocated popular sovereignty through the rule of law. About Spain's colonial administration, Malaspina wrote that "oh, if only by some happy occurrence, amid the terrible convulsions which disturb the human race and combine to impede its increase, it were to be Spain's role to restrain greater spilling of blood and to cauterize some of the present wounds, the consequences of which may perhaps affect adversely the history of the coming century or more in unhappy Europe!" (David et al. 2001–2004, 1:lxxxv). Godoy saw Malaspina as a conspirator, even though both men were advocates of reform, especially in the application of science and rationality in state affairs.

Spain was at war with France as part of the First Coalition of European monarchies that tried to halt the spread of the French Revolution. Malaspina was headstrong and sure of his opinions, and knowing that secret negotiations had begun to strike peace with France, he heedlessly offered his plan for a peace treaty, a plan that Valdés dutifully passed on to Godoy. The plan apparently confirmed Godoy's fear that Malaspina, a well-known and admired officer, was part of a conspiracy to bring him down. Malaspina knew that he was vulnerable, but his egotism drove him closer to eventual ruin. As one of his confidants later wrote, "We all saw with amazement how little circumspection was shown by Malaspina in undertaking to write on such a serious, secret and delicate matter of State" (David et al. 2001–2004, 3:330).

Godoy apparently moved secretly to lay a trap for Malaspina by arranging to intercept messages from him to the king's and queen's religious confessors. The messages appeared to confirm a conspiracy against the state, which Godoy used in a council meeting in November 1795 to obtain a life sentence for Malaspina and the sequestering of the brigadier's papers.

Further Reading

David, Andrew et al., eds. *The Malaspina Expedition, 1789–1794: Journal of the Voyage by Alejandro Malaspina.* 3 vols. London: Hakluyt Society, 2001–2004.

Wagner, Henry R. "Apocryphal Voyages to the Northwest Coast of America." *Proceedings of the American Antiquarian Society* (1932): 179–234.

Wagner, Henry R. "Journal of Tomás de Suría of His Voyage with Malaspina to the Northwest Coast of America in 1791." *Pacific Historical Review* (September 1936): 234–276.

CHAPTER 8

George Vancouver, the Great Survey of the Northwest Coast, 1791–1795

George Vancouver stands in the first rank of British maritime explorers by virtue of his circumnavigation of the world in 1791–1795, a voyage that featured an extensive charting of the North Pacific. The published account of his exploration, *A Voyage of Discovery* (1798), offered exhaustive maps of the Pacific Northwest coastlines, from Baja to the Gulf of Alaska. Not only were his maps replete with detailed charts depicting principal rivers and bays, they also were the first published maps of Puget Sound, Vancouver Island, and the Columbia River. Most important, Vancouver's surveys ended speculation about a Northwest Passage through North America, the long hoped-for waterway between the Atlantic and Pacific Oceans.

This portrait of George Vancouver (1757–1798) is unauthenticated and did not come to light until 1878, but many historians believe it may well be Vancouver because of the setting and his resemblance to images of his brother. (Corbis)

Vancouver grew up with ship commerce as a part of his daily life. He was born on June 22, 1757, in King's Lynn on the Great Ouse River, which emptied into the Great Wash on England's east coast. His father was deputy collector of customs in King's Lynn, where ships and busy wharves were central to the town's life as a major trading port for European countries. When Vancouver was 13 years old, with a grammar school education that included good math instruction and fluency in Latin and French, his father placed him with the navy. At age 14 in January 1772, Vancouver received assignment as midshipman on James Cook's second voyage to the Pacific, which explored the South Pacific and ended speculation about a great southern continent. On Cook's *Resolution,*

Vancouver learned advanced mathematics and astronomical observation from William Wales, one of the creators of Britain's *Naval Almanac* and the officer assigned to test Larcum Kendall's K3 maritime chronometer. Cook admired Vancouver's attention to detail and personal strength and signed him on as senior midshipman for his third voyage in 1776. That voyage took Vancouver to the Northwest Coast for the first time. He was also the midshipman, at the age of 21, who dealt with Hawaiian Natives after Captain Cook was murdered and who brought the commander's body onto the ship.

After the voyage, Vancouver remained in the navy. He was named a lieutenant in 1780 and was on assignment to the West Indies, where he rose through the ranks on three vessels. His training under Wales in navigation, chartmaking, and coastal surveys led to important opportunities. In the West Indies, where Vancouver served as third lieutenant on the *Europa* under Sir Alan Gardner (later on the Board of Admiralty), he impressed his captain with his mastery of navigation technologies, especially the use of maritime chronometers and lunar observation methods for determining longitude. In 1787 Gardner assigned Vancouver and Joseph Whidbey, master of the *Europa,* to chart the harbors of Kingston and Port Royal in Jamaica, valuable experience that helped prepare Vancouver for later tasks on the Northwest Coast. The charts the two made led to promotion for Vancouver, to second lieutenant in 1787 and to first lieutenant the following year, which made him second-in-command to the new captain, James Vashon. On the *Europa,* Vancouver sailed with future *Discovery* crew members, including Joseph Baker and Zachary Mudge, and developed a strong bond with Whidbey and Peter Puget.

After the return of Cook's crew and the publication of his journals, *A Voyage to the Pacific Ocean* (1784), a rush to gather wealth in the Pacific fur trade brought the interest of maritime nations. Britain grew concerned as French and Spanish exploratory missions headed to the Northwest Coast during the late 1780s and worried over other nations' interest in acquiring control of natural resources in the region. In 1789 the Lords Commissioners of the Admiralty decided to mount a follow-up maritime survey of the coast. Although Captain Henry Roberts had been assigned command of the expedition, the Admiralty shuffled commanders when the Nootka Sound Controversy forced Britain to contest Spain's claim of territoriality in the North Pacific. Vancouver was put in charge of the *Discovery,* partly because of his navigational expertise and evident strong personality, and Roberts was reassigned to duty in the West Indies. Vancouver also drew assignment as the crown's diplomatic representative in working out a resolution to the British-Spanish dispute, a task that had been outlined in the 1790 Nootka Convention. The resolution required on-site negotiations between representatives of the two nations in Nootka Sound.

DID YOU KNOW?

In April 1792, George Vancouver encountered the first other ship on the Northwest Coast, Robert Gray and the *Columbia.* Gray told Vancouver about sighting the Columbia River, to which Vancouver scoffed that it was unlikely to be a major river, characterizing it as "inaccessible" and blocked by "reefs and broken water."

On the four-year, 65,000-mile voyage around the world, Vancouver commanded the 330-ton *Discovery,* a refitted merchant ship, and William Broughton captained the 131-ton *Chatham* brig. At the time, it was the longest deepwater sailing voyage on record. The Admiralty gave Vancouver full authority to choose his officers, and he called on his fellow seamen from the *Europa.*

The two ships left Falmouth, England, on April 1, 1791, and headed into the Atlantic bound for the Cape of Good Hope, which they made by early July. After five weeks there, the expedition sailed into the Indian Ocean and on to western Australia by late September, where Vancouver found an expansive and accommodating bay he named King George's Sound Company. By December he had visited New Zealand, named Chatham Island, and arrived in Tahiti. The expedition spent the early weeks of 1792 in Hawaii, leaving for the Northwest Coast in March and arriving at 39°33′ at the present-day northern California state line in mid-April.

The expedition spent most of the surveying season in the Strait of Juan de Fuca, Puget Sound, and the Georgia Strait and on the west coast of Vancouver Island. During their first season in the region they circumnavigated present-day Vancouver Island, disproved the existence of a Northwest Passage, and explored 100 miles of the Columbia River. After spending the 1792 season surveying and negotiating with the Spanish at Nootka Sound on Vancouver Island, they sailed to Hawaii, where they spent the winter refitting the ships and preparing for a second season of surveying the Northwest Coast. Vancouver spent considerable energy and time trying to quell intratribal warfare among Hawaiian chiefs and suggested that ceding the islands to Britain might solve their problem.

DID YOU KNOW?

George Vancouver's midshipman Thomas Manby recorded in his diary on June 5, 1792, in Puget Sound: "The few days the ship passed at this place called Strawberry Bay [named for many wild strawberries onshore], the Brewery was erected on shore which furnished us with twenty puncheons of [spruce] beer. To this beverage we attributed the good health of our Ship's Company; not a sick man on board either Vessel since our first landing in the Straits of Juan De Fuca."

In late March 1793 the expedition left Hawaii for the Northwest Coast, but the voyage east took longer than anticipated, leaving only four months to complete a detailed charting of continental bays, inlets, and harbors north of Vancouver Island. The only conflict with Natives took place in Behm Canal, in the archipelago that forms the present-day Alaskan panhandle, when Tlingit surrounded the boat-based surveyors. The incident ended with gunfire that killed several Tlingits. Vancouver headed to California at the end of the season and then to Hawaii a third time in January 1794.

That year, Vancouver succeeded in negotiating a cession of the Hawaiian Islands to Britain. Kamehameha, a king to his people, agreeably ceded lands to the British crown, but it is doubtful that he perceived how the British understood the agreement. In the end it made little difference, for Britain took no official action to confirm the arrangement. The crown had more difficult foreign policy issues to face by the time Vancouver returned to England in 1795, primarily with revolutionary France, and it is likely that the cession of Hawaii drew little attention.

The last season of surveying on the Northwest Coast began in mid-April 1794 at Cook's River in Alaskan waters, where Vancouver demonstrated that the inlet was a bay, not a river mouth, as Cook had believed in 1778. Vancouver made contact with Russians on the Alaskan coast in mid-May and spent the next few weeks surveying Prince William Sound, sailing south to Yakutat Bay by late June 1794. From the Northwest Coast, he sailed to Mexico and then to Valparaiso, Chile, by March 1795. A five-month voyage from Valparaiso around Cape Horn brought his multiyear circumnavigation to an end in October 1795, with both the *Discovery* and the *Chatham* surviving in good shape to dock at Deptford Yard on the Thames River. In early July 1795 on the way home Vancouver had taken the *Macassar*, a Dutch merchant ship, as a prize at St. Helena in the South Atlantic. For commercial and geopolitical reasons, Britain had forcibly taken Cape Colony from the Dutch that year, and Vancouver took advantage to capture the ship and its value for his country.

Vancouver's return to London did not proceed as he had expected. Unlike Cook, who had been welcomed as a hero on his return from the second voyage in 1771, Vancouver met an indifferent reception and grumblings by some of the ship's crew, who criticized his leadership. Chief among the critics was Thomas Pitt, Lord Camelford and nephew of Prime Minister William Pitt, who charged Vancouver with harsh treatment unbecoming an officer and demanded an official and public redress. Vancouver denied the charges and focused on writing up a thorough account of his four-year expedition. Not only was it important to document the discoveries that he and his men had made on the Northwest Coast, but he also knew that the detailed maps would be extremely valuable to the crown and to merchant traders in any pursuit of wealth and territory in the North Pacific.

Even though Vancouver had considerable materials to create his extensive reports, including nearly all of the logs and journals created by his officers, he had difficulty completing his account of the voyage. In part, the delay was the result of his poor finances. He was on half-pay after the voyage ended, and the Admiralty would not release his portion of prize money from capturing the *Macassar*. His brother John aided him in pursuing his pay and completing the report manuscript, and another brother, Charles, defended him against Pitt's charges and the continuing criticisms that bedeviled the great explorer's last years. Vancouver's declining health complicated everything. He suffered from a range of symptoms that singly or cumulatively led to his death in May 1798, at age 40. Historians have suggested that he may have had tuberculosis, nephritis, or Addison's disease, but it is most likely that he died of kidney failure as a complication of contracting malaria during his West Indies service.

Although Vancouver did not live to see his manuscript in print, there can be no doubt that *A Voyage of Discovery* stands as a masterpiece of reporting and compilation of information that was revelatory about the Northwest Coast. More important in many ways, especially during the decades after publication, were the maps and charts Vancouver generated from his extensive surveys of the North American coast in the North Pacific.

In the final pages of his nearly 500,000-word *Voyage of Discovery*, which John Vancouver prepared for publication, George Vancouver addressed his work succinctly: "Before I bid farewell to the Discovery, I must beg leave to arrest the attention of my readers for a few minutes, for the purpose of taking a short view of the geographical knowledge which had been obtained of the earth, previously to the expedition which I have had the honour to command, and the happiness of bringing thus to a conclusion; and also to notice such parts of the globe as yet remain to be explored to make that species of information complete" (Lamb 1984, 4:1541).

69. Letter of Instruction to George Vancouver, 1791

INTRODUCTION

Vancouver's expedition left England on April 1, 1791, but British interest in undertaking a survey of the Northwest Coast likely originated as early as 1785. Publication of James Cook's A Voyage to the Pacific Ocean *in 1784 had prompted a rush of traders to gather furs on the coast, and Britain knew that other nations sought influence and wealth there. In addition, trader John Meares had appeared before Parliament in April 1790, charging that Spanish com-*

mander Esteban José Martínez at Nootka Sound had captured British ships owned partly by Meares and seized property. War between Britain and Spain loomed, but in October 1790 the nations used diplomacy instead, agreeing to resolve the issue in part by sending representatives to Nootka. Britain sent Vancouver, and Spain sent Juan Francisco de la Bodega y Quadra.

Vancouver's official instructions, composed by the Lords Commissioners of the Admiralty in 1791, included careful explanation of survey objectives and general guidelines for his diplomatic mission at Nootka Sound. Vancouver understood that the principal reason for the voyage was scientific discovery and that settling the dispute with Spain was a secondary purpose. Ironically, though, the Nootka Sound Controversy was the proximate reason for sending the ships to the Northwest Coast in 1791.

Lords of Admiralty to George Vancouver, March 8, 1791

The King having judged it expedient, that an expedition should be immediately undertaken for acquiring more complete knowledge, than has yet been obtained, of the north-west coast of America; and, the sloop you command, together with the Chatham armed tender . . . to proceed, without loss of time . . . to the Sandwich islands in the north pacific ocean, where you are to remain during the next winter; employing yourself very diligently in the examination and survey of the said islands; and, as soon as the weather shall be favorable, (which may be expected to be in February, or at the latest in March, 1792) you are to repair to the north-west coast of America, for the purpose of acquiring a more complete knowledge of it, as above mentioned. . . .

It having been agreed, by the late convention between His Majesty and the Catholic King, . . . you should proceed, in the first instance, to Nootka, or elsewhere, in order to receive, from the Spanish officers, such lands or buildings as are to be restored to the British subjects. . . .

In which examination [of the Northwest Coast] the principal objects which you are to keep in view are,

1st, The acquiring of accurate information with respect to the nature and extent of any water-communication which may tend, in any considerable degree, to facilitate an intercourse, for the purposes of commerce, between the north-west coast, and the country upon the opposite side of the continent. . . .

2dly, The ascertaining . . . of any settlements which have been made within the limits above mentioned, by any European nation, . . . the survey should be so conducted, as not only to ascertain the general line of the sea coast, but also the direction and extent of all such considerable inlets, whether made by arms of the sea, or by the mouths of large rivers, as may be likely to lead to, or facilitate, such communications as is above described. . . .

[T]o avoid any unnecessary loss of time, you should not, and are therefore hereby required and directed not to pursue any inlet or river further than it shall appear to be navigable by vessels of such burthen as might safely navigate the pacific ocean. . . .

[A]nd you are therefore hereby required and directed to pay particular attention to the examination of the supposed straits of Juan de Fuca, said to be situated between 48° and 49° north latitude, and to lead to an opening through which the sloop Washington is reported to have passed in 1789, and to have come out again to the northward of Nootka. The discovery of near communication between any such sea or strait, and any river running into, or from the lake of the woods, would be particularly useful. . . .

In the execution of every part of this service, it is very material that you should use, and you are hereby strictly charged to use every possible care to avoid disputes with the natives of any of the parts where you may touch, and to be particularly attentive to endeavour, by a judicious distribution of the presents, (which have been put on board the sloop and tender under your command by order of Lord Grenville) and by other means, to conciliate their friendship and confidence. Given under our hands 8th of March, 1791.

Source: George Vancouver, *A Voyage to the North Pacific Ocean and Round the World,* 3 vols. (London: G. G. and J. Robinson, and J. Edwards, 1798), 1:xvii–xx, xxii.

ANALYSIS

The *Discovery,* named after one of James Cook's ships, was launched in December 1789, months before the Nootka Sound Controversy erupted. At 330 tons and with three masts, the *Discovery* carried a crew of 100 and armaments of 10 four-pound cannons and 10 swivel guns. Captain Henry Roberts, a veteran of Cook's voyages, had been set to command, but the political crisis at Nootka prompted the Admiralty to boost Vancouver from first lieutenant to captain. That promotion likely came on the advice of Vancouver's former commander in the Caribbean, Sir Alan Gardner, a member of the Board of Admiralty. As in many military assignments in Britain, having well-placed advocates resulted in advancement up the ranks.

The Admiralty initially planned to send "the frame of a small shallop" (Lamb 1984, 1:22), a single-mast fishing boat, with the *Discovery* to conduct a scientific survey of the Northwest Coast. It quickly became evident, however, that the shallop took up too much room and compromised the ship's storage capacity. Instead, a second vessel, the brig *Chatham,* accompanied the *Discovery.* In part, the requirement for cargo space reflected the involvement of Joseph Banks, president of the Royal Society and a veteran of Cook's third voyage, who urged that the work of the expedition include the collection of flora and fauna.

The *Chatham,* named for the first lord of the Admiralty and captained by Lieutenant William Broughton, had a poor reputation with its crew. One called it "our Dung Barge" (Lamb 1984, 1:37). The brig weighed 131 tons, carried a crew of 45, and had armaments of four three-pound cannons and six swivel guns. Several months after the *Discovery* and the *Chatham* sailed from Falmouth, Britain sent a third vessel, the *Daedalus,* which carried provisions and further instructions for Vancouver's negotiations at Nootka. The *Daedalus* arrived on the Northwest Coast in August 1792.

The coastal survey instructions directed Vancouver "to ascertain the general line of the sea coast" from latitude 30° north to the Gulf of Alaska. The Admiralty also expected Vancouver to prove conclusively that there was no Northwest Passage through North America, which he did. They wanted an exhaustive charting of the Strait of Juan de Fuca and the coastal island archipelago, and they especially wanted him to investigate the rumor that American trader Robert Gray had found a passage to the sea east of Nootka Sound. Gray had not discovered a passage, but Vancouver's survey would establish that Nootka was in fact on an island.

The instructions forbade the expedition from pursuing rivers to the interior and directed Vancouver to draw up charts in the interest of maritime utility. The Admiralty wanted Britain's interests advanced, but they also instructed Vancouver that if he encountered any mariners from other nations he was "to treat them in a most friendly manner, and be careful no to do any thing which give any interruption of that peace which now happily subsists between His Majesty and all other powers" (Lamb 1984, 1:285). Near the end of the instructions, the Admiralty cautioned Vancouver about contact with Native people. Cook's murder in Hawaii in 1779 and other conflicts between British mariners and Natives suggested a careful approach, including the beneficial use of trade items to establish friendly relations.

70. First Contact with Northwest Natives, 1792

INTRODUCTION

The diary of Thomas Manby, a midshipman on the Discovery *and master on the* Chatham, *begins in February 1791 and ends in June 1793. Two copies exist, one at the Oregon Historical Society and another at Yale University. Manby intended this personal and secret diary for his Dublin friend, John Lees, warning him that it "is the scribble of a plain blunt Seaman" and not a philosophical account or a naturalist's rendition of the voyage. Manby also kept a log on the* Discovery *and the* Chatham *that detailed his observations on six survey excursions. That log is among 20 from the voyage archived in the Public Record Office in London. After the voyage, Manby was one of the crew who was critical of Vancouver for the captain's harsh treatment of seamen and his use of violent language. There is no evidence, however, that Vancouver disliked or distrusted Manby. Vancouver promoted Manby to master of the* Chatham *in June, only weeks after publicly dressing him down.*

This section of Manby's diary, when he was master mate on the Discovery, *describes the expedition's first contact with Native people on the North American continent, at 39°33′ north. The ships had left Hawaii in March 1792 to begin surveying the Northwest Coast, and Manby recorded his first view of the coast on April 17, 1792: "Few Parks in England could equal this Natural Landscape only inhabited by Indians and Wild beasts" (Manby Diary).*

Thomas Manby Diary

25th [April] At 1 P.M. it fell little wind, we found the influence of the current setting the ship fast towards the Rocks. The anchor was dropped in 37 fathoms, and soon after the Chatham followed our example. The calm lasted all the afternoon, which permitted two canoes with six Indians in each to visit us without hesitation and came on board with little Invitation, being the first Inhabitants of the Continent we had seen. . . . Each man had his quiver hung round him well fitted with arrows, the points of them were either of Bones or Stone, placed in with great ingenuity and neatness. . . . The yellow buttons on our cloaths fixt their attention, for there them they bartered every thing they possessed. We procured from them some curious flint knives about four inches in the blade and some skins of Bears, Deer, Foxes, Racoons and other small Animals. . . . We found red Ochre in their possession. After mixing it with grease or train oil they smear the face, hair and body over with it and to make themselves more beautiful in their own Opinion, a shinning mineral not much unlike writing sand is laid plentifully on the Eyebrows, Nose & Chin. It was difficult to read their countenances, so much were they disfigured by their odious fashions; the Septum of the Nose was perforated and a piece of bone four inches long pierced through it lying athwart the face. By constantly living in the smoak their eyes have acquired a loathsome redness which makes them appear the most furious of the furies. When spoken to or given any thing they constantly repeated the words Slaw, Slock, Slee; this we imagined to be a term of friendship, as they gave evident marks of satisfaction when we repeated it after them. Having spent two hours on board, they returned to the shore with their finery suspended to the Ears and Nose. The canoes were very ill formed, made from the pine and hollowed out by fire, about twelve feet in length and four or five in breadth; in their appearance they much resembled a large Butchers' Tray.

Source: Manby Diary, Oregon Historical Society, MS 2389.

ANALYSIS

Only a few of the men on the expedition had been to the Pacific, so their contact with Natives during the voyage created new and important impressions of people much different from themselves in appearance, behavior, and culture. By the time Vancouver reached the Northwest Coast, the expedition had encountered several communities of Natives in the Pacific. From England, the expedition had sailed around the Cape of Good Hope to Australia, New Zealand, and Tahiti. Among seamen, the reputation of Tahiti focused on the beauty of the women, their willingness in intimate relations, and the islands' spectacular landscape. The expedition spent four weeks in Tahiti, interacting with Natives mostly on good terms, but perceived thefts by Natives created unanticipated friction and near conflict before Vancouver left for Hawaii in January 1792. Vancouver approached Hawaii with some concern, for he had been there in 1779 when Natives had murdered his commander, James Cook, and the ships' crews knew the story. During three weeks in the Hawaiian Islands, Vancouver quickly discovered that Natives would not trade for anything but arms and gunpowder, quite a change from his experience in 1779.

Manby and other seamen were struck by Native adornments, especially the cosmetic application of grease, tinctures, and glitter, which underscored how distinctive and foreign the people of Tahiti and Hawaii appeared to the sailors. They recorded derogatory evaluations of the Natives, in contrast to their appreciation of Tahitian beauty. While there were some disagreements, generally good relations prevailed, and Vancouver set sail for the Northwest Coast on March 16.

On April 18, 115 miles north of present-day San Francisco, a seaman on the *Discovery*'s topmast sighted land, and Vancouver's Northwest Coast survey began. Of the Northwest Natives, Manby noted that "the eye received them with disgust; in stature small and in person filthy and stinking, we considered them the nastiest race of people under the Sun" (Manby Diary). The term "train oil" refers to whale oil and is from *traan,* the Dutch word for "tear," because rendering of whale oil by boiling produces tear-like droplets. The "shining material" that Natives applied to their faces was likely silica, which flashes reflectively in direct sunlight.

Manby's focus on Native clothing and artifacts was less ethnographic than commercial. Along with other expedition members, he traded expedition stores for furs and "curiosities." The instructions to Vancouver had been explicit that a "judicious distribution of presents" should be used to establish friendly relations, and he openly allowed crew members to conduct trade with Natives. The men found that they had difficulty with Northwest Coast languages, which included glottal stops and unfamiliar sounds, unlike the easily learned South Pacific languages that they had picked up in a few days.

71. Near-Conflict in Puget Sound, 1792

INTRODUCTION

Vancouver began surveying Puget Sound in May 1792, after entering and sailing east through the Strait of Juan de Fuca. His survey had several objectives, including an exacting search for any indication of a Northwest Passage. Vancouver reported in his account that they "completely explored every turning" of Puget Sound by simply and thoroughly exploring every mile of continental coastline. It was an enormous task considering that Puget Sound, which stretches about 100 miles north to south, encompasses more than 1,000 square miles of water and over 1,300 miles of coastline.

The expedition proceeded into Puget Sound, anchoring south of present-day Bainbridge Island and exploring in boats to near present-day Tacoma, including circumnavigation of Vashon Island. Vancouver put names on his charts as a claim on explored lands and for British naval personnel or lords to use in acknowledgment of imperial support for exploration. For boat surveys, Vancouver gave chief responsibility to his close friend, 35-year-old Joseph Whidbey, and 22-year-old lieutenant Peter Puget. Vancouver used Puget's journal to compose the section of A Voyage of Discovery *on Puget Sound. The events he described took place at present-day Carr Inlet, an exploration that Puget undertook after surveying the west side of Vashon Island.*

Captain George Vancouver, A Voyage of Discovery, May 1792

Mr. Puget had little more to communicate respecting his late expedition than what had fallen under my own observation, excepting the disorderly behaviour of an Indian tribe . . . whose conduct had materially differed from that of the natives in general; and in particular from that of a party consisting of about twenty natives whom they had before seen in that route. . . . About eight in the evening, attended by some of the natives in two canoes, they landed for the night. These people could not be invited nearer our party than about an hundred yards, where they remained attentive to all the operations until the tents were pitched, when it became necessary to discharge some loaded muskets, the noise of which they heard without any apparent surprize, and exclaimed *poo!* after every report. They soon afterwards paddled away to the westward. The next morning . . . they saw a canoe making towards them, on which they rested on their oars to wait its approach. The canoe suddenly stopped, and no offers of presents, nor signs of friendly inclinations, could induce the Indians to venture near the boat. In order to remove their apprehensions, Mr. Puget fastened some medals, copper, and trinkets, to a piece of wood which he left floating on the water; and when the boat was at a sufficient distance, the Indians picked it up. . . . [M]ost of them had lost their right eye, and were much pitted with the small-pox. They now attended the boats for a short time, and having received some additional presents, returned to the shore. . . . They seemed well acquainted with the value of iron and copper, but would not dispose of their weapons, or any other article in exchange for either. . . . [S]ix canoes were seen paddling hastily round the point of the cove they were in, and directing their course towards the boats . . . on whose approach, a line on the beach was drawn, to separate the two parties from each other; which was readily understood, and obeyed. They now divided their numbers into two sets, one remaining on shore with their bows and quivers, and other retiring to their canoes, where they quietly seated themselves. . . .

On a seventh canoe joining the Indian party, those on the beach immediately embarked; and the whole number, amounting to twenty four persons, evidently entered into a consultation. . . . This conduct tended to increase the suspicious that their inclinations were otherwise than friendly. . . . At this time, an eighth canoe joined the party; on which all of them paddled to the beach, jumped on shore, and strung their bows. This was manifestly preparing for an attack, as they had not ever been seen, on any former occasion, with their bows strung. . . . Mr. Puget thought it might answer a good purpose to fire a swivel, shotted; the effect of which, might teach them to respect, hereafter, our powers of defence, and induce them, on the present occasion, to prefer a pacific deportment. . . . Although, on the report of the gun, or the distant effect of the shot, which was fired over the water, not the least visible astonishment or apprehension was expressed, yet, the measure was almost instantly attended with every expected good consequence. Their bows were soon unstrung; and instead of their menacing a combat, their weapons became articles of traffic, in common with other trifles they had to dispose of, for copper, buttons, knives, beads, and other ornaments; in which friendly intercourse, they accompanied the boats until towards the evening, when they peaceably took their leave, and returned to their home.

Source: George Vancouver, *A Voyage to the North Pacific Ocean and Round the World,* 3 vols. (London: G. G. and J. Robinson, and J. Edwards, 1798), 1:275–278.

ANALYSIS

Vancouver had a broad range of experience with Native people in the South Pacific and the Pacific Northwest. Perhaps most traumatic for him was the murder of James Cook on the sands of Kealakekua Bay on the big island of Hawaii in February 1779. In the aftermath of Cook's death, Vancouver had personally helped diminish tensions between Hawaiians and crew members. On his own expedition to the Northwest Coast in 1792, Vancouver faced no threats from Natives, principally because he had been instructed to conduct friendly relations with them. He understood that without tacit approval from local people, the expedition's boat-based exploring parties would be vulnerable. Natives also had good reasons to avoid confrontations. They desired trade with the Europeans, and Vancouver's men often traded for furs during their surveys. Nonetheless, Vancouver's exploratory crews did not escape all friction with Natives, including when the expedition was far north of Puget Sound in 1794 and trade goods on the *Discovery* ran low.

Vancouver's exploration method, which limited how long the *Discovery* and the *Chatham* remained in any anchorage, meant that the expedition's connection with Native people was usually brief and connected mostly to trade. As Puget put it in the fall of 1792, his contact with Natives was "more from conjecture and appearance than Real Information" (Frost et al. 1999, 155). Although he lamented his lack of information about Native people, Puget recorded more ethnographic details and his representations contain more dispassionate descriptions of Native culture than any other expedition journal keeper. In describing the people he encountered near present-day Quilcene Bay on Hood Canal, for example, Puget wrote that "the Conduct of these People impressed me with an high Idea of their Honesty, for whatever they had to barter, was suffered to be taken away, without an Exchange & it would be sometimes ten Minutes or a Quarter of an Hour before the person returned from the Boats with the Things he intended to give. . . . Surely then, if these People behave with such Confidence to Strangers, may we not infer, that Innate Principles of Honesty actuated their Conduct on this Occasion?" (Blumenthal 2007, 23).

When Puget decided to fire his swivel gun loaded with grapeshot he hoped that it would "cower" the Natives, but they paid little mind. Perhaps, he surmised, they knew all along the explorers' intentions. "I now felt Real Satisfaction in not having carried Matters to an Extremity," he wrote. "But I still am of Opinion, that the Desire with which we entreated the first Canoe off the Village to come alongside & the Presents made him, had impressed them with an Idea we were purchasing their Friendship out of fear of the Numbers" (Blumenthal 2007, 35). In *A Voyage of Discovery,* Vancouver wrote that the Natives' behavior on Hood Canal was "certainly very mysterious, and renders the foregoing an extraordinary circumstance, for which it is difficult to account" (Vancouver 1798, 1:275).

72. Sounding and Surveying Puget Sound, 1792

INTRODUCTION

Before making their way to Nootka Sound on the west coast of present-day Vancouver Island, the Discovery *and the* Chatham *coasted north to the Strait of Juan de Fuca from where they intercepted the coast of North America at about 39° north. They entered the strait, sailed east, and began exploring Puget Sound by mid-May. Edward Bell, clerk on the* Chatham, *kept a journal throughout most of the expedition. The journal was undiscovered until the early 20th century, when historian Edmond S. Meany (1915) purchased the portion covering Puget Sound from a book dealer in London. Meany figured out that Bell was the author, although the book dealer remained unconvinced. Meany verified his findings and published an edited version of the journal in 1915. The section on Bell's stay in Hawaii was published in the* Honolulu Mercury *(1929), and the* Oregon Historical Quarterly *(1932) published the section on his trip up the Columbia with Lieutenant Broughton in October 1792. Bell's manuscript journal for the South Pacific portion of the expedition is in the Turnbull Library in Wellington, New Zealand. Bell often gave more space in his journal to social interactions between expedition members and Natives, but he also recorded major events during the survey of the Strait of Juan de Fuca and Puget Sound.*

Edward Bell Journal, 1792

As the Discovery had only been waiting for us here [Strawberry Bay on Cypress Island in San Juan Islands] we left it the following day [June 6, 1792] and steered for a very extensive opening trending about N. and came to anchor in a very pleasant Bay which was called Birch Bay [near present-day Blaine, Washington]. From this place two Boat expeditions were undertaken [June 12, 1792] one by Captain Vancouver and the other by Mr. Whidby. In the meantime the Observatory was set up for the purpose of regulating the watches and Spruce Beer brew'd for the Ships Companies. Our operations on shore were carried on in a very convenient place there being a fine Grass plot of nearly a mile in length with a fine fresh water River at the back of it. Capt. Vancouver set out with his two Boats and 10 days provisions on the 12th to the Westward and Mr. Whidby with two Boats and a weeks provisions towards an opening to the Eastward of us. . . . [On being surprised by two ships near them they proceeded] [w]e soon came up with them and they hoisted Spanish Colours. A Boat with an officer was sent on board the Brig where he was very politely received by the

Commander. They proved to be his Catholic Majesty's Brig Soutile commanded by Don Dionisio Galiano and the Schooner Mexicana, Don Cayetano Valdez, Commander; both Captains of Frigates in the royal Navy of Spain and employed in surveying these Streights to complete the parts left unfinished by Seigr. Malespini with whom these two gentlemen had been Lieutenants. . . . Don Galiano offered us every information & civility in his power and sent on board some milk & cabbages that he had brought from Nootka. . . . In this dreary place [Desolation Reach] (the first place that deserves that name that we had been since we entered De Fuca's Streights) we lay about three weeks in the course of which time no less than three Boat expeditions were undertaken from us and two by the Spaniards. In the last of ours by Mr Johnstone a passage to sea was discover'd by an extensive Arm that led into Queen Charlotte's Sound and to which the continent had been carried. Mr Johnstone's situation in this Arm of the Sound was once or twice rather critical, for coming into it unexpectedly he was surprised to find himself among several villages, populously inhabited and well arm'd with Musquets, and they had endeavoured to decoy him to a place where he observed, as he proceeded on, several large canoes well mann'd, he however did not go near them, and prevented them from following him.

Source: Edmond S. Meany, ed., *A New Vancouver Journal on the Discovery of Puget Sound by a Member of the Chatham's Crew* (Seattle, 1915), 12–13.

ANALYSIS

Joseph Banks, president of the Royal Society, had drawn up much of the specific instructions for Vancouver's coastal survey, which in Puget Sound required using small boats to navigate narrow inlets and to thoroughly view bays, islands, and river mouths. Vancouver had left port with two advanced maritime chronometers, Larcum Kendall's K3 on the *Discovery* and John Arnold's 82 on the *Chatham.* The mariners used the chronometers shipboard to measure time against Greenwich Mean Time, thereby establishing longitude in degrees and minutes distance from England. At times, they also used a less accurate and more time-consuming lunar-sight method to calculate approximate longitude. While Vancouver consistently placed his position at sea several degrees east of his actual location, chronometers were a revolutionary technological advancement for mariners.

On land, surveyors Peter Puget, Joseph Whidbey, and James Johnstone used several methods to record the coastline precisely. They set up observatory tents, where they established location by using sextants, theodolites (an instrument to measure horizontal and vertical angles), and station pointers (devices that aided in plotting positions on charts). In most cases, the surveyors fixed positions through triangulation with landmarks or stations that had known locations. Using small boats, they applied triangulation methods, depth soundings, comparative compass readings, and rough sketches to draw up the outline of the coastline. They compiled the information and matched, compared, and combined time and distance statistics to create charts.

Vancouver moved his ships from one anchorage to another and sent survey boats out to investigate the multiple inlets and bays in the sound. The survey crew included Peter Puget and Thomas Manby, sailing a pinnace and a launch boat. They headed east and north, exploring Burrard Inlet, the present-day harbor entrance to Vancouver, British Columbia. Puget sighted the opening of the Strait of Georgia, the waterway around Vancouver Island. It was a significant finding. When Vancouver and Robert Gray had talked in April 1792, Gray denied that he had sailed around Vancouver Island, as John Meares claimed in 1790. Nonetheless, Puget's sighting made it clear that Nootka Sound was on an island.

Vancouver and Puget, in the pinnace, investigated Howe Sound and Jervis Inlet, while Manby returned to the *Discovery* in the launch. Whidbey and Johnstone explored west and encountered the *Sutil* and the *Mexicana,* Spanish ships making their own exploration of the Northwest Coast. Vancouver learned from the captain of the *Sutil,* Dionisio Alcalá Galiano, that Juan Francisco de la Bodega y Quadra was at Nootka as Spain's representative for the Nootka Convention negotiations.

Vancouver's instructions included sharing information with exploring parties of other nations, because this promoted good relations and benefited everyone in charting a complex region with hundreds of bays, inlets, shoals, and river openings. After some informative discussions, he proposed that the Spanish join with the British to do a survey. As a result, both expeditions proceeded north to Desolation Sound, a place that Bell described as "dreary." The joint enterprise would not be without friction, but their explorations included identifying Johnstone Strait and the strong water flow from the north as the sailing channel from the inland waters to the ocean and Haida Gwaii (Queen Charlotte Islands).

73. Negotiating the Nootka Convention, 1792

INTRODUCTION

Conflict between Spain and Britain over the Northwest Coast erupted in 1790, when British trader John Meares charged that the Spanish commandant at Nootka Sound, Esteban Martínez, had illegally captured his ships and confiscated property at Nootka in 1789. Meares sought £469,865 in compensation. Alexander von Humboldt, the great explorer of South America, later characterized the incident as a tempest about "a few sheds erected on the coast, and a miserable bastion defended by swivel guns, and a few cabbages planted within an enclosure." It was much more, however, and portentous for both nations.

Spain had claimed hegemony in the Pacific based on treaties with Portugal in the 15th and 16th centuries, which had split control of the world's oceans between the two countries. It was unlikely that Spain was prepared to defend this claim, but Spain nonetheless worried about British interests and approved the occupation of Nootka in 1789. Commanding the Spanish fort at Nootka, Martínez enforced Spanish claims by holding British ships in port and purportedly confiscating an outpost building and other property. British prime minister William Pitt had been under pressure from commercial and military interests to challenge Spain in the Pacific, and he threatened war while also pressing for a diplomatic solution. In Madrid in late October 1790 the two nations agreed to the Nootka Convention, which eliminated Spain's claim of sovereignty in the Pacific and returned British property but left open how the two countries would accommodate their competing interests in the region. In 1792, Vancouver and Bodega met in Nootka to represent their nations' interests.

Letters between George Vancouver and Juan de la Bodega y Quadra, Nootka Sound, September 1792

J. F. de la Bodega y Quadra to George Vancouver, September 2, 1792

According to the 1st Article of the Convention, and the Royal order which I have, I am only to restore to His Britannic Majesty, the Edifices, districts, or portions of land, which in April 1789 were taken from His subjects.

I have been able to justify that the small hut they had did not exist on the arrival of Martinez, nor is our establishment in the place they had theirs.

In answer to your attentive and esteemed favor of the 1st current, you say you do not consider yourself authorised to agree upon those points which exist in my Commission. Each of us will lay before our respective Courts what we think most just, that from thence they may be able to resolve; and you from hence to be in possession of what Mears occupied; and at your command the houses, gardens, and offices, which we have, and I shall retire till the decision which ought to result after clear and authentic informations.

I return you many thanks for the Kind offer of your works, &c. I shall have the greatest satisfaction in accompanying you in those which remain, & in occasions where I may more and more shew how sincerely I am, &c.

George Vancouver to Bodega y Quadra, September 10, 1792

I am favored with your very civil and obliging letter of the 2nd instant, and as it was verbally translated to me I did not conceive it required any immediate reply: the translation however appearing in writing seems still to contain a retrospective discussion of right of the respective sovereigns in the territories here in question; which I have already observed doth not appertain to our present business, and must, therefore beg leave to decline entering into it.

You may depend on the justness of my representation to the British Court in every thing that has come within my observation (or at least so far as my abilities will allow me) respecting our transactions here.

I have therefore only again to observe that I am ready to be put into possession of the said territories, agreeable to the 1st article of the Convention &c. &c. so soon as it may, in every respect, be compatible with your situation and convenience; taking this opportunity of returning you my best thanks for the Charts you have favored me with; and to which, in two or three days, I shall have the honor of making a similar return. I beg leave to assure you I have the honor to be, &c.

P.S. I trust you will excuse the lateness of this reply, as Mr. Dobson's arm not allowing him to write prevented the translation taking place before.

Source: W. Kaye Lamb, ed., *George Vancouver; A Voyage of Discovery to the North Pacific Ocean and Round the World, 1791–1795,* Vol. 4 (London: Hakluyt Society, 1984), 1572–1573. Used by permission of the Hakluyt Society.

ANALYSIS

In 1768 to support their exploration of the Northwest Coast, Spain established the Naval Department of San Blas, Pacific, in present-day Mexico west of Guadalajara. Twenty-one years later, Spain sent Juan Revillagigedo to Mexico City as viceroy and Juan Francisco de la Bodega y Quadra as commandant at San Blas to deal with the Nootka Sound Controversy. Revillagigedo maintained that Meares had never built a structure at Nootka and, further, that James Cook's 1778 landing at Nootka had been preceded by Juan Pérez in the *Santiago* in July 1775. Revillagigedo considered Nootka "barren and rugged" land, of little value to Spain, and recommended abandoning it. Nonetheless, the Spanish crown commissioned Bodega to reestablish a Spanish presence at Nootka while securing specific concessions from the British in the process of negotiations.

Spain had full confidence in Bodega, who came from a titled Basque family and had taken part in Northwest Coast explorations in 1775 and 1779. His instructions accepted the Nootka Convention's first article, which called for the return of all British possessions sequestered in April 1789. In addition, Bodega was instructed to negotiate two other agreements: a division of the region at the Strait of Juan de Fuca—with Britain in control to the north and Spain in control to the south—and an east-west division of interest north of the strait on a line running from the entrance north to 60°, giving Spain control of bays and land east of the line, including the continental interior, and giving Britain hegemony west of the line. Bodega pressed for the north-south division of interest, but he did not pursue the east-west cleaving of the region. He knew it was unlikely that Vancouver would agree to losing nearly the entire coastline.

The negotiations between Bodega and Vancouver took place almost entirely by letters, 13 in all. In face-to-face discussions they clarified statements, articulated disagreements, and tried to avoid conflict. Midshipman Thomas Dobson, a seaman on the *Daedalus* who could read and write Spanish, served as the official interpreter. In these exchanges, Vancouver felt disadvantaged. His original instructions referred only to the convention's first article and included no authority to negotiate regions of interest on the Northwest Coast. Additional Admiralty instructions that came in August with the supply ship *Daedalus* provided little help. Vancouver was frustrated as he tried to respond to Bodega's request for the territorial cessions and limitations. Writing to the Admiralty in January 1793 after concluding the negotiations, he complained that his instructions left him at a disadvantage. "No there can be little doubt I should either proved myself a most consummate fool or a traitor," he wrote, if he had given in to Bodega's suggested division of the region. Almost whining, Vancouver complained that "*I was still left totally in the dark what measures to take*" (Tovell 2008, 264).

By terms of agreement in the Third Nootka Convention in 1795, Spain eventually gave up on its plan to claim hegemony at Nootka. Nonetheless, the negotiations between the two countries at Nootka Sound marked a turning point in the larger contest over the Northwest Coast. Vancouver's steadfast refusal to accede to Bodega's insistence on dividing the coast at the Strait of Juan de Fuca effectively blunted Spain's last opportunity to hold on to a portion of its theoretical control of the region's sea-lanes. Britain emerged the winner, with Spain subsequently focused on competing for the sea otter trade and protecting its missions in California.

74. Maquinna's Feast at Tahsis, 1792

INTRODUCTION

In early September 1792, Juan Francisco de la Bodega y Quadra, George Vancouver, and more than 80 British and Spanish officers and seamen visited Maquinna's winter lodge at Tahsis Inlet in Nootka Sound. Bodega had suggested to Vancouver that they pay an official visit at Maquinna's lodge to consecrate their nations' relationships with this important leader. Maquinna had long been an important chief among the Mowachaht at Nootka Sound, where he controlled an extensive area on the northwest side of the sound, including three prominent villages. For Maquinna, the meeting surely enhanced his local prestige, while the British got Maquinna's attention and the promise of enduring relations that might allow them to establish a base at Friendly Cove near the village of Yuquot in Nootka Sound.

Maquinna and other Mowachaht entertained the European guests with elaborate dancing. A sketch made at the event by Atanasio Echeverria, an assistant artist from the Academia de San

Carlos in Mexico City, has been lost. Jose Maria Velazquez, a Mexico City art student, painted the scene based on Echeverria's drawing. In the painting, drying fish hang from poles suspended from ceiling beams, and a large totem carving dominates the space. The painted scene also highlights Maquinna in the midst of a dance for Bodega, Vancouver, and their naval officers while dressed in a magnificent sea otter fur cloak.

Captain George Vancouver, September 1792

Agreeably to appointment, about eight in the morning of Tuesday the 4th, Senor Quadra accompanied me in the discovery's yawl, which, with our own and a Spanish launch, and the Chatham's cutter, containing as many Spanish and English officers as could be taken, we departed for Tasheis. . . . [W]e reached Tahsheis about two in the afternoon: Maquinna received us with great pleasure and approbation, and it was evident that his pride was not a little indulged by our shewing him this attention. He conducted us through the village, where we appeared to be welcome guests, in consequence perhaps of the presents that were distributed amongst the inhabitants who all conducted themselves in the most civil and orderly manner. After visiting most of the houses, we arrived at Maquinna's residence, which was one of the largest. . . . [H]ere we found seated in some kind of form, Maquinna's daughter, who not long before had been publicly and with great ceremony proclaimed sole heiress to all his property, power, and dominion. Near her were seated three of his wives, and a numerous tribe of relations. The young princess was of low stature, very plump, and with a round face, and small features; her skin was clean, and being nearly white, her person altogether, though without any pretensions to beauty, could not be considered as disagreeable. To her and to her father I made presents suitable to the occasion. . . . These ceremonies being ended, a most excellent dinner was served, which Senor Quadra had provided. . . . After dinner Maquinna entertained us with a representation of their warlike achievements. A dozen men first appeared, armed with muskets, and equipped with all their appendages, who took their post in a very orderly manner within the entrance of the house, where they remained stationary, and were followed by eighteen very stout men, each bearing a spear or lance sixteen or eighteen feet in length, proportionably strong, and pointed with a long flat piece of iron, which seemed to be sharp on both edges, and was highly polished. . . . The men made several movements in imitation of attack and defence, singing at the same time several war songs. . . . I was presented with two small sea-otter skins, and the warriors having laid by their arms, performed a mask dance, which was ridiculously laughable, particularly on the part of Maquinna, who took considerable share in the representation. We were not backward in contributing to the amusements of the day, some songs were sung which the natives seemed much to admire, and being provided with drums and fifes, our sailers concluded the afternoon's diversion with reels and country dances. In the evening we took leave of Maquinna, who was scarcely able to express the satisfaction he had experienced . . . and that he would in a few days return us the compliment; on which he was given to understand, he should be entertained in the European fashion.

Source: George Vancouver, *A voyage of Discovery to the North Pacific Ocean,* Vol. 1 (London: G. G. and J. Robinson, Paternoster-Row, and J. Edwards, 1798), 395–396.

ANALYSIS

In the late summer and early fall of 1792 at Nootka Sound on the west coast of present-day Vancouver Island, 49-year-old Bodega and 33-year-old Vancouver represented Spain and

Britain in working out details for the two nations' claims on the Northwest Coast. In August Bodega awaited Vancouver's arrival, but the British captain delayed, hoping that the supply ship *Daedalus* would arrive with new instructions. When word came that the ship had arrived, Vancouver proceeded to Nootka on August 28. The following day, Thomas Manby (Manby Diary) reported that "the Spanish colours were saluted with 13 guns and the same number instantly returned from a brig that bore Don Quadra's masthead pendant." Vancouver's naturalist, Archibald Menzies described the Spanish "fort" at Nootka as merely some cannon secured on a platform, but the commandant's house was substantial: "We found him [Bodega] on shore at a decent house two story high, built of Planks with a Balcony in the front of the Upper Story after the manner of the Spanish Houses. One end of the ground floor was occupied as a Guard Room, & the other as a Kitchen & Servants' Hall, while the Upper Story was divided into small apartments & occupied by the Governor & his Officers, who were separated by a large Hall in the middle where they commonly dined" (Tovell 2008, 204–205). Bodega had prepared for his British guests by gathering fresh meats and foods and had brought his own silver service and preserved delicacies. He kept "a sumptuous table," Manby wrote, "served on plate." Vancouver and his officers dined with Bodega nearly every day during the three-week negotiations.

Bodega assiduously pandered to Maquinna, the acknowledged leader of the Mowachaht at Nootka. Maquinna had become chief at the onset of the fur trade and quickly had mastered the trade with Europeans and controlled access by other Native groups who competed for trade goods. Above all, Maquinna determined to acquire firearms and often withheld his best furs from trade unless he received them. He gained prestige and power among other Native leaders because of his hospitable and remunerative relations with Bodega and Vancouver.

On September 4, 1792, Maquinna welcomed the two captains to his village, Tahsis, 15 miles up an inlet from Nootka. Bodega thought it important to take "the opportunity of imposing on the minds of the natives the highest and most favourable opinion of our little squadron; and the more effectually to insure a good understanding in the future" (Tovell 2008, 243). He and Vancouver offered gifts of copper sheets, cloth, blankets, and beads, and Maquinna reciprocated with furs. In the chief's large plankhouse, which featured beams 32 feet long and 12 feet in diameter, men conducted dances. "They were all of the Fierce & Warlike style and subject," the *Chatham*'s clerk, Edward Bell, recorded, "and one of two of them ended with a frightful Yell that to a strangers ear was truly terrific. Maquinna, dancing, now entered, dressed in a very rich garment of Otter skins with a round Black Hat, and a Mask on, and with a fanciful petticoat or apron, around which was suspended hollow tubes of Copper and Brass and which as he danced, by striking against each other made a wonderfully tingling noise." Vancouver responded by getting a few seamen "to dance a reel or two to the fife" (Meany 1915, 20–22). Vancouver left Tahsis fully convinced that he had impressed Maquinna and had achieved a diplomatic success.

75. Menzies at Friendly Cove, 1792

INTRODUCTION

Born in 1754, Archibald Menzies grew up in a family of botanists, studied at the Royal Botanic Garden in Edinburgh, trained in surgery and apothecary, and served in the Royal Navy as an assistant surgeon. He drew the notice of Joseph Banks, a wealthy patron of science and a member of James Cook's second voyage in 1768–1771, who became president of the Royal Society in 1778. Banks sponsored Menzies as botanist on James Colnett's circumnavigation in 1776–1789 and pressed his appointment to Vancouver's expedition in 1790. Throughout the voyage, Menzies collected plants for the Royal Botanic Garden and corresponded with Banks. Menzies discovered several Northwest species, including the madrone (Arbutus menziesii) *and Douglas-fir* (Pseudotsuga menziesii), *and he surveyed Northwest Coast landscapes for antiscorbutic plants to combat scurvy.*

Banks's political influence and manipulative role in planning the expedition made George Vancouver suspicious of Menzies, and friction between the two men boiled up late in the expedition. Menzies charged Vancouver with destroying some collected plant species and, according to Vancouver, "behaved to me on the quarter-deck, with great contempt" (Lamb 1984, 1:218). Vancouver briefly put Menzies under arrest, but relations between them improved by the end of the voyage. Nonetheless, Menzies was the only journal keeper on the Discovery *who did not surrender his log to Vancouver, a requirement the captain enforced to facilitate writing an account of the voyage.*

Archibald Menzies Journal of Vancouver's Voyage, April–October, 1792

The 23d [September 1792] & three following days we had fair pleasant weather but cloudy with a moderate breeze of wind. . . .

On the 27th very strong gales set in from the Southward & Southeast quarter accompanied with thick gloomy weather & almost incessant heavy rain, so that we had very boisterous & unsettled weather for the remainder of this month.

In the evening of the 28th one of the Chatham's Cables parted in the nip of the clinch in a violent squall & in swinging round to the other Anchor she took the ground on the North West side of the Cove but was soon hove off again by the other Cable without having received any apparent injury. . . .

Early on the 1st of October the *Fenis* Brig saild for *Macao* & in her Mr. Mudge first Lieutenant of the Discovery went home with dispatches for Government by the way of China, this occasiond a Vacancy which Mr. Swain who had been lately made Master of the Chatham was appointed to fill & he was succeeded in the Chatham by Mr. Manley.

By Mr. Mudge I sent home a collection of Seeds adressd to Sir Jos: Banks Bt for his Majesty's Garden & which I was afterwards happy to find that Mr. Mudge had taken great care in their preservation, by which some valuable Plants were added to the great collection at Kew through the uncommon skill & industry of Mr. Acton in rearing them. . . .

Maquinna & his Brother Wagh-elas-opulth visited the Cove this day together with another young Chief named *Nannacoos* who was a very great favourite with Sr Quadra & all his Officers, & they were very much hurt indeed at not being receivd as usual at the Governor's house,—for while Sr Fidalgo was station'd at De Fuca's entrance his Pilot strolled one day into the Woods & was murderd in a most shocking manner by the Natives, in consequence of which Sr Fidalgo was resolved to keep the Chiefs & all the rest of the Natives at due distance & not suffer any of them to approach the Settlement without being minutely watched. The Nootka Chief could not bear the idea of suspicion & left the Cove in a huff. . . .

There were now seven English Vessels, a Spanish Frigate & two American Vessels riding at Anchor in the Cove, besides the two small ones that were building on Shore. . . . Capt Vancouver & Mr. Broughton waited on Sr Fidalgo in the forenoon to pay their respects to him before their departure, & it is but justice to add that he was very obliging & ready in facilitating our operations by every aid in his power, & in the evening he returnd the visit by coming on board the Discovery to take leave of us. It was but natural to feel some reluctance at parting as during our stay at Nootka the Spanish Officers & we lived on the most amicable footing. Our frequent & social meetings at Sr Quadra's hospitable mansion afforded constant opportunity of testifying our mutual regard & friendship for each other, by that harmony & good understanding which always marked our convivial hours. In short even in this distant sequesterd gloomy region we passed our time together chearfully & happy.

Source: C. F. Newcombe, M.D., ed., *Menzies' Journal of Vancouver's Voyage, April to October, 1792* (Victoria: British Columbia Legislative Assembly, 1923), 125–127, 129–131.

ANALYSIS

In 1790, Archibald Menzies appealed to Joseph Banks to join Vancouver's expedition. In part, Menzies's desire came from his earlier visit to the Northwest Coast with Captain Colnett. At Nootka Sound in 1789, Menzies had taken Chief Maquinna's daughter-in-law on plant-collecting forays, gaining considerable information about the use of native plants from her. Menzies knew the potential for scientific discovery, if he could only join Vancouver. Banks got Menzies assigned to the *Discovery* and also drew up instructions that detailed the specifications Menzies should use for collecting seeds and plants, including "an investigation of the whole of the Natural History of the Countries you are to visit as well as an enquiry into the present State & comparative degree of civilization of the inhabitants you will meet with" (Lamb 1984, 1:43). Banks also made several impositions on Vancouver, including a requirement that he maintain plant frames for Menzies's specimens and provide regular watering by an assistant.

At the end of the 1792 survey season, Menzies sent substantial specimens from Nootka to London by consigning them to the care of the first lieutenant of the *Discovery,* Zachary Mudge, who took passage on the Portuguese brig *Fenis* to China. Mudge made it to London in June 1793, with the specimens in good order. He also delivered papers from Vancouver. "To this gentleman," Vancouver wrote of Mudge, "I intrusted extracts from the most important parts of my journal, with a copy of our survey of this coast" (Lamb 1984, 2:683). Menzies credited William Aiton with maintaining his valuable collections at Kew Gardens, where Aiton had been in charge since the Kew's creation in 1759 to study plants collected by British explorers.

In September 1792 Salvador Fidalgo, commander of the *Princessa,* a 189-ton frigate, arrived at Nootka to take charge of the harbor. He had surveyed the Northwest Coast for Spain in 1790 and had contributed to charts that Bodega passed on to Vancouver. Fidalgo had negative and suspicious views of all Natives, a consequence of the recent murder of his officer Antonio Serantes in June at Neah Bay, at the mouth of the Strait of Juan de Fuca. Spain had sent Fidalgo to the strait to establish Nuñez Gaona, a prospective colonial property that was the first white settlement west of the Rocky Mountains, north of San Francisco, and south of Alaska. Spain would abandon the effort by 1796 and retreat from the Northwest Coast.

At Nootka, although Maquinna and other Natives had poor relations with some of the Spanish men, in Bodega they found an ingratiating officer. Bodega had been particularly hospitable to Maquinna at his Nootka residence, a situation that gave the chief prestige in his relations with other Native leaders. The men also had good relations because naturalist Jose Mozino had grasped the Native language and facilitated conversation between them.

In his journal, Menzies listed more than 30 vessels from Britain, America, Spain, Portugal, and France in Nootka Sound during 1792. It was a busy place, and mariners of all nations saw benefit in aiding other vessels, even though Britain and Spain had nearly gone to war over control of the region. Menzies's comments on the collegiality among Europeans at Nootka—a region he described as "gloomy"—underscore the distance between them and Natives. It was a distance that could be bridged most often when they found mutual benefits in shared information, trade, and political relations among Northwest Coast peoples.

76. Broughton's Map of the Columbia River, 1798

INTRODUCTION

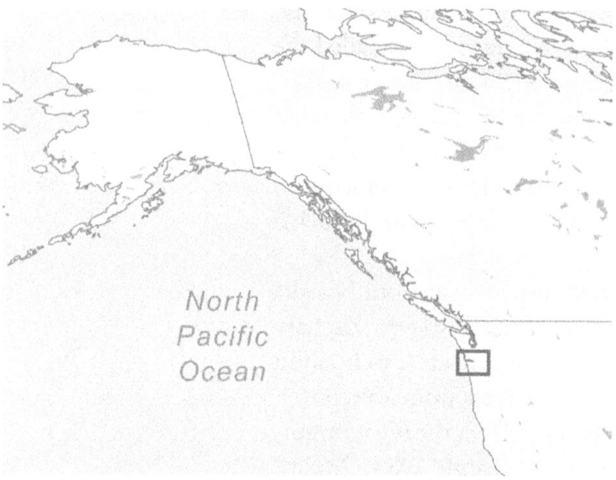

London mapmaker and publisher Aaron Arrowsmith engraved this map of the Columbia River (River Oregan) in the same year that George Vancouver's narrative and atlas of his expedition to the Pacific Northwest were published. Arrowsmith often had access to the latest information from British naval and commercial explorations in North America, and his maps of the continent were the most authoritative of the time. For this map, he faithfully reconstructed the extensive survey of the river by Lieutenant William Broughton, commander of the Chatham *on Vancouver's expedition. Broughton and his men had entered the Columbia River on October 20, 1792, and left on November 10. During that time, they ascended and mapped the Columbia in two small boats for an estimated 100 miles upstream. Broughton's manuscript chart is preserved in the British Columbia Archives and Records Service, and its survival affords a unique opportunity to compare the field map with the engraved copy by Arrowsmith.*

Broughton asserted in his account that he did not believe that Robert Gray had actually entered the river. Broughton believed that the river originated farther upstream than Gray had gone. His own survey and eventual possession-taking ritual convinced him that the true right of discovery of the river fell to the British. Even though Broughton knew that the name Columbia likely would be associated with an American identity, he titled his manuscript A Sketch of the Columbia River Explored in His Majesty's Arm'd Brig Chatham—Lieut. Broughton—Commander in October 1792. *His use of that toponymy was testimony to Vancouver's and Broughton's integrity in respect to Gray's primacy of "discovery."*

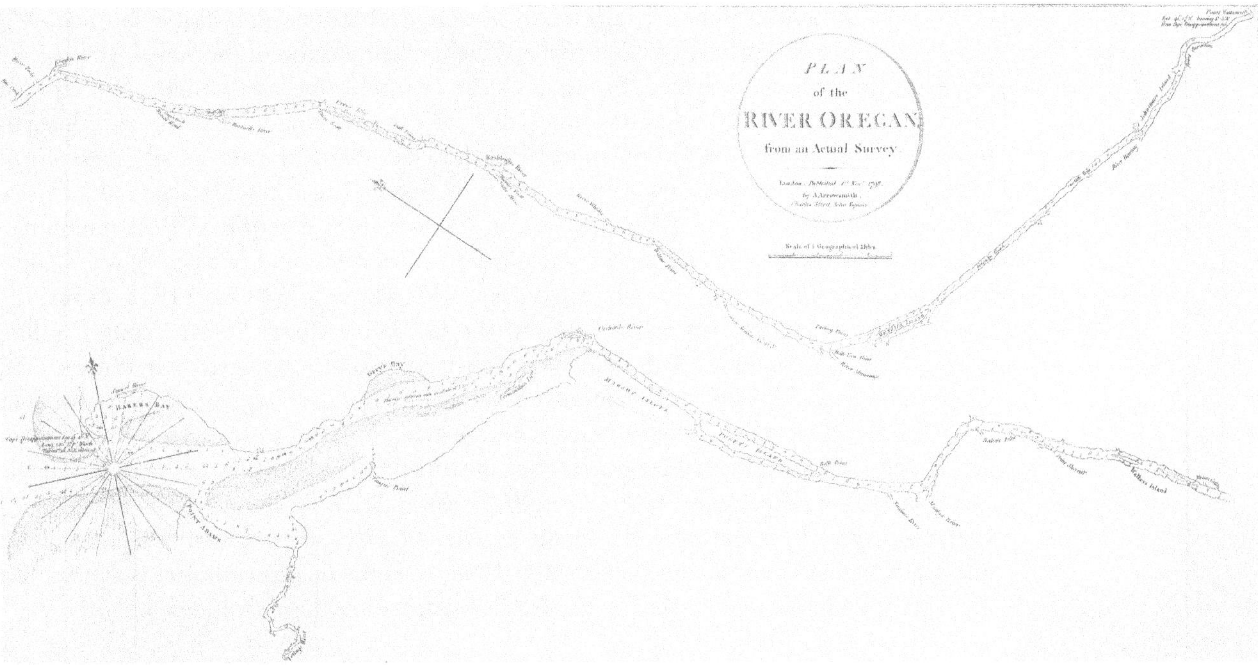

"Plan of the River Oregan from an Actual Survey." (Courtesy of the Royal BC Museum, BC Archives)

ANALYSIS

Lieutenant William Robert Broughton, a British Royal Navy officer who participated in the battle of Bunker Hill and was a prisoner of war during the American Revolution, explored the Columbia River in 1792. He surveyed nearly 100 miles upriver from the mouth to a location he called Point Vancouver near present-day Cottonwood Point, upstream of Washougal, Washington. There on October 30, Broughton "formally took possession of the river, and the country in its vicinity, in His Britannic Majesty's name, having every reason to believe, that the subjects of no other civilized nation or state had ever entered the river before; in this opinion he was confirmed by Mr. Gray's sketch [of the river], in which it does not appear that Mr. Gray either saw, or was even within five leagues of its entrance." His account appeared in the second volume of Vancouver's *Voyage of Discovery to the North Pacific Ocean, and Round the World in the Years 1791–1795.* Many of Broughton's placenames on his manuscript chart—and hence on Arrowsmith's map—survive, including Baker's Bay, Young's River, Tongue Point, Gray's Bay, Pillar Rock, Walker's Island, Mount Coffin, and Willow Point.

For nearly 300 years, mapmakers represented ideas about conceptual and undocumented navigable water passages above or through the North American continent. The River of the West, one possible passage, supposedly originated west of the Mississippi River and met the Pacific somewhere between ~45° and 49°. In 1778, Englishman Jonathan Carver applied the word "Origan" or "Oregon" to this conceptual River of the West. His book and map were influential, and mapmakers before and after Vancouver's voyage often prominently depicted the present-day Columbia River as "River Oregon."

In 1792, American Robert Gray's report of a large estuary on the coast at latitude ~46° north raised British expectations that perhaps the western portion of the fabled River of the West had finally been identified. Broughton's survey soon dispelled that idea. He recorded in his account that at 100 miles upstream "thus far the river could hardly be considered as navigable for shipping." Nevertheless, many of those familiar with conceptual geography of the day retained the idea of a navigable river called Oregon. Even President Thomas Jefferson gave Meriwether Lewis and William Clark an instruction to determine if "the Columbia, Oregon, Colorado or any other river may offer the most direct and practicable water communication across this continent for the purposes of commerce" (Jackson 1978, 1:61).

Thus, Arrowsmith's map of the Oregan (the Columbia River) linked a long-standing geographical concept with a recent discovery. In what could be interpreted as a strategic bias, Arrowsmith copied Broughton's Columbia River manuscript chart, except for the name, and in 1802 he used the name River Oregan on a large map of North America. In 1814, William Clark used Columbia R. on his map that accompanied the publication of the results of the Lewis and Clark expedition. On Arrowsmith's revised edition of his map of North America in the same year, he adopted Clark's toponymy. Arrowsmith's map of River Oregan and the maps in Vancouver's atlas, both published in 1798, were the first to authoritatively establish the reality of a major river in the Pacific Northwest at midnorthern latitudes.

77. Vancouver's Great Map of the Coast, 1798

INTRODUCTION

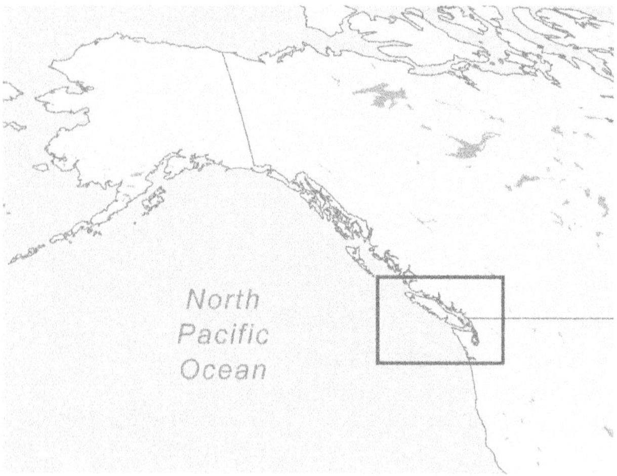

Vancouver's three-volume account of his voyage was published in London in 1798 with the title A Voyage of Discovery to the North Pacific Ocean and Round the World in Which the Coast of North-West America Has Been Carefully Examined and Accurately Surveyed. *This elegant copper-plate-engraved map was one of eight charts of the Northwest Coast in an accompanying untitled atlas. It illustrates the section of the coast from approximately Cape Lookout at 45°30′ north to Cape Swaine at 52°15′. One of three inset maps, "Entrance of the Columbia River" recorded part of the survey of Lieutenant William Broughton, who entered, ascended, and surveyed the river for 11 days in October and November 1792. Vancouver compiled his information for the finished maps from the results of his surveys of the coast from Cape Mendocino to the head of Cook Inlet (~61°), which he had visited from April 1792 to November 1794. He also incorporated information he had acquired from Russian maps and the most recent Spanish surveys. Two of Vancouver's officers prepared summary charts from more detailed manuscript maps made during the voyage. Four highly skilled engravers from the British Admiralty office engraved the plates for publication.*

Despite declining health, Vancouver prepared most of the account and oversaw preparation of the charts. He died on May 10, 1798, shortly before completing the work, which was published within a few months. A second English edition with just one chart was published in 1801, but

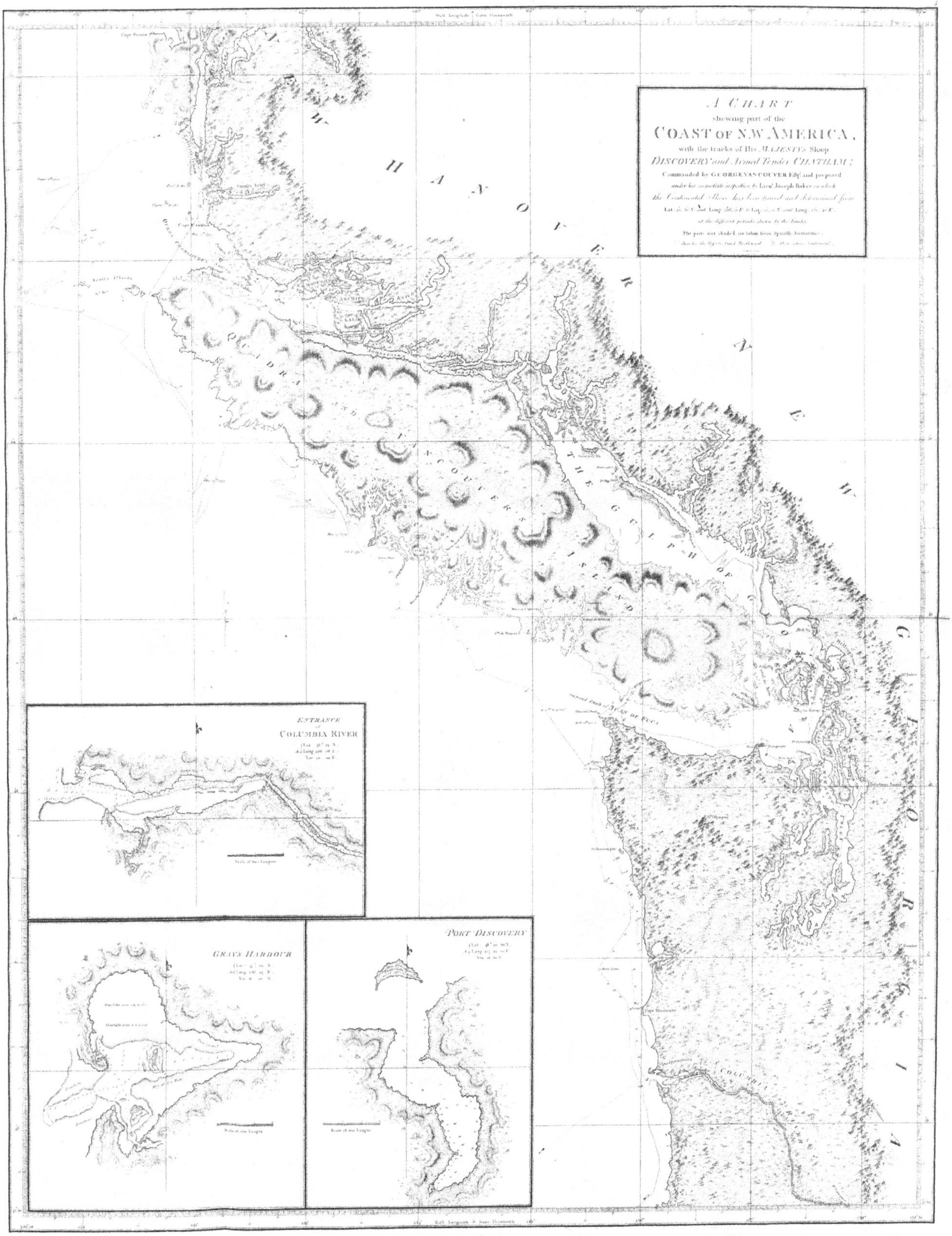

"A Chart shewing part of the Coast of N.W. America with the tracks of his Majesty's Sloop *Discovery* and Armed Tender *Chatham*." (Copyright © 2000 by Cartography Associates)

the original copperplates had been stolen, so the maps had to be reengraved by different craftsmen. Other editions were published in French, German, Swedish, and Russian.

ANALYSIS

From both a geographical and a geopolitical perspective, Vancouver's charts were landmarks. They recorded a meticulous and extended English survey of a vast and complicated coastline and archipelago. When combined with information from Russian maps and recent Spanish exploration, for practical purposes mapping of the North Pacific from latitudes 30° to 61° north—from Mexico to the Alaskan panhandle—was complete.

Vancouver failed to see the estuary of the Fraser River near present-day Vancouver, British Columbia, and by modern standards his longitude measurements were somewhat imprecise. Nevertheless, his charts were useful for navigation purposes well into the 19th century. They ended the long searches for the apocryphal Northwest Passage that purportedly existed between latitudes 43° and 53° and had eluded Pacific explorers for nearly 60 years.

Vancouver's sectional chart of the Northwest Coast was the first to depict from direct observation the complicated waterways of Puget Sound and Admiralty Inlet, the insularity of Vancouver Island, and much of the hydrography of the Columbia River. Lewis and Clark carried sketches of parts of Vancouver's charts on their expedition, and their journal entries in November 1805 expressed their relief as they approached the mouth of the Columbia and began to recognize some of the features illustrated on Vancouver's charts. From a strictly geographical perspective, Vancouver's charts appear authoritatively modern in their unembellished style, superb engraving, meticulous details of soundings, and rendering of ship's tracks. Readers were assured on the title page that the coast on the maps had been "Accurately Surveyed [and] Undertaken by His Majesty's Command."

Vancouver's charts also conveyed nongeographic information that was even more powerful than the precision of his surveying measurements (Clayton 2000). For example, the use of an infrastructure of a grid of longitude and latitude and skillfully applied hachure lines (short spaced lines used to indicate topography) were part of long-established West European mapping traditions. Ships' tracks and small anchor images were suggestive of only a European presence. The charts were also tools of nation building. Most conspicuously, the use of non-Native names was a compelling and imperialist cartographic method of representing the Northwest Coast.

In the interior regions of this map, Mt. Rainier, Mt. St. Helens, Point Vancouver, and Mt. Hood were all named for British officials or naval associates. The place-names Puget's Sound, Johnstone's Strait, Strait of Georgia, Queen Charlotte's Sound, and Broughton's Archipelago also had English origins. The journals of expedition participants often recorded the names that Native peoples used for geographical features, but few of them were used on Vancouver's charts. Furthermore, these English names largely replaced previously applied Spanish names that would have been known to Vancouver through his exchanges with the Spanish captain Bodega. In some cases, Vancouver retained Russian and Spanish names in places he had not explored. This also occurred when he left the name Columbia on the river that American Robert Gray entered just four months before Lieutenant Broughton surveyed it upstream for 100 miles. Perhaps the most obvious example of appropriation of identity by use of names was the toponym Quadra and Vancouver Island, which Vancouver prominently engraved over present-day Vancouver Island in commemoration of the two commanders' mutual respect. The name appeared on most British Admiralty and Spanish charts until the Hudson's Bay Company officially eliminated Quadra's name in 1825.

In May 1792, George Vancouver named Mount Rainier for Peter Rainier, later to become Rear Admiral. The image depicts the mountain from Admiralty Inlet in the Strait of Juan de Fuca. (Getty Images)

Finally, the concept of using names to convey dominion was nowhere more evident than on Vancouver's small-scale chart of the Pacific Coast. There, the toponyms New Albion and New Georgia (after King George) were applied in large type from latitude 35° to 50°. Commercial mapmakers such as Aaron Arrowsmith quickly adapted and widely disseminated the names on Vancouver's charts. Arrowsmith published a new edition of his popular map of North America in 1802, which was updated with Vancouver's information.

78. The Public Humiliation of George Vancouver, 1795–1798

INTRODUCTION

When George Vancouver returned to England in 1795, he and his social and political friends were disappointed in the public's reaction to his voyage. Unlike James Cook, who had been posthumously lauded by nearly all Englishmen for his unprecedented accomplishments in exploration,

"The caneing in Conduit Street. Dedicated to the flag officers of the British Navy." Engraving by James Gillray. Published by H. Humphrey, October 1, 1796. (Library of Congress)

Vancouver's accomplishments met lackluster interest. His health had deteriorated during the voyage home, and he required rest before setting to his main task: writing a full account of his explorations using his and his crew members' journals, which he had demanded from his officers.

Vancouver's shipboard discipline of Thomas Pitt, Lord Camelford, a member of the politically powerful and wealthy Pitt family (his uncle William Pitt was prime minister in 1783–1801) left Vancouver, a commoner, vulnerable to Pitt's obsessive desire to publicly have vengeance. Pitt famously attacked Vancouver with a cane on Conduit Street in London as Vancouver and his brother Charles were on their way to meet with Lord Grenville, the secretary of state, to get his protection from Pitt's stalking. The incident received immediate press coverage. This scathing cartoon by caricaturist James Gillray was published by print seller Hannah Humphrey on October 1, 1796.

ANALYSIS

George Vancouver's life after his return from his triumphal voyage around the world could not have been more troublesome. Suffering from declining health caused by the onset of kidney disease, his symptoms included significant weight gain, episodic stomach pains and vomiting, and a general loss of strength. His condition made it difficult for him to gather materials and write an official account of his great voyage, a document that the Lords Commissioners of the Admiralty and the crown fully expected. Most galling and confusing to Vancouver, though, was the animosity directed at him by the *Discovery's* officers and politicians. Thomas Pitt, Lord Camelford, stirred up criticism of his captain by spreading stories of his brutality at sea. Vancouver had disciplined Pitt for destroying equipment and for rank disobedience and three times had ordered the officer lashed on deck. Pitt had finally been sent home as an unfit seaman on the supply ship *Daedalus* in 1794. Vancouver's relations with other officers had also deteriorated, notably Archibald Menzies and Joseph Baker. In addition, although Vancouver had interceded on behalf of several crew members who suffered financial difficulties or petitioned for promotion in rank after the voyage, he could not secure his own pay and benefits. Once the Admiralty decommissioned the *Discovery*, Vancouver lost his command and went on half-pay. For unknown reasons, they withheld Vancouver's just reward for capturing the Dutch ship *Macassar* at St. Helena in May 1795, during the *Discovery*'s return voyage to England.

Pitt, knowing Vancouver's routine, jumped him and his brother Charles in Conduit Street in early September 1796, when they were walking to meet the lord chancellor to file an injunction against Pitt's harassment. In good measure, the difficulties that Vancouver encountered had class origins. His middle-class background left him without protection from the disdain that Lord Camelford and other naval officers harbored toward lower-class government officials. The cartoon played off the class divisions. The balloon script attached to Pitt, who is being restrained by Charles Vancouver, reads "give me Satisfaction Rascal. . . . Draw your Sword, Coward! What you won't? Well, then take that Lubber & that & that." Vancouver replies "Murder, Murder . . . Watch Constable! Keep him off Brother, while I run to my Lord chancellor for protection! Murder, Murder." From Vancouver's neck hangs a scarf imprinted with "Present from King of Owyhee to George III," a reference to Vancouver's successful negotiation of the cession of Hawaii to Britain in 1794. A trailing roll of paper draping from Vancouver's pocket reads "List of those degraded during the voyage," and the title over the building door—"The south Sea Fur Warehouse from China . . . No contraband goods sold here"—refers to the perception that Vancouver made profit from trading furs on the Northwest Coast. The two boys observing the scene communicate the ridicule that the cartoonist hoped readers would adopt toward Vancouver. Gillray largely succeeded. Nearly everywhere Vancouver went he felt scorned, and none of his earlier sponsors, including Sir Alan Gardner at the Admiralty, came to his defense. Vancouver died in 1798 with his report on the voyage incomplete, suffering from professional slights and with the knowledge that he had not overcome a damaged public image.

Further Reading

Blumenthal, Richard. *With Vancouver in Inland Waters: Journals of 12 Crewmen, April–June 1792.* Jefferson, NC: McFarland, 2007.

Clayton, Daniel W. *Islands of Truth: The Imperial Fashioning of Vancouver Island.* Vancouver: University of British Columbia Press, 2000.

Frost, Alan, et al. *Pacific Empires: Essays in Honour of Glyndwr Williams.* Vancouver: University of British Columbia Press, 1999.

Jackson, Donald, ed. *Letters of the Lewis and Clark Expedition, with Related Documents, 1783–1854.* 2nd ed. 2 vols. Urbana: University of Illinois Press, 1978.

Lamb, W. Kaye, ed. *George Vancouver; A Voyage of Discovery to the North Pacific Ocean and Round the World, 1791–1795.* 4 vols. London: Hakluyt Society, 1984.

Manby Diary, MS 2389, Oregon Historical Society, Portland, Oregon.

Meany, Edmond S., ed. *A New Vancouver Journal on the Discovery of Puget Sound by a Member of the Chatham's Crew.* Seattle: n.p., 1915.

Newcombe, C. F., M.D., ed. *Menzies' Journal of Vancouver's Voyage, April to October, 1792.* Victoria: British Columbia Legislative Assembly, 1923.

Tovell, Freeman. *At the Far Reaches of Empire: The Life of Juan Francisco de la Bodega y Quadra.* Vancouver: University of British Columbia Press, 2008.

Vancouver, George. *A Voyage of Discovery to the North Pacific Ocean and Round the World.* 3 vols. London: G. G. and J. Robinson, and J. Edwards, 1798.

Expeditions of Robert Gray

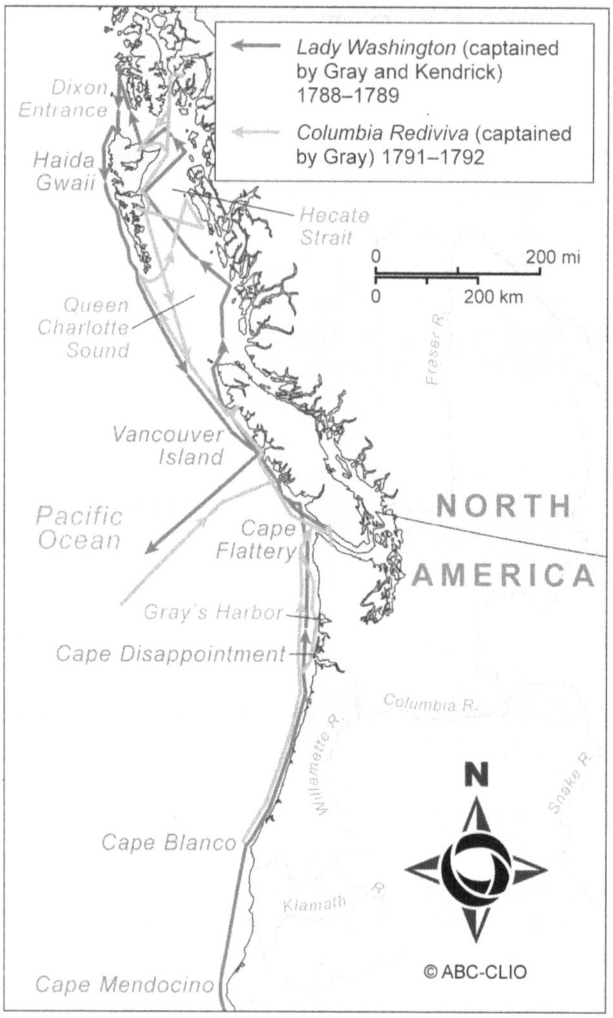

CHAPTER 9

Robert Gray, First American Commercial Northwest Coast Voyages, 1787–1793

Robert Gray is best known in the history of the Pacific Northwest for piloting his ship, the *Columbia Rediviva,* into the Columbia River on May 11, 1792, and naming it Columbia's River after his vessel. It is unclear whether "Rediviva" in the ship's name refers to Christopher Columbus or to America as a country, but the implication either way is that the name conveyed the idea of revival or revitalization. Gray was also the first American mariner to skipper American ships around the world by first sailing the *Washington* from Boston in 1787 to the Northwest Coast and then returning the *Columbia* from those waters back to its home port three years later.

Gray sailed as a commercial ship captain, so he was not a representative of the U.S. government in the 1790s, nor were his voyages assigned official status or supported with government funds. Nonetheless, by naming and drawing a chart of the Columbia River estuary in 1792, Gray documented American presence at that important location. More than half a century later when the United States and Great Britain negotiated their claims to the Oregon Country, Gray's actions inadvertently offered the nation the claim of discovery to the river and the coastal region. Although his documentation of the Columbia River did not figure prominently in the Oregon Treaty agreement of 1846, his bold sailing prompted George Vancouver to send Lieutenant William Broughton up the river in October and November 1792. The results of that exploration became known to the world in the 1798 publication of Vancouver's journals.

Despite the importance of his activities to U.S. and Oregon history, surprisingly little is known about Gray's life before the voyages of the *Columbia.* Even the place, date, and cause of his death are uncertain. The only extant portrait of Gray is undocumented and was likely composed years after his death, because it does not depict him wearing an eye patch or with a missing eye, even though several who knew him mentioned it. What we know is that he was born on May 17, 1755, in Tiverton, Rhode Island, and that he apparently went to sea at a young age. During the American Revolutionary War he may have served in the Continental Navy, as claimed by his widow in 1846, but no records of his service have been found. By the 1780s, however, he was captain of the *Pacific,* a ship in the coastal trade that sailed from Boston to Charleston, South Carolina. Presumably, he had learned seafaring

This 1847 artistic rendition of the *Columbia* in battle against Natives likely depicts the scene in May 1792, when Gray ordered cannon fire at canoes he thought meant harm to his ship. (Herbert Orth/The LIFE Picture Collection/Getty Images)

during the 1770s, probably on cargo vessels. Crowell Hatch and Samuel Brown, two of the partners in the Columbia expedition, owned the *Pacific,* and likely they recommended that Gray captain the *Washington* on his first voyage to the Northwest Coast.

Joseph Barrell, a prosperous Boston merchant, and five business partners organized Gray's two Pacific voyages. Gray was their employee on the first voyage in 1787–1790, and he became an investor and commander on the second voyage in 1790–1793. John Kendrick, 15 years older than Gray, commanded the first voyage as captain of the *Columbia.* Gray captained the *Washington* and dutifully followed Kendrick's lead, although the logs of his officers make it clear that Gray held Kendrick in low regard and thought that he was a timid sailor and slow to make command decisions.

The principal goal of the voyages to the Pacific was commercial gain, and that focus affected and determined most of the decisions Kendrick and Gray made. Because they did not have the benefit of governmental authority, they avoided the burden of completing political missions, establishing American claims to territory, or creating international relationships. Nonetheless, Barrell had instructed his captains to purchase land if the opportunity arose as a means to secure trade operations in the Northwest. Kendrick claimed that he bought land from Natives, and Gray may have done the same. But while there are documents to support Kendrick's claims, there are none to support Gray's. Barrell also gave Kendrick and Gray explicit instructions to work cooperatively with other nations' mariners on the Northwest Coast and to facilitate commercial ties with Native peoples.

The American-based trade on the Northwest Coast was financed principally by Boston merchants. They took advantage of Britain's trade-crippling mercantilist policy that had awarded monopolies to the South Sea Company in 1711, which controlled trade on the Pacific Coast by the late 18th century, and to the East India Company in 1600, which had complete authority for purchasing or selling items to China. American captains faced no such restrictions. They were individual entrepreneurs or representatives of private trading companies who could strike unfettered deals on the Northwest Coast and in China. Within 15 years of James Cook's last voyage, American ships had begun to challenge many British traders on the Northwest Coast in the pelagic fur trade. In 1787 five British ships competed with Gray; by 1793 six American and six British ships were trading on the Northwest Coast (Gibson 1992, 299–300).

On his two voyages to the Northwest Coast, Gray approached the trade in a single-minded fashion, not unlike others in the same business. He meant to purchase sea otters from Natives at the lowest possible price and trade them to Chinese merchants for trade goods that could bring high prices in Boston or other markets. The cost of purchasing furs, however, included the original investment in trade goods plus the cost of inducements offered to Natives and the costs incurred through loss of goods and materials to Native theft. It was a dynamic trade in which Native desires could change rapidly. They wanted iron, sheets of copper, cutting tools such as knives and axes, and chisels plus cloth, glass, armaments, and liquor. Which items Natives preferred at any one time might pivot on what they had already purchased or the prices offered by the traders. As *Columbia*'s master Joseph Ingraham put it, their offered trade goods often left Natives so uninterested that he felt they had arrived "the day after the fair." The trade flourished for just over three decades and was an "ephemeral and strangely diversified trade," in historian F. W. Howay's words, that "was, in reality, merely a looting of the coast" (Howay 1932, 7). Nonetheless, the trade had consequences, for it began an engagement with Native peoples that brought great social and economic change. Not the least effect was the trade's capriciousness in the hands of merchant captains such as Gray, who often displayed little patience with Native traders. For the traders, all costs were threats to their voyage's monetary return. For his part, Gray acted unsentimentally and sometimes vengefully in his relationships with Natives.

DID YOU KNOW?

Although Robert Gray was the first American to circumnavigate the world, he incurred the ire of his superior, Captain John Kendrick, on the Northwest Coast for sailing dangerously close to shore and damaging both ships, the *Lady Washington* and the *Columbia Rediviva*.

On the first voyage, Gray took the *Washington* into bays and inlets on present-day Vancouver Island and north to Haida Gwaii (Queen Charlotte Islands). Although the *Washington* was a small boat and could maneuver in shallow waters, its large mainsail and single mast sometimes left it unable to navigate in difficult conditions. Gray apparently ignored the *Washington*'s vulnerabilities and sailed too tight to coastlines, which led him to damage the ship on the first voyage. He spent most of his time at anchor in Clayoquot Sound and Nootka Sound on Vancouver Island, where he made contact with powerful trading chiefs, Wickananish at Clayoquot and Maquinna at Nootka. In July, Kendrick ordered Gray to take command of the *Columbia* and sail to China with the furs they had collected and then to continue back to Boston. The change in plans was a result of Gray damaging the *Washington* two months earlier by sailing too close to the Alaskan coast, which forced him to repair the ship at Nootka. Kendrick remained at Nootka to repair the *Washington,* while Gray made his way to Canton to trade before returning to Boston in August 1790. Meanwhile, Kendrick traded on the Northwest Coast and by early 1790 sailed to Canton in the repaired *Washington* to sell additional furs. For reasons that are not entirely clear, Kendrick decided to conduct trading on his own and made two trips from China to the Northwest Coast, one in 1791 and a second in 1793–1794. Without authorization, he "sold" the *Washington* to himself. He never returned to Boston and died in an accident in Hawaii in December 1794.

Gray returned to Boston as a hero in 1790. He was the first American to circumnavigate the globe—sailing the *Washington* to the Northwest Coast and returning in the *Columbia* via the Cape of Good Hope—and the *Columbia* became a symbol of the nation's new strength and ambitions in seafaring and trade. The partners, however, had realized almost no profit from the voyage, and the *Washington* and its captain were still on the Northwest Coast with unknown prospects. Rather than abandon the enterprise, they decided to finance a second voyage to the Pacific and chose Gray to captain the *Columbia.* In a demonstration of their determination to succeed, they assigned John Hoskins, as supercargo, the task of monitoring Gray to ensure that he made businesslike decisions. It was an arrangement that created friction and made Gray's relationship with Hoskins difficult. Nonetheless, the second voyage proved to be more efficient than the first, in part because Gray advised the partners about which trade goods to carry on the voyage. The partners also decided to load special materials sufficient to build a small craft once the *Columbia* arrived in the Pacific Northwest, a boat that could navigate in small inlets. Gray made Clayoquot Sound in June 1791 and built the vessel, the *Adventure,* in the spring of 1792. They used the *Adventure* to advantage while on the coast, but in the end Gray sold the vessel to Juan Francisco de la Bodega y Quadra, the Spanish commander at Nootka Sound, before returning to Boston in 1793.

DID YOU KNOW?

Gray was a hard-nosed trader, but his investors hedged their confidence in his abilities by assigning John Hoskins as supercargo on the *Columbia* in 1790 to monitor Gray's trading and ensure an accurate accounting of all financial dealings on the voyage.

Gray conducted a strong trade on Vancouver Island, in Haida Gwaii (Queen Charlotte Islands), and along the Washington coastline. His impatience with Natives, however, was frequent and sometimes intense, and he was quick to use violence to protect the *Columbia* and his trading options, in one case destroying an entire village out of vengeance. Nonetheless, he kept to the trade and took the opportunity to enter the Columbia River, thereby achieving an important discovery that benefited American geopolitical interests. His tendency to sail boldly in dangerous waters, though, led to several accidents that brought damage to the *Columbia.* By July 1792 the ship's condition forced him to rely on Spanish hospitality and materials at Nootka Sound, where Gray oversaw a complete overhaul of the ship. The repairs required hauling the *Columbia* ashore and laying a repaired keel, work that required help from Spanish and British ships' carpenters. This episode and others document Gray's diplomatic skills, but he could not overcome the trading obstacles at Canton. When he arrived there in December 1792, more than 50 European and American ships had furs to sell, and the fees and other costs of trade with the Chinese sharply limited his returns. This meant that the value of the tea and other goods that Gray took back to Boston fell below expectations.

Gray arrived home in July 1793 after sailing around the Cape of Good Hope and across the Atlantic. In 1794, he married Martha Howland Atkins; they would have four daughters. Gray remained active in merchant shipping, commanding vessels based in Boston. In 1798 he headed back to the Pacific Northwest on a trading voyage in command of the *Alert,* but his ship fell into the hands of French privateers who forced Gray to sail the ship to Montevideo as a prize. The following year Gray was captain of an American privateer, the *Lucy,* during the Quasi-War with France (1798–1800). The date of Gray's death is uncertain, but it is likely that he died of yellow fever in 1806 on a voyage to South Carolina. He was 51 years old. Long after his death, in 1846 Martha Gray petitioned the U.S. Congress for compensation in payment for his discovery of the Columbia River and his purported purchase of land near the mouth of the river. By 1848 when she petitioned Congress a second time, surveys of the recently acquired Oregon Country by the U.S.-Britain Oregon Treaty of 1846 were beginning, and she hoped to receive something from the government for her husband's historic voyage and discovery. Over the years she continued to petition for compensation, but Congress regularly denied her claims.

79. The Partners' Instructions for the First Voyage, 1787

INTRODUCTION

Publication of James Cook's and James King's A Voyage to the Pacific Ocean *in 1784 and John Ledyard's* A Journal of Captain Cook's Last Voyage *in 1783 made the world aware of the potential for trade in sea otter pelts from the Northwest Coast to China. In the United States, Boston became the new nation's center for launching trading ships bound for the North Pacific in search of prime sources of pelagic furs. Before the American Revolution, New England had developed a strong shipping industry in more than a dozen ports that specialized in North Atlantic whaling. Boston also competed with New York and Philadelphia for merchant shipping cargoes to all destinations, even though they operated under British trade restrictions. In the Pacific, for example, Britain's South Sea Company controlled commerce on the western coast of North America and South America, while the East India Company had a monopoly on British trade in Asia. The American Revolution, however, changed their world, as they became untethered from the constraints of British trade regulations and sought ways to use a fleet of ships that had been expanded by the war. But American merchants also faced a serious post-Revolution economic recession and the challenge of developing new trade relationships. In January 1787, a group of Boston merchants recognized the potential for new trade business and formed a partnership to enter the North Pacific fur trade. The partners set about securing two ships, the* Columbia Rediviva *and the* Lady Washington. *The merchant group's instructions to John Kendrick, captain of the* Columbia *and head of the expedition, reflected the partners' ambition, the times, and their concerns about the voyage.*

Orders Given Captain John Kendrick of the Ship *Columbia* for a Voyage to the Pacific Ocean, 1787

The ship Columbia and sloop Washington being completely equipped for a voyage to the Pacific ocean and China, we place such confidence in you as to give you the entire command of this enterprise. It would be impossible upon a voyage of this nature to give with propriety very binding instructions, and such is our reliance on your honor, integrity and good conduct, that it would be needless at any time. You will be on the spot, and as circumstances turn up you must improve them; but we cannot forbear to impress on your mind our wish and expectation that the most inviolable harmony and friendship may be cultivated between you and the natives, and that no advantages may be taken of them in trading, but that you endeavor by honest conduct to impress upon their minds a friendship for Americans. . . .

If you make any fort or improvement of land upon the coast, be sure you purchase the soil of the natives, and it would not be amiss if you purchased some advantageous tract of land in the name of the owners, if you should let the instrument of conveyance bear every authentic mark the circumstances will admit of. . . . You will constantly bear in mind that no trade is to be allowed on the coast, on any pretence whatever, but for the benefit of the owners. . . . You are strictly enjoined not to touch at any part of the Spanish dominions on the western continent of America, unless driven there by some unavoidable accident, in which case you will stay no longer than is absolutely necessary, and while there be careful to give no offence to any of the subjects of his Catholic Majesty, and if you meet with any subjects of any European prince, you are to treat them with friendship and civility.

The certificate you have from the French and Dutch consuls you will make use of if you meet with any ships of those nations, and you will pay them every respect that is due to them.

The sea letters from Congress and this State you will also show on every proper occasion; and although we expect you will treat all nations with respect and civility, yet we depend you will suffer insult and injury from none without showing that spirit which will ever become A FREE AND INDEPENDENT AMERICAN.

Source: Frederic W. Howay, ed., *Voyages of the "Columbia" to the Northwest Coast, 1787–1790 and 1790–1793* (Portland: Oregon Historical Society Press, 1990), 111–112.

ANALYSIS

As these orders make clear, the voyage of the *Columbia* and the *Washington* to the Northwest Coast was a private commercial effort. The formation of the partnership to fund the voyage took place in January 1787 in Charles Bulfinch's home on Bowdoin Square in Boston. A budding architect, Bulfinch had recently returned from France with an idea he had for a commercial venture while talking to Thomas Jefferson and John Ledyard: a fur-trading expedition to the Northwest Coast. The principal funding came from Joseph Barrell, a merchant grocer who had grown wealthy during the American Revolution by supplying the French fleet in Boston Harbor with provisions. In addition to Barrell, five men took shares in the prospective voyage: Charles Bulfinch, John Derby, Crowell Hatch, Samuel Brown, and John Pintard. Barrell took two shares, each valued at $50,000, while the others took one share apiece. Brown and Hatch had made their money in the slave trade and as privateers during the American Revolutionary War. Derby was a member of a prominent seafaring family in Salem, Massachusetts, and had been one of the first Americans to develop the maritime trade with China in the 1780s. Pintard, a wealthy member of a New York trading family, had a hefty inheritance and wanted to invest in the China trade.

In choosing the ships and captains for the trading voyage to the Pacific, the partners selected substantial ships they could readily purchase and sea captains they knew and trusted. They found a ship that met their needs under construction in Plymouth, Massachusetts, the

Columbia Rediviva—212 tons and 83 feet long with a 24-foot beam. For captain of the *Columbia,* they chose John Kendrick, a 47-year-old sea captain who had a reputation for steady seamanship and grit as commander of privateers during the war. The expedition's second vessel, the sloop *Lady Washington,* was likely a Boston-based privateer whose part owner during the 1770s was Samuel Brown. The *Washington* displaced 90 tons, stretched 60 feet in length, and was sloop-rigged, with a single mast, a large mainsail, and a sizable jib sail. The partners hired 32-year-old Robert Gray as the *Washington*'s captain. Gray had sailed for Brown and Hatch in the coastal trade from Boston to the Carolinas in the 1770s, and the two partners evidently saw him as a sailor who had good seafaring judgment.

Only one crew member of the *Columbia* or the *Washington* had ever been to the Pacific. Kendrick's first mate, Simeon Woodruff, had been a gunner's mate on the *Discovery* on James Cook's third voyage and knew seafaring. On the *Columbia,* however, he crossed Kendrick and was forced to leave the voyage at St. Jago in the Cape Verde Islands in early November 1787. Woodruff may have been too British in demeanor to suit Kendrick, for the partners made the *Columbia* expedition a patriotic enterprise, one that would carry the American flag into the Pacific and around the world. As captains, Kendrick and Gray carried sea letters from the U.S. Congress certifying that the vessels were legitimate commercial ships that would abide by accepted sea laws and conventions and should be accorded due respect and consideration. Nonetheless, the partners instructed the captains to avoid Spanish ports, knowing that Spain jealously guarded its claims to land in the Pacific. The instructions made no mention of mapping the region, presumably because the purpose of the expedition was business and the partners decided not to invest in personnel who were capable of astronomical calculations and chartmaking. It is also likely that Barrell relied on Kendrick to have adequate knowledge of the region based on Cook's descriptions and maps of his third voyage, published in 1784.

80. Kendrick's Sailing Instructions, 1788

INTRODUCTION

The Columbia *and* Washington *left Boston on September 30, 1787, with the* Columbia*'s captain John Kendrick in command. Kendrick sailed at a modest pace to the Cape Verde Islands, a Portuguese colony about 400 miles off the west coast of Africa. They arrived in early November and set about gathering provisions, likely the best place to restock before the ships took on the passage around Cape Horn to the Pacific. They added 142 goats, 2 bulls, 1 cow, 3 hogs, and 3 sheep to the* Columbia*'s crowded deck. One of the female goats, Nancy, became a minor celebrity, for she made the full voyage back to Boston and also shipped out on the second voyage, only to die on board near the Falkland Islands in April 1790.*

At the Cape Verde Islands, Kendrick took considerable time filling water casks and reorganizing cargo holds on the Columbia*. All was not cheery, though, as third mate Robert Haswell recorded: "Much discord (from what cause they arose I will not pretend to say) subsisted in our Ship ever since our departure from Boston, and now prevailed in a more vilent degree" (Howay 1990, 7). Conflicts between the first mate, Simeon Woodruff, and Kendrick ended with the captain ordering Woodruff off the ship on December 7 to find his way back to Boston. A few days later the ship's surgeon asked the Portuguese governor of the islands to grant him protection from*

Kendrick's anger, which prompted the captain to put the doctor ashore. Kendrick never explained his actions to the partners in Boston. Third mate Robert Haswell criticized Kendrick for dilatory seamanship and for being a poor manager of the crew. After he had a physical confrontation with Kendrick on the quarterdeck, when the captain threatened to shoot him, Haswell requested that he be released from the voyage. Instead, the captain allowed Haswell to move to the Washington *once the ships paused in the Falkland Islands off the coast of Argentina, where they arrived in February 1788.*

John Kendrick's Instructions to Robert Gray, on Board Ship *Columbia* Lying in Britts [Bretts] Harbor in Falklands Islands, February 1788

Capt. Robart Gray as you have the Command of the Sloop Washington my orders is that you Sail with the Columbia and do all in your power to Keep Company with her But Should you get Seperated from her By Bad weather or any other axerdent What Ever you air to proseed on your voige Round Cape horn in to the pacific ocan and then Stand to the Northward as far as the Lattd. 49°36′ North and Longitude of 126°42 West from the Meridion of London. their you will find a harbour By the Name of Nootka Sound on the West Side of North america. . . . But take great Care not to put your Self in the power of the Spaniads If possibal to avoide it. When you Stand to the Northward from these [Juan Fernandez] Islands you must Not Come nigher to the Continant of america then Seventy or Eighty Leagues till you get thrue the trade Wind into the varibel on the Coast of California on account of the indraft of Wind that Blowes in on the Coast. . . . Surplys may Be got after your arrival at Nootka Sound[.] I would have you purchase as much furs as possibal of the Natives and in perticuler the Sea otters But not Refuse any[.] Should you git five hundered Sea otters Skins and other furs in perpotion the first Season I would have you proseed from thence to Macao in China and their inquire for Letters Directed to me or your Self and then follow the orders you Receve from them Butt Should you not git the Number of Skins mentined I would have you Stay at Nootka or Sum other harbour on the Coast of america all winter By all means and in the Spring If you find the trade not good at Nootka or their abouts to proseed to the Northward as far as Cook-River But I would have you vissit the Coast as you pass a Long in Ever place whare you think it Safe and try for trade[.] I would have you treet the Natives with Respect whare Ever you go Cultivate frind Ship with them as much as possibel and take Nothing from them But what you pay them for according to a fair agreement and not Suffer your peopel to affrom them or treet them Ill and allways Remember that their is no trade to Be made Butt for the Benifet of the owners and Should it Be in your power to Catch Sea otters with the helpe of your traps or any orther way it must Be for the Bennifit of the owner of Ship Columbia and Sloop Washington.

Source: Frederic W. Howay, ed., *Voyages of the "Columbia" to the Northwest Coast, 1787–1790 and 1790–1793* (Portland: Oregon Historical Society Press, 1990), 114–115.

ANALYSIS

While in the Falklands, Kendrick planned the expedition's route to the Pacific. Most captains avoided Ferdinand Magellan's difficult and narrow passage through the archipelago of islands at the tip of South America, choosing instead to sail south and face the fierce currents and gales that made the route around Cape Horn among the most feared maritime passages on the globe. By this time in the voyage, Kendrick also recognized the differences in his sailing style and Gray's. Gray was much less timid than Kendrick, choosing to sail tight to coastlines, while Kendrick traveled at a slower rate, often used less sail, and generally stood off the land. It is likely that Kendrick wanted to pause in the Falklands for two months or more, but he met with a quick response from the crew. "Every Offiser immedeately expressed there disgust at the Very Idea," Haswell recorded, "and for my part I informed him that was I in his case the Commander of the Ship, after spending so much time in the Cape de Verds and making such long passages I would go round the Cape or loos the Vessells in the attempt. . . . Captain Gray urged moast strongly the danger of staying on this side the Continent" (Howay 1990, 14–15).

With that in mind, Kendrick seemed to expect that he and Gray would lose touch on their way to the Pacific Ocean, but he also wanted Gray's strict adherence to the expedition's general procedures and goals. Gray pushed the *Washington* to its limit against gales and rough seas to round the cape by early April, and both ships made their way to the western end of the Straits of Magellan. There the *Washington* lost contact with the *Columbia.* Rather than follow Kendrick's recommendation to head west from South America to catch strong trade winds that would propel him to the Northwest Coast, Gray kept the *Washington* close to the continent and headed directly north. He sailed by the Galapagos Islands without seeing them and bypassed nearly every opportunity to put ashore and gather fresh provisions for men and animals, dangerously inviting scurvy aboard his ship.

Kendrick's instructions posed two challenges. On the one hand, he emphasized that Gray avoid Spanish ships, ports, and colonies, while on the other hand he made it clear that Gray was to collect as many skins as possible before heading to China. This meant that the *Washington* had to find sources for supplies and trading opportunities that kept the American trading venture free of conflict with Spain or traders from other nations. In fundamental ways, Kendrick's instructions largely determined how Gray managed his trading on the Northwest Coast in 1788–1789, and it is likely that Kendrick's knowledge of conditions in the North Pacific came from Cook's journals of his third voyage, which had been published in 1784.

81. At Murderers Harbor, 1788

INTRODUCTION

In late April 1788 after rounding Cape Horn, Gray arrived in the San Fernandez Islands off the coast of Chile, intent on replenishing supplies. Spanish mariners had been the first Europeans to encounter the islands, and it was there that Alexander Selkirk had been marooned in 1704–1709, an experience that Daniel Defoe fictionalized in Robinson Crusoe *in 1719. The Spanish in Chile had also used the San Fernandez Islands as a penal colony. Mindful of Kendrick's instructions to avoid Spanish forces and finding high surf at potential landing areas, Gray decided to bypass the islands and sail directly for North America. He caught favorable trade winds that brought the Washington to the continent near present-day Cape Blanco, at 42° north and 124° west, on August 2, 1788. Gray coasted north, unsuccessfully looking for an appealing harbor. He and his crew made contact with Natives who came to the* Washington *by canoe nearly every day, with large numbers encountered near the present-day Alsea River. On August 12 Gray encountered a bay near Tillamook Head, where he sent a longboat to replenish the wood supply. The following day, August 13, the crew dropped anchor in present-day Tillamook Bay, their first landing since leaving the Falkland Islands in February. In his log, second mate Robert Haswell describes a violent encounter with the Tillamook who lived on the bay.*

Robert Haswell, A Voyage Round the World Onboard the Ship *Columbia-Rediviva* and Sloop *Washington,* August 16, 1788

Earley in the Afternoon I accompaneyed Mr. [R. David] Coolidge onshore in the long Boat to amuse ourselves in taking a walk while our boat was loaded with grass and shrubbs for our stock[.] we took all the people in the Boat who were affected by the Scurvey our number in all amounted to seven. . . . we had two Muskets and three or four Cutlaces[.] we boath took our swoards and each of us a pistol on our first landing. we visited there Houses. . . .

They [Indians] then amused us shewing there dexterity with there arrows and spears[.] they then began a war dance[.] it was long and hedious accompaneyed with frightfull howlings[.] indeed there was somthing more horrid in there song and the jestures which accompanied it than I am capable of Discribing[.] it Chilled the bludd in my vains. The dance over we left the natives. . . . we went past the boat a little way but within call to a small sand flatt in hopes to find some clams[.] while we were digging for these shell fish a young Black man Marcus Lopius a native of the Cape de Verd Islands and who had shiped [as] Captain Grays servant at St. Jago's . . . had carelessly stuck his Cutlas in the sand[.] one of the natives seeing this took a favourable oppertunity to snatch it. . . . one of the people observing him before he was quite out of sight called vehemonantly thretening to shoot him in hopes he would abandon the stoln goods . . . I had given positive orders to our people not to fier but in cases of the most absolute emergence when for self defence it might be nececery.

'Twas the hollowing of our people that first roused our attention and we immediately flew to know the cause, we were informed . . . that the Black boy had followed him in spite of every thing they could say to the contorary. . . .

[W]e met several Chiefs persons whose friendship we had taken every oppertunity to obtain by kind youseage and liberal preasants. . . . Mr. Coolidge offered several articles to them of great value to bring back the man unhurt, this they refused intimating there wish for us to seek him ourselvs. I now remarked to Mr. Coolidge that all the natives we saw were unusualy well armed having with them there bows arrows and spears however we proceeded still further and on turning a clump of trees that obstructed our prospect the first thing which presented itself to our view was a very large groop of the natives among the midst of which was the poor black with the thief by the colour [collar] loudly calling for assistance saying he had cought the thief, when we were observed by the main boddy of the Natives to haistily aproach them they instantly drenched there knives and spears with savage feury in the boddy of the unfortunate youth. He quited his hold and stumbled but rose again and stagered towards us but having a flight of arrows thrown into his back and he fell within fifteen yards of me and instantly expiered while they mangled his lifeless corse. . . .

Our only method was to get to the boat as fast as posable for this purpos we turned leaving the dead body. . . . we made the Best of our way for the Boat assaulted on all sides by showers of arrows and spears. . . . we jumped into the Boat and pushed of and were soon out of arrow shot. . . .

I must confess I should not have lett them enjoy there festervile so peasabley had I been Cap. Gray but his humanity was commendable.

Murderers Harbour for so it was named is I supposed the enterence of the river of the West[.] it is by no meens a safe place for aney but a very small vessel to Enter[.]

Source: Frederic W. Howay, ed., *Voyages of the "Columbia" to the Northwest Coast, 1787–1790 and 1790–1793* (Portland: Oregon Historical Society Press, 1990), 37–39.

ANALYSIS

Robert Haswell was only 19 years old when he signed on with the *Columbia* as third mate. His conflict with first mate Simeon Woodruff and Captain Kendrick forced him off the ship in the Falkland Islands, where he became Gray's second mate on the *Washington*. The son of William Haswell, a lieutenant in the British Royal Navy, and a child of a Loyalist family during the American Revolution, Haswell had moved with his family to Nova Scotia in 1778 and then to Yorkshire, England, before returning to Massachusetts sometime before 1787. He was the only crew member to keep logs of both the 1787–1790 and 1790–1793 voyages. On the second voyage, he was the *Columbia*'s first officer.

Gray kept the *Washington* close to shore as he sailed up the coast, so close that crew members often saw Natives onshore and in small villages. He had two objectives: to find a good harbor to replenish water and secure fresh food for his scurvy-affected crew and to collect as many skins as possible through trade. Gray and his crew had few interactions with Natives, although many approached the ship in canoes. It was Gray's first trading experiences on the North American coast, and he was cautious, allowing only headmen to come aboard the *Washington*. He handed out gifts to encourage trade and acquired a few high-quality pelts, but he worried at the Indians' seemingly aggressive attitude and was concerned that some were clad in clamons, thick elk-hide breastplates that could repel arrows. Haswell noted that all of the men in canoes were armed with arrows and spears and "both Iron and stone knives which they allways kept in there hands uplifted in readiness to strike" (Howay 1990, 33).

After 10 days of searching for a good harbor, Gray sailed into present-day Tillamook Bay, which had an easy entrance and good anchorages. Trade ensued, with the Americans' knives and axes traded for good-quality otter skins, although Haswell noted that they could have acquired many more skins if they had copper sheets for trade. The *Washington*'s anchorage, however, was some distance from good water sources, so Gray planned to leave and find a better harbor.

On August 16, Gray sent out a boat to gather wood and forage for the ship's animals. The landing party included Markus Lopius, the first documented black in present-day Oregon. Lopius may have been enslaved or free, and he may have been African or born in Cape Verde, a Portuguese colony where the sugar industry had relied on African slave labor from the 15th century. Captain Gray may have purchased Lopius as a slave or hired him as a personal servant. Four other Cape Verdeans had joined the crew of the *Columbia*, suggesting that all may have been free blacks.

The landing party from the *Washington* apparently went about gathering wood and forage without much fear, although they were startled by a warlike dance at a Tillamook village. Tensions increased, however, and during a violent encounter Markus Lopius was murdered. It is telling that Gray did not attempt to retrieve Lopius's body, something that would have been a high priority if he had been white. Haswell was angry at the killing and wanted vengeance, but it is likely that Gray saw their situation as too weak to contest the Natives onshore.

Haswell's reference to Tillamook Bay as a potential entrance to the River of the West reflected his knowledge of James Cook's search for the Northwest Passage. If Haswell believed it to be the fabled opening to the passage, however, he appeared to warn others to avoid Murderers Harbor.

82. Gray Takes Command of the *Columbia,* 1789

INTRODUCTION

Gray sailed north from Murderers Harbor for nine days, reaching Barkley Sound on the west coast of Vancouver Island on August 26, 1788, and later Clayoquot Sound on September 10. There he met Native traders and their principal chief, Wickananish, who held sway in trading with Euro-American fur traders in the region and surprised Gray by listing off the names of British ship captains who had already visited. Gray sailed north to Nootka Sound, where he met British traders John Meares, captain of the Felice Adventurer, *and William Douglas of the* Iphigenia Nubiana, *ships more than double the size of the* Washington. *Meares had been at Nootka for four months and had established trade with Maquinna, the local trade broker among Natives at Nootka.*

Kendrick arrived by September 23, but he had a scurvy-ridden crew that needed considerable recovery time. In the spring of 1789 while Kendrick continued to refurbish the Columbia, *Gray itched to gather skins and head to China. First he went back to Clayoquot Sound, 49° north and 126° west, and then north to Haida Gwaii (Queen Charlotte Islands). During the tour, Gray once again sailed too close to shore and put the* Washington *on the rocks. Barely saving the vessel, he had to return to Nootka for repairs. When he arrived on June 17, 1789, Kendrick surveyed the damage and decided to swap roles with Gray, sending him as captain of the* Columbia *to China and then to Boston while Kendrick took over the* Washington *and remained on the Northwest Coast to trade.*

Robert Gray to Joseph Barrell, Canton, December 18, 1789

I have the pleasure to inform you of my safe arrival at Canton in the Ship Columbia from the NW Coast of America in the Lattd. 49°9′ N. the 30th July, where Captn. Kendrick thought best to change Vessells and take all the property on board the Sloop Washington and Cruize in her himself on the Coast, and for me to make the best of my way to the Sandwhich Islands, there to procure sufficient provisions to carry us to Canton, our provisions on the Coast were not sufficient for both Vessells to Cruize the Season out was the reason of our seperation.

I had four weeks passage from the Coast to the Sandwhich Islands, where I lay three weeks and salted five Punchions of Pork and took on deck One hundred and fifty live hogs—our passage from the Islands to Canton was nine weeks, where I arrived on the 17th Novr. found fourteen Americans and about seventy Ships of other Nations—I received Instuctions from Mr. Howe (whome Captn. Kendrick appointed to assist me) to value on Messrs. Shaw & Randall which we have done on Mr. Randall Mr. Shaw being absent.

The Cargo consists of seven hundred skins and three hundred pieces which is not yet disposed of, but expect will be in a few days—I am getting the Ship in readiness in order to receive a Cargo of Bohea teas for Boston, and expect to Sail by the last of Jany [January].

The Ship is much out of repair and the expenses will I fear exceed your expectation but rest assured that nothing Shall be wanting in attention to render the expences as light as possible—when I parted with Captn. Kendrick he was well and all on board the Washington the Sloop in good order and better found than when She saild from Boston, having on board twenty men, and am in hopes they will make a good Cruize, having provisions to

Cruize two months—but Captn. Kendrick never informd wether he intended to come to Canton or not in case he meet with no accident, which makes it a matter entirely uncertain, in case he does not arrive this month I shall conclude he means to Stay another season or has meet with some accident.

The Spaniards have taken possession of all the Coast and fortyfiing in many places, they have taken possession of a Snow and Schoo[ne]r under portuguese Colours and have captured a Brig and Sloop belonging to English merchants in this place and sent them to St. Blass for tryal.

Our expedition Gentlemen will not be equal to your expectations nor is there any American Ships here but will make bad Voyages—I have only to add that Mr. Howe joins me in best wishes for your health and prosperity.

Source: Frederic W. Howay, ed., *Voyages of the "Columbia" to the Northwest Coast, 1787–1790 and 1790–1793* (Portland: Oregon Historical Society Press, 1990), 128–130.

ANALYSIS

Before Gray left Nootka for China, he had taken note of Spanish activities in Friendly Cove, which included building a fort. The newly arrived commander, Esteban Martínez, came to enforce Spain's claim of authority in the Pacific, which he did by capturing two British trading vessels and eventually taking them as prizes to San Blas. His actions led to what became known as the Nootka Sound Controversy, which brought Britain and Spain to the edge of war. Despite this conflict, the Americans maintained good relations with Martínez from their anchorage in Marvinas Bay, north of Friendly Cove. The commander accepted shipbuilding materials from *Columbia*'s stores and requested that the Americans transport Spanish-owned furs to China, along with captured British crew from the *North-West America,* the schooner that John Meares had built at Friendly Cove in 1788.

Gray left Nootka in late July 1789, sailing to Kealakekua Bay in Hawaii, where he purchased provisions and added a personal cabin boy, Attoo, and first mate Joseph Ingraham brought on board a servant, Kalehua. Both Hawaiian youths went on the *Columbia* to Boston. Kalehua would sail again with Ingraham on the *Hope* in 1790, and Attoo would board the *Columbia* on the second voyage in 1790. After leaving Hawaii in September and sailing west for just over two months, Gray arrived in November at Macao, a Portuguese trading port. A letter from Joseph Barrell awaited him, with instructions on the sale of furs and the potential sale of the *Washington.* Barrell had warned Gray not to sell furs at Canton, located on the Pearl River more than 100 miles by water from Macao, believing that trade with the Portuguese would incur fewer port costs and bribes. Why Gray ignored Barrell's instructions and went to Canton is unclear, but likely he thought that he could effect a better trade with Chinese merchants, especially with help from resident American trade agents Samuel Shaw and Thomas Randall. Shaw and Randall had first been to Canton in 1785 aboard the *Empress of China,* the first American trade ship to that country, but had returned to make a business of the China trade. At Canton, traders had to meet specific requirements laid down by Chinese authorities in the 13 factories authorized to trade with foreigners. Seeing the other ships waiting to trade and facing the necessity to pay bribes and other fees, Gray warned Barrell that the sale price for the 700 skins he transported on the *Columbia* likely would not be sufficient to bring the partners a return on their investment.

Gray worked through Shaw and Randall with the aid of Richard Howe, who had sailed from Boston as *Columbia*'s supercargo, the officer responsible for acquiring and selling cargoes. Nonetheless, the trading at Canton turned out to be less straightforward than Gray's letter to Barrell suggested. On January 4, 1790, Gray traded 1,215 skins, which included the 700 reported to Barrell plus 96 that Martínez had given Kendrick to pay for transport of the *New-Northwest* crew and 24 that Gray withheld for his own pay. That accounting still left 395 skins untallied, and it is likely that Gray and other crew secreted them for their own gain.

On January 12, Kendrick arrived at Macao on the *Washington.* In letters exchanged between the two captains, Gray warned Kendrick that Shaw and Randall would not take the furs on the *Washington* and that Kendrick would have to sell them at Macao. For his part, Kendrick wrote that he could not "at present determin whether I shall dispose of my Vessell, take a Cargo for America—or return again to the *Coast*" (Howay 1990, 136). Gray left Canton in February for Boston, while Kendrick returned to the Northwest Coast. The voyage of the *Columbia* expedition had turned out quite differently than the partners' expectations.

83. The Partners' Instructions for the Second Voyage, 1790

INTRODUCTION

Joseph Barrell and his partners realized little or no return on their investment in the first trading voyage to the Pacific. There were questions about the sale of furs in Canton, because the returns had fallen far short of the partners' expectations—about $75 per skin—and repairs to the Columbia *had added unanticipated costs. While Gray had received only about $30 per traded skin, he had been successful in his trading cruises on the Northwest Coast, had willingly taken over the* Columbia *on Kendrick's order, and had brought the ship home to Boston. Gray's descriptions of trading conditions and opportunities argued for another voyage.*

The investors in the second voyage in 1790 were all Bostonians (John Pintard from New York and John Derby of Salem left the partnership). Barrell took a stronger lead, with 5 shares of a total 14. Samuel Brown held 3 shares, while Crowell Hatch and Dr. Thomas Bulfinch (who replaced his son Charles) held 2 each. The last 2 shares were held jointly by Robert Gray and two merchants, Rufus Davenport and John McLean. Gray's inclusion likely came after his appointment as captain of the Columbia, *although it is unclear how he funded his share. It is also unclear why Barrell preferred Gray as commander. Joseph Ingraham, Kendrick's first officer, likely would have had preference, but he signed with a rival merchant company as captain of the brig* Hope, *which would sail on its trading voyage to the North Pacific two weeks before the* Columbia. *Ingraham's defection so angered Barrell that he forbade the* Columbia *from carrying any letters or cargo destined for the* Hope.

Joseph Barrell's Instructions to Robert Gray, Boston, September 25, 1790

The ship Columbia being completely equipped for a voyage to the northwest coast of America, China, and elsewhere, we place such confidence in you as to give you the command.

In all matters of traffic on the northwest coast of America, China, or elsewhere, you will consult with Mr. John Hoskins, who is with you, in whose industry, integrity, and honor we place the utmost confidence; we therefore expect the most perfect harmony to subsist between you, your officers, and him.

It is of importance, and we therefore enjoin it upon you, that the most friendly intercourse be observed in all your traffic with the natives, and no unjust advantage taken of them in trade, but endeavor, by your honorable conduct, to impress them with a respect for Americans.

You will touch at the several ports on the coast, as you shall think best; but you must bear in mind that there are several American vessels on the trade, and you will therefore make such despatch on your voyage as shall prevent your being overtaken by any of them; and to reach the coast before the brig, which has already sailed, as it must be of great importance.

At any time, at the close of a season, when you have furs sufficient, you may dispatch the sloop (which we call the Nootka) with them to market, if you should think it best under all circumstances; but you will bear in mind, if the sloop should be lost, the injury would be irreparable, and, if you should send her, be sure you give such orders, that whoever goes may avoid the excessive charges of going up to Canton; and, from the experience we have had, it appears plainly a much higher price may be obtained at the mouth of the river of Canton than in the city; we would therefore advise your selling at the mouth of the river.

If you should send the sloop at any time, we would have Mr. Hoskins go with her, and he will apply to such persons as will assist him in disposing of the furs to the best advantage for the interest of the concerned.

We depend that the ship remains on the coast until your provisions are consumed, and longer if you find the trade advantageous, and can get the supplies you want.

When you leave the coast in the ship, we advise to your trying what can be done on the coast of Japan and Pekin, where, if you find safety in anchoring, and can dispose of your furs to advantage, we would advise you by all means to do it.

You are strictly enjoined not to touch at any port of the Spanish dominions on the western continent of America, unless driven there by unavoidable accident, in which case you will stay no longer than is necessary, and while there, be careful to give no offence to any of the subjects of his Catholic Majesty, nor on any pretence whatever trade for a single farthing, unless for the supplies you may want, and let that be done under the immediate eye and sanction of authority.

Your sea-letters, and letters from this State, you will produce upon all proper occasions, and treat the subjects of every nation you meet with civility and respect, offering no insult to any, nor receiving any without showing the becoming spirit of a free, independent American.

Source: Frederic W. Howay, ed., *Voyages of the "Columbia" to the Northwest Coast, 1787–1790 and 1790–1793* (Portland: Oregon Historical Society Press, 1990), 443–445.

ANALYSIS

Joseph Barrell and his partners were determined to realize a return from the second voyage. To ensure a careful attention to costs and oversight of the trading in China, Barrell insisted that John Hoskins serve as supercargo for the *Columbia.* Gray was unhappy with having a monitor on board, but considering that he owned a portion of a share in the voyage and had a vote on such matters, he must have gone along with the majority of the partners and agreed to the arrangement. Barrell was a close friend of Hoskins's father, a prominent Boston import-export merchant and a member of the U.S. government's Treasury Board in 1779. John was just 18 years old when his father died in 1786, but Barrell readily took him into his household and employed him in his office. Hoskins acted as Barrell's advocate on board the *Columbia,* and in his log—one of the few that covered the voyage—he did not withhold criticism of Gray where he thought it was warranted.

Although the partners knew that John Kendrick had remained on the Northwest Coast with the *Washington,* which he had refitted in China to become a two-masted ship, they sent materials on the *Columbia* sufficient to build a small trading vessel that could navigate shallow bays and inlets. In other words, Barrell expected Gray to follow the trading methods he had used on the first voyage and to visit many villages rather than remain in one location and attract Native traders. Barrell identified the sloop as "Nootka," but Gray christened it the *Adventure* when it was built in March 1792 in Clayoquot Sound. His instructions also encouraged Gray to be competitive with Americans and traders from other nations, including beating the Boston competitor's brig *Hope* to the Northwest Coast at nearly any cost. More important, considering Gray's apparent impatience with Natives, Barrell emphasized that he deal with Indian traders in "friendly intercourse" and with "honorable conduct." Barrell had two goals in mind: creating a genuine and ongoing trading relationship with Natives and establishing a preference among them for American traders.

Barrell's instructions also included patriotic references that had been part of the guidelines for the first voyage. Gray should stay clear of Spanish locations, and he should wave his sea letters before representatives of other nations to underscore the competitive spirit and strength of the United States. He carried letters from President George Washington and Governor John Hancock of Massachusetts affirming that his voyage and activities were lawful and had government support. Beyond patriotic purpose, though, the letters were meant to facilitate the trading enterprise. On that topic, Barrell made it clear that Gray should base his decisions on securing the highest returns for the partners, including ensuring that John Hoskins made it to China to oversee the sale of furs.

84. At Clayoquot Sound, 1792

INTRODUCTION

After reaching landfall at Barkley Sound in June 1791, Gray sailed a few miles north to Clayoquot Sound, following his instructions to avoid the Spanish at Nootka Sound. After trading north of Vancouver Island and losing a sailor to Native attack in Cholmondeley Sound on Prince of Wales Island, Gray returned to Clayoquot Sound for the winter of 1791–1792 to repair the Columbia *and build the sloop* Adventure *from materials they had carried on the voyage. The crew erected a blacksmith forge and a makeshift boatyard on present-day Meares Island in the sound, a location they called Adventure Cove. Gray also ordered that cannon and some other arms be stored nearby the boatyard. Beginning in September 1791, they worked on the vessels while trading with Wickananish and other Native parties, who watched the building of the* Adventure *with much interest. The Indians' curiosity led to some friction, but Wickananish and other chiefs maintained good trading relations. Nonetheless, Gray and his crew remained on guard and occasionally fired muskets in the air to keep Natives at a distance. In January 1792, rumors that Wickananish planned an attack on the* Columbia *put Gray on high alert and fed his anxiety, which eventually led him to level a harsh punitive action on March 26 against the village of Opitsat, on the south end of Meares Island. John Boit was the only crew member to document the event; Gray, John Hoskins, and Robert Haswell left it out of their accounts. Note: Boit's dates are advanced one day from the actual day of occurrence.*

John Boit, Remarks on the Ship *Columbia*'s Voyage from Boston, January 18–March 27, 1792

January 18

The situation of the Ship at this period was very favorable to their [Tla-o-qui-aht's] views, and must have encouraged them with the hope of destroying the whole of us, without the loss of a man on their side. . . .

But fortunately, in the evening, the Sandwich Island lad [Attoo] made a Confession to his Master. . . . He said *Tatochkasettle,* (*the Cheif*) told him, that Wickananish was about to take the Ship and Massacree all the Crew, and said he shou'd be a great man. . . .

It does not appear that *Wikananish* wish'd to conquer a part of us, as he had frequent opportunitys to have accomplish'd itt for two or three times a week a boat was down at the *Village,* generally with an officer and four *sailors,* but I suppose he very prudently thought that shou'd he cut a boats Crew off, there was still enough left for to destroy his Villages. The Cheifs had been telling us for some time that they was going to war with a distant tribe and wish'd for us to lend them Musketts and Ammunition, which *some* of these fellows used as well as ourselves. We had observed of late that they did not seem so cheerfull as common, but seem'd to be deeply wrapt in thought. After this no more of the Natives visited Adventure Cove except some old women and a young girls, who brought us berries and fish, and most probable they was sent as *Spies*. . . .

March 4. This day the Ship [*Columbia*] was completely rig'd, Hold stowed, and in every respect in readiness for sea. She look'd like a *fiddle*! . . .

[March] 22. Launch'd the Sloop Adventure. She went off admirably. [T]ook a Hawser and got her along side the Ship [*Columbia*], and soon had her rig'd.

[March] 24. The Sloop Adventure is ready for sea. Capt. Haswell, 1st mate of ship, went on board and took charge. . . .

[March] 25. . . . During our long tarry in Adventure Cove, we all enjoy'd good health, although the Crew was at times very much exposed. . . . We pick'd Whurtle and Blue Berries, throughout the winter, which was very fine, and Whurtle Berry pudings was quite common with us. We kept the Crew continually supplied with Spruce beer. . . . perhaps this method kept the *Scurvey* off. . . .

[March] 27. I am sorry to be under the nessescity of remarking that this day I was *sent* with three boats, all well man'd and arm'd, to destroy the Village of Opitsatah[.] it was a Command I was no ways tenacious off, and am grieved to think Capt. Gray shou'd let his passions go so far. *This* Village was about half a mile in Diameter, and Contained upwards of 200 Houses, generally well built for *Indians*[.] ev'ry door that you enter'd was in resemblance to an human and Beasts head, the passage being through the mouth, besides which there was much more rude carved work about the dwellings some of which was by no means *innelegant.* This fine Village, the Work of Ages, was in a short time totally destroy'd.

Source: Frederic W. Howay, ed., *Voyages of the "Columbia" to the Northwest Coast, 1787–1790 and 1790–1793* (Portland: Oregon Historical Society Press, 1990), 388–391.

ANALYSIS

The captain's passions, as Boit refers to Gray's anger at Native attitudes and behaviors, had increased during the winter stay at Adventure Cove. The friction originated in June 1791 during the *Columbia*'s first trading stop. Attoo, a Hawaiian teenager who had accompanied Gray on the *Columbia*'s first voyage back to Boston and joined the second voyage to return to Hawaii, left the ship to be among the Tla-o-qui-ahts. An angry Gray decided to hold Chief Tootiscosettle, Wickananish's older brother, as a hostage on the deck of the *Columbia,* threatening to carry him out to sea if his tribesmen did not return Attoo. Tootiscosettle complied and ordered that Attoo be fetched. When the Hawaiian appeared, Gray had him tied to a frame on the *Columbia* and put to the lash, forcing the chief and other Natives to view the spectacle. Tootiscosettle carried the humiliation with him and months later threatened Gray with the story of an imminent attack on the *Columbia.*

Gray became convinced that an attack would come on March 21, but at noon that day the aged father of Wickananish and Tootiscosettle came with presents and evidently tried to calm the situation. But as John Hoskins recorded in his journal, "their countenances betrayed the most evident signs of fear[.] [T]he old gentleman came on board but Tootiscoosettle would not." After they presented a gift of two otter skins, Gray ordered Wickananish and his father "to depart and never to return again on pain of death" (Howay 1990, 275).

Regardless of where Gray had decided to pause to build the *Adventure* and repair the *Columbia* for the 1792 trading season, his crew would be exposed to potential raids or attacks from Indians, who would be tempted by the copious iron and other materials used in the boat building. To guard against this threat he chose a small bay, "so small," Boit recorded, "that when the Ship was moor'd, you might throw a Stone upon the beach in any direction. [T]he passages in was not to exceed 100 feet" (Howay 1990, 381). Gray also built a substantial reinforced log structure—they called it Fort Defiance—to house materials, tools, and arms, including a ship's cannon. The two-story structure was 36 feet long and 18 feet wide with two ports for cannon and openings for small arms. From Boston, they had carried substantial beams for the stern and stem of the hull and some deck timbers, but the masts, yardarms, fittings, and other wood components had to be made from local sources. When completed, the 50-foot-long single-masted *Adventure* displaced 45 tons and could accommodate 12 crew. Robert Haswell became captain during the 1792 trading season, sailing as far north as present-day Sitka, Alaska. In September, Gray sold the vessel to the Spanish commander at Nootka, Juan Francisco Bodega y Quadra, for more than 70 otter skins.

On leaving Clayoquot Sound, Gray took vengeance on Wickananish and Tootiscosettle by ordering the destruction of Opitsat, a large summer village that had few winter inhabitants, who had fled in fear of conflict with the Americans. The village houses that Boit described each accommodated several families and had an open interior design, cedar-plank sides, and a shed roof, with openings for entryways that usually featured carved animal figures on structural posts and beams. Twenty to 30 people lived in each house, which included thousands of board feet of timber. Only chiefs had the authority to construct and own houses. Houses and seagoing canoes constituted the Natives' principal investment of labor and materials.

85. Conflict at Gray's Harbor, 1792

INTRODUCTION

After the successful launch in March 1792 of the sloop Adventure *in Clayoquot Sound, Robert Haswell took the vessel north to trade in early April. Gray also left the sound in April and coasted south in the* Columbia. *On parting, the two agreed to meet months later in Columbia Cove, an inlet north of Clayoquot Sound. Gray sailed south past the Strait of Juan de Fuca down the coastlines of present-day Washington and Oregon and stood off several bays but could not find any agreeable anchorage. By April 11 the ships were off the mouth of the Umpqua River on Oregon's south coast, but they returned north to the Strait of Juan de Fuca, where they encountered George Vancouver. The two captains discussed their impressions of the Columbia River, and when Gray left Vancouver he sailed close to the shore to near Tatoosh Island off Cape Flattery. There, John Boit noted in his log, "many Indians came off, and brought plenty of furs" (Howay 1990, 393). Additional trading took place between the* Columbia *and Natives who came to the ship from villages near the present-day village of La Push, at the mouth of the Quillayute River. The trade was "brisk," as Boit put it, but they acquired few skins in the exchanges, which brought them many fish. Gray and his crew, eager to make profitable contact with Natives in a location that promised more sustained trade, finally spotted a welcoming bay on May 7.*

John Boit, Remarks on the Ship *Columbia*'s Voyage from Boston, at Anchor at Gray's Harbour, May 7, 1792

[May] 7. N. Latt. 46°58′. Saw an inlett in the land, which had the appearance of an harbour, sent the Cutter under charge of 2d Officer to examine it. Laying too strong Current with squally weather. The Boat returned, and the Officer reported that he cou'd find nothing but breakers at the entrance, but farther in itt had the appearance of a good harbour. This appearance being so flattering, Capt. Gray was determin'd not to give itt up. Therefore ordering the boat a head to sound with nessescary signalls the Ship stood in for the weather bar and we soon see from the Mast head a passage in between the breakers.

Bore off and run in NE *b[y]* E, having from 4 to 9 fathom sand, an excellent strong tide setting out. the boat having made a signall for anchorage and a good harbour we continued to stretch on till completely within the shoals when we anchor'd in 5 f[atho]m In an excellent harbour. Vast many canoes came off, full of Indians. they appear'd to be a savage sett and was well arm'd, every man having his Quiver and Bow slung over his shoulder. Without doubt we are the first Civilized people that ever visited this port, and these poor fellows view'd us and the Ship with the greatest astonishment. their language was different from any we have yet heard. The Men were entirely naked, and the Women, except a small apron before, made of *Rushes,* was also in a state of Nature. they was stout made, and very ugly. Their Canoes was from the Logs rudely cut out, with upright ends. We purchas'd many furs and fish. . . .

[May 8.] We was fearfull to send a Boat on discovery but I've no doubt we was at the Entrance of some great river, as the water was brackish and the tide set out half the time. This evening heard the hooting of Indians, all hands was immediately under arms, sevrall canoes was seen passing near the Ship, but was dispers'd by firing a few Musketts over their heads. At Midnight we heard them again, and soon after as 'twas bright Moon light, we see the Canoes approaching to the Ship. We fir'd sevrall cannon over them, but still persisted to advance with the war Hoop. At length a large Canoe with at least 20 Men in her got within ½ pistol shot of the quarter, and with a Nine pounder, loaded with langerege [iron shrapnel] and about 10 Musketts, loaded with Buck shot, we dash'd her all to pieces, and no doubt kill'd every soul in her. The rest soon made a retreat. I do not think that they had aney Conception of the power of Artillery. But they was too near us for to admit of any hesitation how to proceed. . . .

[May 9.] I am sorry we was oblig'd to kill the poor Divells, but it cou'd not with safety be avoided. These Natives brought us some fine Salmon and plenty of Beaver Skins, with some Otters, and I believe had we staid longer among them we shou'd have done well.

Source: Frederic W. Howay, ed., *Voyages of the "Columbia" to the Northwest Coast, 1787–1790 and 1790–1793* (Portland: Oregon Historical Society Press, 1990), 394–396.

ANALYSIS

Gray called the harbor he entered on May 7 Bulfinch Harbor, later called Gray's Harbor. In the only fragment we have of Gray's journal, he described their search that day for a sufficient anchorage: "lowered away the jolly-boat, and went in search of an anchoring-place, the ship standing to and fro, with a very strong weather current. At one, P.M., the boat returned, having found no place where the ship could anchor with safety; made sail on the ship; stood in for the shore. We soon saw, from our mast-head, a passage in between the sandbars. At half past three, bore away, and ran in north-east by east . . . and, as we drew in nearer between the bars, had from ten to thirteen fathoms, having a very strong tide of ebb to stem. Many canoes came alongside. At five, P.M., came to in five fathoms water, sandy bottom, in a safe harbor, well sheltered from the sea by long sand-bars and spits" (Howay 1990, 435).

Boit reported that many canoes came quickly up to the ship in an aggressive manner, and he concluded that the *Columbia* crew was the "first Civilized people that ever visited this port." He may have been correct, but Natives in the area had already seen and traded with British trader Charles William Barkley in 1787 at the mouth of the Quinault River, not far from Gray's Harbor. Boit also noted that the Native language was new to the crew, not surprising because it was their first encounter with Chinookan-speaking people. They sold Gray "many furs and fish" in exchange for "Blanketts and Iron," references to square cuts of blue-, green-, or red-dyed canvaslike cloth and 5.5-pound wedge-shaped iron chisels. Gray had noticed on the first voyage that above all else Natives desired cutting tools made of iron, and they snapped up chisels whenever they could. With that in mind, Barrell had loaded the *Columbia* with nearly 6,800 pounds of chisels for the second voyage.

It is unclear who manned the 20 canoes that appeared to challenge the *Columbia* in the early morning hours of the next day. Gray had interpreted the "hooting of Indians" at night as preparation for some kind of attack, but when canoes approached the ship and seemingly ignored a warning shot from the ship's cannon, he decided to level a devastating blast. If the Natives were an attacking party, it is likely that they were from villages on rivers that drained into Gray's Harbor, because by morning there were several trading parties from the interior who traded skins and communicated easily with Indians at Gray's Harbor. In his log Gray made no mention of the fusillade he had directed at the Natives, but he may have gained some information about the existence and location of the Columbia River from the Indian traders. When Gray left the harbor on the night of May 10, he sailed swiftly in a direct course to the mouth of the Columbia River, which he entered on May 11.

86. Gray on the Columbia River, 1792

INTRODUCTION

On April 29, 1792, Captain Gray encountered George Vancouver offshore, near Cape Flattery at the Strait of Juan de Fuca. The Columbia *was the first ship Vancouver had met in coastal waters on his voyage to the Northwest Coast. The two captains exchanged information, and Gray was surprised at news that John Meares had reported that Gray had circumnavigated Vancouver Island in the* Washington. *In fact, he had not accomplished the feat and had gone only a short way into the Strait of Juan de Fuca. But Gray did mention that he had discovered a river mouth at about 46° north in August 1788 and had spent a few days in the* Washington *trying to enter the river under poor conditions. Vancouver had also seen evidence of the river two days before but doubted that it was significant, commenting that "if any inlet or river should be found, it must be a very intricate one, inaccessible to vessels of our burden" (Lamb 1984, 1:78). Gray disagreed and told Vancouver that he intended to investigate the opening again. Two weeks later on May 11 (May 12 in Boit's log), Gray stood off the entrance to the Columbia River and evaluated crossing the tempestuous river bar, a stretch of sandy barrier shoals three miles wide by six miles long that created dangerous surf conditions and a threat of grounding. The river mouth is one of the most treacherous on the Pacific Ocean, the so-called Graveyard of the Pacific that even today requires bar pilots to safely guide ships from ocean to river.*

John Boit, Remarks on the Ship *Columbia*'s Voyage from Boston, at Anchor in Columbia's River, May 12–14, 1792

This day [May 12] saw an appearance of a spacious harbour abrest the Ship, haul'd our wind for itt, observ'd two sand bars making off, with a passage between them to a fine river. Out pinnace and sent her in ahead and followed with the Ship under short sail, carried in from ½ three to 7 f[atho]m, and when over the bar had 10 f[atho]m Water quite fresh. the River extended to the NE as far as eye cou'd reach, and water fit to drink as far down as the *Bars,* at the entrance. we directed our course up this noble *river* in search of a Village. The beach was lin'd with Natives, who ran along shore following the Ship. Soon after above 20 Canoes came off, and brought a good lot of Furs and Salmon, which last they sold two for a board Nail. the furs we likewise bought cheap, for Copper and Cloth. they appear'd to view the Ship with the greatest astonishment and no doubt we was the first civilized people that they ever saw. We observ'd some of the same people we had before seen at Gray's harbor, and perhaps that was a branch of this same River. at length we arriv'd opposite to a large village, situate on the North side of the river about 5 leagues from the entrance, came too in 10 f[atho]m sand, about ¼ mile from shore. The river at this place was about 4 miles over. We purchas'd 4 Otter Skins for a Sheet of Copper, Beaver Skins, 2 Spikes each, and other land furs, 1 Spike each.

We lay in this place till the 20th May, during which time we put the Ship in good order and fill'd up all the *water* casks along side itt being very good. These Natives talk'd the same language as those farther South, but we cou'd not learn itt. Observ'd that the Canoes that came from down river brought no Otter Skins, and I beleive the Otter constantly keeps in Salt water. they however always came well stocked with land furs and capitall Salmon. The tide sett down the whole time and was rapid. Whole trees sometimes come down with the *Stream.* The Indians inform'd us there was 50 Villages on the banks of this river. . . .

On the 15th took up the Anchor, and stood up River but soon found water to be shoal so that the Ship took the Ground, after proceeding 7 or 8 miles from our 1st station, however soon got off again. Sent the Cutter and found the main Channel was on the South side, and that there was a sand bank [present-day Sand Island] in the middle, as we did not expect to procure Otter furs at any distance from the Sea, we contented ourselves in our present position which was a very pleasant one. I landed abrest the Ship with Capt. Gray to view the Country and take possession, leaving charge with the 2d Officer. Found much clear ground, fit for Cultivation, and the woods mostly clear from Underbrush. none of the Natives come near us.

18. Shifted the Ship's birth to her Old Station abrest the Village *Chinoak,* command'd by a cheif name *Polack.* Vast many Canoes full of Indians from different parts of the river where constantly along side. Capt. Grays named this river *Columbia's,* and the North entrance Cape Hancock, and South Point *Adams.* This River in my opinion, wou'd be a fine place for to sett up a *Factory.*

Source: Frederic W. Howay, ed., *Voyages of the "Columbia" to the Northwest Coast, 1787–1790 and 1790–1793* (Portland: Oregon Historical Society Press, 1990), 396–399.

ANALYSIS

John Boit was only 16 years old when he signed on as fifth mate aboard the *Columbia* in 1790, and his journal is the only extant record of the full course of the *Columbia*'s second voyage to the Northwest Coast. It is likely that Boit's father, a Boston merchant, had some role in his son joining Gray's crew, and it is noteworthy that after returning to Boston in 1793, Boit went back to sea and remained in the maritime business as ship captain and merchant until his death in 1829.

The *Columbia* had a draft of from six to eight feet, deep enough to caution any captain from entering a shallow bay or river. Nonetheless, even though the surf at the bar indicated shoals, Gray sent out the pinnace, the ship's boat, to take soundings. He then crossed the bar in the *Columbia* on the morning of May 11, 1792, becoming the first ship to enter the river. His purpose was trade, not exploration, so it is not surprising that Gray did not emphasize his discovery and drew only a rough sketch of the river estuary. In addition to noting Gray's naming the river Columbia's River after his ship, Boit appears to document a claim of possession. It turns out, though, that the phrase was added to the log in a different color and texture of ink much later, likely during the negotiations between the United States and Great Britain over the division of the Oregon Country. In 1826 when the negotiations began, Britain discounted Gray's discovery as a "lucky adventure" and pointed to Britain's official exploration of the river in October–November 1792 by Lieutenant William Broughton as a true claim to discovery (Clayton 2000, 213). British negotiators may have had a point, but in the end Britain agreed in 1846 to abandon its claim to the Oregon Country south of the 49th parallel, including the Columbia River.

Even though Boit realized the river's importance when he described it as at least four miles wide and populated by an estimated 50 villages, neither he nor Gray made any allusion to the Columbia as the River of the West. Trading opportunities were more important. Boit reported minor fur collecting during the first few days, but trading increased once they moved upriver. When they returned to their first anchorage he reported a "vast many Canoes full of Indians from different parts of the river where constantly along side." He suggested that a fur factory, or post, be built on the lower Columbia, and he purposefully noted the potential for agricultural cultivation. Boit was in error, though, when he thought he saw Gray's harbor people on the Columbia River and guessed that the river might connect to that bay, more than 100 miles to the north. If Boit accurately reported their experience on the Columbia, it seems likely that Gray left to seek more trading opportunities once he had gleaned enough furs. Gray had such insufficient concern about his discovery of the river as a national claim that he later passed along his sketch to Juan Francisco de la Bodega y Quadra at Nootka. Bodega obligingly handed it off to Vancouver, who decided to investigate the river himself.

87. Gray's Columbia River Sketch, May 1792

INTRODUCTION

During the last decade of the 18th century several fur traders published accounts and maps of their voyages. But in the surviving written accounts of Robert Gray's second voyage, there is no mention that he drew any maps of the Northwest Coast, much less a map of his nine-day stay in May 1792 in the river he named Columbia's River. At some point, however, Gray drew a sketch map of the river, because it is mentioned later in the historical record. Gray's map then disappeared until 1961, when a copy was discovered in the expedition papers of George Vancouver in the archives of the Public Record Office in London. A description of this map was first published by Paul Rigdon in 1975.

The story of this map includes elements of serendipity and irony. The map itself is not particularly well drawn, since Gray was not a well-trained cartographer and the primary purpose of his voyage was not to create detailed maps of his discoveries. As the surviving accounts of his time on the Columbia River record, Gray's most important activity there was trade. The simple title of the map is deceptively important. Gray chose the possessive form for the toponym, Columbia's River, which imparted an understanding that the river was associated with an American-sponsored vessel.

ANALYSIS

We can reasonably reconstruct a timeline of what happened to Gray's map from the extant diaries of John Boit and Robert Haswell, a remnant log perhaps written by Gray himself, and the 1798 published account of George Vancouver. On May 11, the day after his 37th birthday, Gray crossed the bar at the entrance to the Columbia River. At some point near the end of or shortly after his nine-day sojourn on the river, Gray probably completed his sketch map. On May 19, he christened the river Columbia's River. He left the river the next day, sailing north until, on June 27, the *Columbia* struck rocks and underwent significant damage. On July 24 the ship was in such bad repair that Gray entered Nootka Sound, where he received assistance from the Spanish commandant Juan de la Bodega y Quadra. During the next few weeks, Gray shared a copy of his map of Columbia's River with Bodega. The information was likely of considerable interest to Bodega, since he had sailed with Bruno Hezeta on the second Bucareli expedition when Hezeta first documented the presence of what was to become known on Spanish maps as the Entrada descubierta por Dn Bruno Ezeta.

Gray left Nootka on August 24. Four days later, George Vancouver arrived at Nootka Sound to conduct negotiations with Bodega on the Nootka Convention. Over the next few weeks the two exchanged geographical information, and Bodega shared a copy of Gray's map with Vancouver. Subsequently Gray returned to Nootka on September 21, when he met with Vancouver, and as Haswell recorded, they "mutually informed each other of our discoveries." Haswell then noted that "Captain Vancouver told me it was his intention to visit Columbia's River" (Howay 1990, 353–354). Gray left Nootka on September 22.

On September 30–October 1, Vancouver sent his lieutenant, Zachary Mudge, to England with dispatches and other information related to both his explorations and his diplomatic mission in the Pacific Northwest. Mudge may have carried a copy of Gray's map with him. If so, Vancouver retained a copy. Vancouver noted that on October 21 when Lieutenant William Broughton entered the Columbia River, "Mr. Broughton had, for his guidance thus far up the inlet, a chart by Mr. Gray, who had commanded the American ship Columbia, but it did not much resemble what it purported to represent" (Vancouver 1798, 1:53). Broughton was dismissive of Gray's chart, principally because his own soundings and location of treacherous sandbars did not agree with Gray's. Gray, however, had recorded his observations five months earlier, during the spring floods. Despite Broughton's objections that Gray undeservedly claimed credit for discovering this major river, his own much larger (four feet by two feet) and detailed manuscript, *A Sketch of the River Columbia,* and the engraved map published in 1798 in the atlas to Vancouver's *A Voyage of Discovery* bore the place-name Columbia River. That designation had the potential to promote the case for American sovereignty in the area, an ironic point considering that Britain and the United States would argue over who had the greater claim to the Oregon Country during the 1830s and 1840s.

Ultimately, it is unclear how many copies of Gray's map were made in 1792, but its reappearance nearly 170 years later allows us to supplement the existing journals with a graphic representation of what Gray observed during nine days in May. The map, which is undated, measures 16 by 10 inches. Based on the drawn scale, the extent of Gray's reconnaissance was 12 leagues, or approximately 36 nautical miles, from the entrance to the farthest point east reached by the ship. Gray's latitude coordinate of 46°10′ north closely corresponds with today's given latitude of 46°14′. His longitude of 122°42′ was almost 1.5 degrees off from today's 124°3′ west. For purposes of annotating the rest of the map, the following letters have been added:

A. Gray named this cape at the northern entrance Cape Hancock, although he did not include the name on his sketch. This is today's Cape Disappointment, a name given by John Meares in 1788 and adopted by Vancouver.
B. Gray named this cape Point Adams in honor of President John Adams. The cape was first described by Bruno Hezeta, who named it Cabo Frondoso. Between these two capes, Gray noted that "this shore composed of Sunken Rocks and Overfalls." He used stippling to indicate the extensive sandbar that allowed navigation only through a northern channel, where extensive soundings in fathoms marked the course of the river upstream.

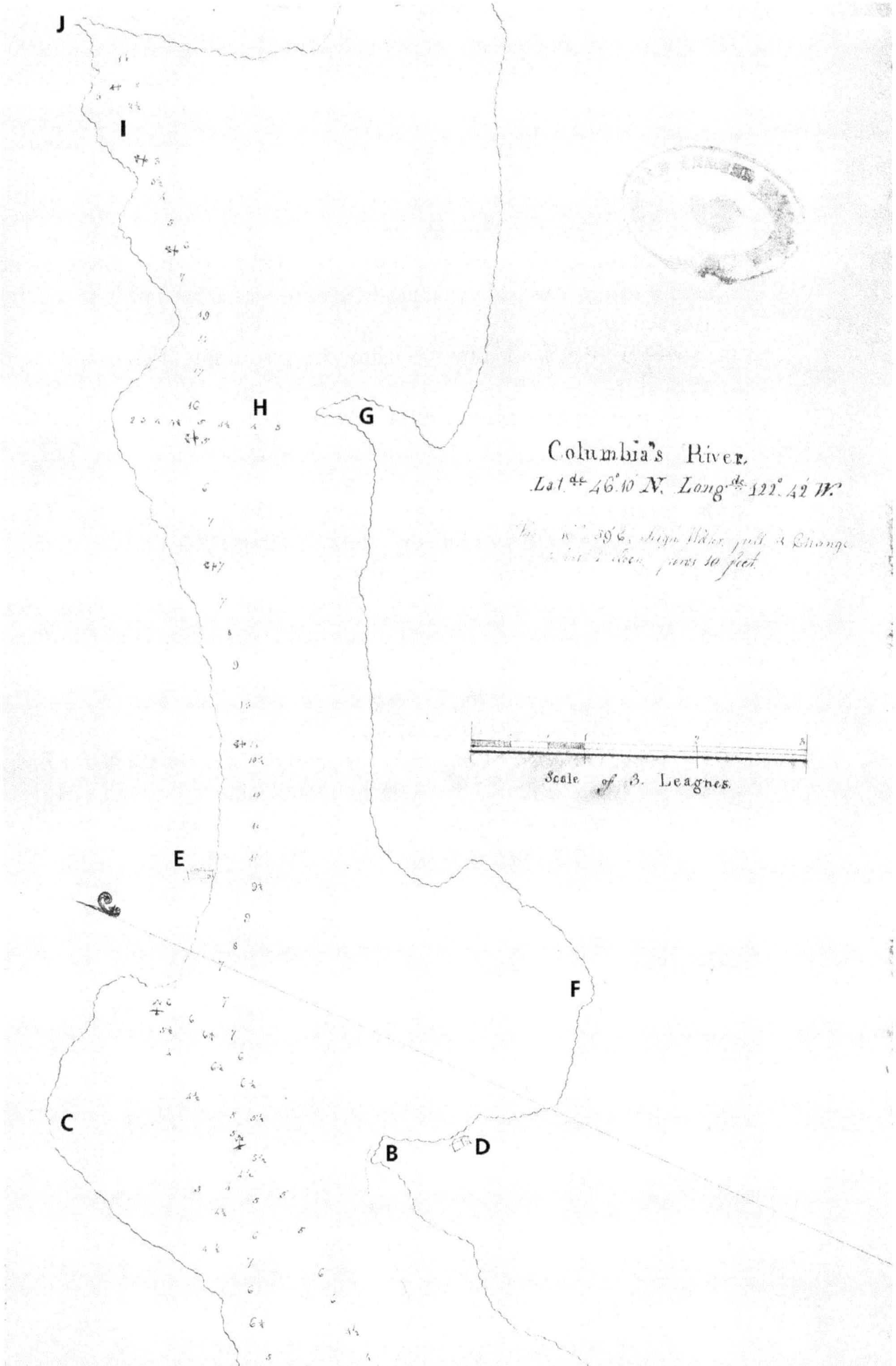

"Columbia's River." (British National Archives)

C. Bakers Bay, named by Broughton.
D. This small structure probably represented a Clatsop village.
E. This drawing undoubtedly represents the Chinook village that was led by the chief whom Boit identified as Polack. The *Columbia* spent a considerable amount of time at this site, trading for sea otter, beaver, and other furs. The slightly elongated building may suggest the architecture of a Native longhouse.
F. Young's Bay, named by Broughton.
G. Tongue Point, named by Broughton.
H. This line of soundings, extending north and south, probably represent Gray's imprecise recording of the navigable channel that meanders from north to south approaching Tongue Point.
I. Today's Gray's Bay, named by Broughton in recognition of Gray and the farthest point he reached.
J. Probably the mouth of present-day Gray's River.

Further Reading

Clayton, Daniel. *Islands of Truth: The Imperial Fashioning of Vancouver Island.* Vancouver and Toronto: University of British Columbia Press, 2000.

Gibson, James R. *Otter Skins, Boston Ships, and China Goods: The Maritime Fur Trade of the Northwest Coast, 1785–1841.* Seattle: University of Washington Press, 1992.

Howay, F. W. "Perkins and Company, Canton 1803–1827." *Bulletin of the Business Historical Society* (March 1932): 1–5.

Howay, F. W., ed. *Voyages of the Columbia to the Northwest Coast, 1787–1790 and 1790–1793.* Portland: Oregon Historical Society Press, 1990.

Lamb, W. Kaye, ed. *George Vancouver; Voyage of Discovery to the North Pacific Ocean and Round the World, 1791–1795.* 4 vols. London: Hakluyt Society, 1984.

Vancouver, George. *A Voyage to the North Pacific Ocean and Round the World.* 3 vols. London: G. G. and J. Robinson, and J. Edwards, 1798.

Expedition of Galiano and Valdés, 1792

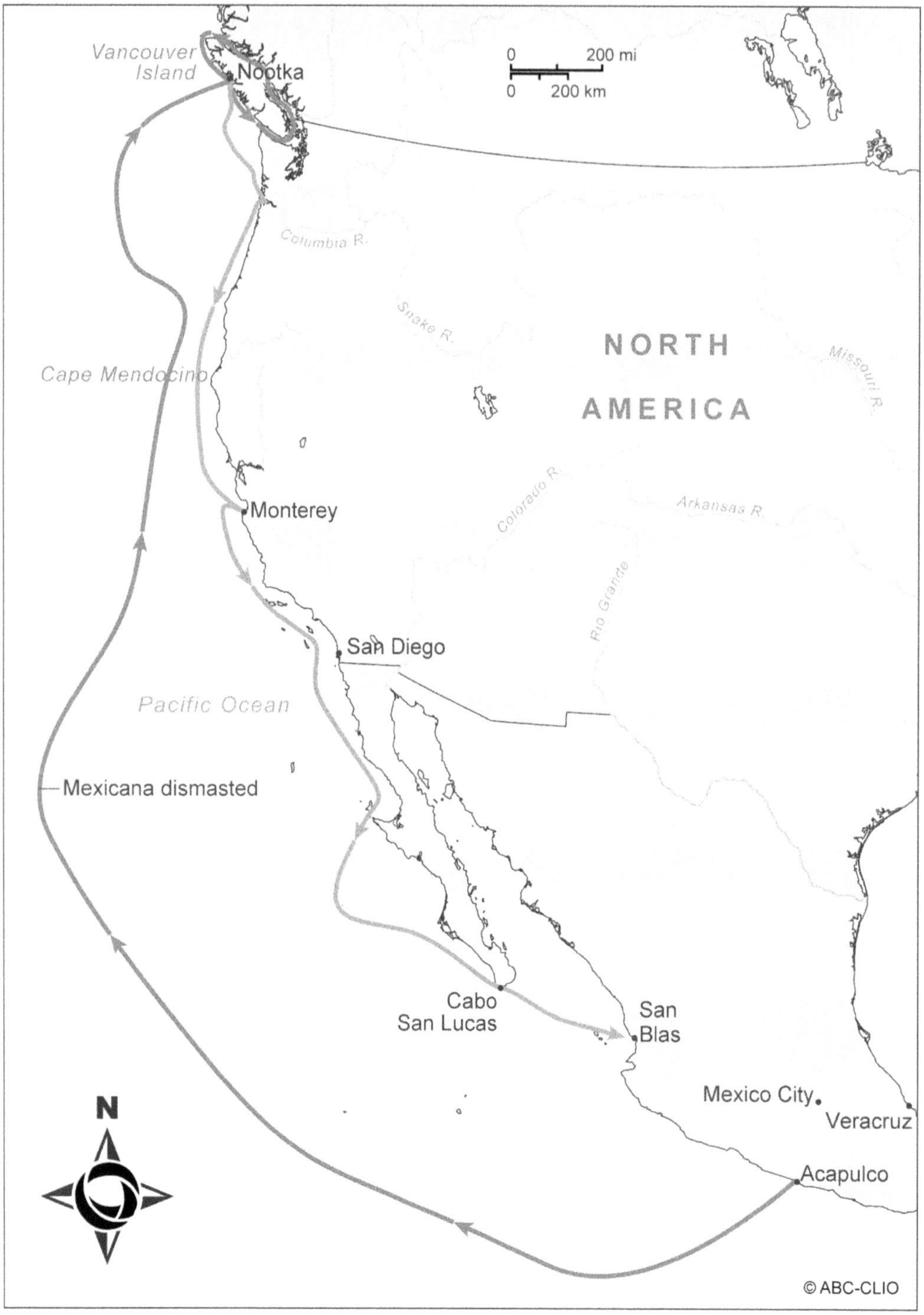

CHAPTER 10

Dionisio Alcalá Galiano, Exploration of the Strait of Juan de Fuca, 1792

Dionisio Alcalá Galiano was a Spanish naval hero, explorer, and cartographer whose career as a mariner encompassed both the apogee of Spain's territorial dominion on the Northwest Coast and the beginning of that country's withdrawal from the region. In 1791, Alejandro Malaspina appointed Galiano commander of an exploratory voyage beyond the Strait of Juan de Fuca. By the turn of the 19th century Galiano's published maps, along with those of George Vancouver, had contributed to a modern view of the coast from Acapulco to the Gulf of Alaska, and Galiano had helped dispel the centuries-old idea of a transcontinental water passage from the Atlantic Ocean to the Pacific. Ironically, his reputation was related directly to Malaspina's disgrace in 1795, which delayed the publication of his own account of the voyage until the late 19th century, well after Galiano's narrative had been published.

Galiano was born in Cabra, Córdoba, Spain, in 1760. At age 11 he entered the Spanish Royal Navy. He began his formal education in 1775 at the prestigious naval academy, Real Colegio de Guardiamarinas in Cádiz, at a time when the international focus on the Pacific Northwest Coast of North America was increasing. Carlos III had ascended the throne in 1759, intent on making Spain equal in arts, sciences, and naval education to other European powers, especially Britain and France. Two years later, the Spanish ambassador to St. Petersburg sent the first report to Madrid on Russian activities in the North Pacific. Spain's navy has been characterized as La Marina Illustrada (the Enlightened Navy), and the purpose of the Real Colegio, as Felipe V (1700–1724) declared in 1748, was to "educate and endow the noble youth of my kingdoms in the sciences most useful to the navy" (qtd. in Tovell 2008, 8).

Courses at the academy included mathematics, cosmography, cartography, and instruction in the use of the latest navigational instruments, which often were purchased in London or Paris. Galiano entered the academy months after Juan Francisco de la Bodega y Quadra had sailed for San Blas, Mexico, where he was sent to explore the North Pacific Coast. Bodega had spent a year at the academy studying the latest techniques in navigation, including the use of newly acquired chronometers under the guidance of master cartographer Vicente Tofiño. Also in San Blas in 1774 was Juan Pérez, who had departed to the Pacific

The expedition of Galiano and Valdés (above) was one of the last Spanish voyages of exploration on the North Pacific coast in the 18th century. From June through August 1792, these intrepid officers conducted the first true circumnavigation of Vancouver Island and surveyed part of the Strait of Juan de Fuca simultaneously with George Vancouver. The Spanish and English captains exchanged charts, but only a few modern place names attest to the presence of Galiano and Valdés in 1792. (Album/Art Resource)

Northwest on the first Spanish voyage of exploration to the region since 1602. By the time of Galiano's graduation from the academy in 1779, he would have learned the latest in naval science and technology and would have understood Spanish imperial interests in North America.

Galiano honed his skills as a junior officer on hydrographical surveys, including tours in South America on the River Plata and in the Falkland Islands and in a survey of the Strait of Magellan. He first met Malaspina in 1783, when the two studied together under Admiral Tofiño at the Royal Observatory at Cádiz. Tofiño was so impressed with Galiano that he selected him to help create the *Atlas Maritimo de España,* the great national atlas of the coasts of Spain, published in 1789.

Malaspina selected the 29-year-old sublieutenant to be a hydrographer and cartographer on his 1789–1794 scientific voyage of exploration in the Pacific, including the Northwest Coast. On the voyage, which left from Cádiz in July 1789, Galiano served under José de Bustamante y Guerra on the *Atrevida,* helping chart the coast of South America from Patagonia to Mexico and recording astronomical observations and navigational measurements. When the Spanish crown directed Malaspina to explore Alaska and Prince William Sound two years later, he left Galiano in Mexico to prepare the charts and observations for publication. Galiano pursued his task using all available colonial documents relating to New Spain and prior explorations to the Northwest Coast.

After Malaspina's return to Acapulco in September 1791, he interceded with the viceroy, who was making plans to explore the Strait of Juan de Fuca, and arranged for Galiano and Cayetano Valdés y Flores (who had been with Malaspina on the voyage to Alaska) to command the two ships of the expedition. Malaspina recommended that both officers be promoted to frigate captain and gave them many of his instruments, including several of his best chronometers. Malaspina also transferred a young artist from his command, Manuel José Cardero Antonio, to Galiano. Cardero's portraits of Native people and his sketches of scenes, artifacts, and events are valuable records of this period.

Galiano's orders anticipated the need for close-to-shore navigation and surveying, which required two specially constructed ships. The ships' construction began in San Blas, but when they arrived in Acapulco, Galiano and Valdés ordered that they be remodeled to increase stability and below-deck capacity for provisions. With the repairs completed and the ships rerigged, the expedition departed on March 8, 1792. Still, the two small ships struggled under challenging ocean conditions. When George Vancouver examined the ships in June, he was "astonished at the vessels in which they [Galiano and Valdés] were employed to execute a service of such a nature. They were each of about forty-five tons burthen . . .

and were navigated by twenty-four men, bearing one lieutenant, without a single inferior officer. Their apartments just allowed room for sleeping places on each side, with a table in the intermediate space, at which four persons, with some difficulty, could sit, and were, in all other respects, the most ill calculated and unfit vessels that could possibly be imagined for such an expedition" (Lamb 1984, 1792).

DID YOU KNOW?

In June 1792 while surveying Bellingham Bay, Dionisio Galiano and Cayetano Valdés observed the eruption of Mount Baker, previously named La Gran Montaña del Carmelo based on the resemblance of the snow-capped peak to white-robed Carmelite monks.

On May 13, 1792, the expedition arrived at Nootka Sound, where the officers and men spent three weeks repairing and resupplying the ships, making astronomical observations, and recalibrating their chronometers. Galiano met frequently with Commandant Bodega, who had arrived two weeks earlier to prepare for negotiations with Vancouver over articles of the Nootka Convention. At Nootka, Galiano recorded his observations about prominent Nuu-chah-nulth people, including his interpretations of Native hierarchical relationships, customs and beliefs, and changes in Native-European trade relationships, including the Europeans' occasional resort to violence. The expedition left Nootka on June 5 and visited Puerto de Nuñez Gaona (present-day Neah Bay) on the south shore of the Strait of Juan de Fuca, the newly established second Spanish settlement north of San Francisco. Galiano's observations on the suitability of this post as a fallback location for a permanent Spanish settlement in the Pacific Northwest would prove useful within the year.

With the Makah chief Tatacus aboard, Galiano left Gaona on June 8. For the next 84 days, he explored and mapped the complicated hydrography and archipelagoes between present-day Vancouver Island and mainland British Columbia before returning to Nootka on August 31. Although some information on Galiano's and Valdés subsequent chart came from George Vancouver, their map was a singular achievement, illustrating the highest skills in navigation, surveying, and cartography. Although Vancouver reached Nootka before Galiano did, Galiano earned the distinction of being the first to circumnavigate Vancouver Island as his was the only exploration to start and end at Nootka. While at Nootka, Galiano learned from Bodega (who had learned the news from Vancouver) that Robert Gray had entered the Entrada de Ezeta and renamed it Columbia's River. Galiano left Nootka within 24 hours of hearing the news. He sighted the Columbia on September 8 on the way to San Blas, and fixed its location but did not enter it. The expedition reached San Blas on November 23.

DID YOU KNOW?

In September 1792, Galiano and Valdés used a Spanish copy of Robert Gray's map of the Columbia River to confirm the location of the river that they still called Entrada de Ezeta.

At San Blas and later in Mexico City, Galiano prepared his journal and maps, which were initially intended for publication as part of the larger account of the Malaspina expedition. Viceroy Juan Vicente de Revillagigedo (1789–1794) and Commandant Bodega sought Galiano's opinion on several pending policy issues, including whether to abandon San Blas in favor of a single naval base at Acapulco, the suitability of Puerto de Núñez Gaona as a permanent settlement, and whether further explorations north of Vancouver Island should be undertaken.

In December 1793 after serving the Spanish Royal Navy for nearly three years in Mexico and the Pacific Northwest, Galiano left for Spain. The following August, he presented his report of the expedition to the king and continued to work on preparing the information from the five-year expedition for publication. Two engraved maps derived from Galiano's field maps were published in 1795, but Malaspina's arrest and imprisonment in late 1795 for his public criticism of Spain's colonial policies delayed publication of the larger work. Galiano apparently escaped ensnarlment in Malaspina's case and was awarded membership in the Order of Alcántara, a prestigious military and religious organization.

Galiano continued his cartographic activities until he was called into combat duty when war broke out with England in 1796. He commanded the frigate *Vencedor* during the Spanish defeat at the Battle of Cape Saint Vincent in 1797. In 1802 while commanding a naval vessel escorting treasure ships carrying gold and silver from South America to fund the war, Galiano was blockaded in Havana Harbor. Also that year, the official account of the 1792 voyage, *Relación del viage hecho por las goletas Sutil y Mexicana en el año de 1792,* was published in Madrid. The author of the account is still debated by scholars, but Galiano was certainly a principal contributor. In late 1804, Spain and Britain again went to war. The 45-year-old Galiano, now a brigadier, commanded the 76-gun ship the *Bahama* at the Battle of Trafalgar. On October 21, 1805, he was killed by a cannonball. He was buried at sea, and his ship was taken to Gibraltar as a prize.

Galiano's remarkable training and service to his country during his 34-year career are reason enough to memorialize the intrepid naval officer, but his principal legacy is the narration of his exploration in the Strait of Juan de Fuca during 84 days in 1792. Topographical reminders of his presence in the Pacific Northwest are Galiano Island in the Strait of Georgia, Alcalá Point and Dionisio Point on the northern end of that island, and Galiano Gallery, a scenic natural formation on present-day Gabriola Island.

88. In Search of Spanish Dominion in the Strait of Juan de Fuca, 1792

INTRODUCTION

Spanish authorities initiated the Galiano-Valdés expedition of 1792 with a different perspective of the strategic importance of the Pacific Northwest. Through the terms of the First Nootka Convention of October 1790, Britain and Spain averted war over confrontations at Nootka Sound between an aggressive Spanish naval officer and British fur traders in mid-1789. The articles of the convention clearly challenged Spanish hegemony in the Pacific Northwest and required restitution of the British property the Spanish had seized at Nootka as well as reinstitution of British commerce in the region. Although Spanish authorities may have ceded sole control of the Nootka region, they claimed that they maintained sovereignty by right of possession or first exploration elsewhere in the Pacific Northwest.

The royal order of December 25, 1790, from the Spanish court to the viceroy of New Spain was written just two months after the Nootka Convention was signed. The order suggests that within the limits of the convention, boundaries of demarcation could be placed around British penetration to the south and east into the North American continent.

Nootka Sound Convention Articles

Articles of the First Nootka Sound Convention, Signed on October 28, 1790

Article I. It is agreed that the buildings and tracts of land situated on the Northwest Coast of the continent of North America, or on islands adjacent to that continent, of which the subjects of His Britannic Majesty were dispossessed about the month of April, 1789, by a Spanish officer, shall be restored to the said British subjects.

Article II. Further, a just reparation shall be made, according to the nature of the case, for every act of violence or hostility which may have been committed since the said month of April, 1789, by the subjects of either of the contending parties against the subjects of the other; and in case any of the respective subjects shall, since the same period, have been forcibly dispossessed of their lands, buildings, vessels, merchandise, or any other objects of property on the said continent or on the seas or islands adjacent, they shall be replaced in possession of them or a just compensation shall be made to them for the losses which they have sustained.

Article III. And in order to strengthen the bonds of friendship and to preserve in the future a perfect harmony and good understanding between the two contracting parties, it is agreed that their respective subjects shall not be disturbed or molested either in navigating or carrying on their fisheries in the Pacific Ocean or in the South Seas, or in landing on the coasts of those seas in places not already occupied, for the purpose of carrying on their commerce with the natives of the country or of making establishments there; the whole subject, nevertheless, to their restrictions and provisions which shall be specified in the three following articles.

Article IV. His Britannic Majesty engages to employ the most effective measures to prevent the navigation and fishery of his subjects in the Pacific Ocean or in the South Seas from being made a pretext for illicit trade with the Spanish settlements; and with this in view it is moreover expressly stipulated that British subjects shall not navigate nor carry on their fishery in the said seas within the distance of 10 maritime leagues from any part of the coast already occupied by Spain.

Article V. It is agreed that as well in the places which are to be restored to British subjects by virtue of the first article as in all other parts of the Northwest Coast of North America or of the islands adjacent, situated to the north of the parts of the said coast already occupied by Spain, wherever the subjects of either of the two powers shall have made settlements since the month of April, 1789, or shall hereafter make any, the subjects of the other shall have free access and shall carry on their commerce without disturbance or molestation.

Royal Order of December 25, 1790

The last establishment of Spain is the port of San Francisco, at about 38° latitude, and its dependencies and missions ascend in the interior to more than 45°; and going even farther inland are near 55°; thus it would be advisable to fix the point of division [between Spanish and British dominions] on the coast at 48°, for that would take into account all districts purchased by the English, leaving in common the trade, navigation, and establishments of both nations from the said 48° northward. Thus there will be a neutral area between the

Spanish port of San Francisco and the new establishments that will aid all parties in avoiding transgressions and discords.

It would also be advisable to establish a boundary to the interior of these lands so that English traders do not penetrate to the missions [to the south]; and it should run [from 48°] as far north as Prince William and Bucareli, at 55° and 60°, to avoid the turbulence, jealousies, and suspicions that those traders might introduce into New Mexico and its missions.

Source: William Ray Manning, "The Nootka Sound Controversy," in *Annual Report of the American Historical Association* (Washington, DC: U.S. Government Printing Office, 1905), 454–455; quoted in Warren L. Cook, *Flood Tide of Empire: Spain and the Pacific Northwest, 1543–1819* (New Haven, CT: Yale University Press, 1973), 248.

ANALYSIS

In the late 18th century, Nootka Sound was the center of both Spanish naval activity and the burgeoning maritime fur industry. Royal Navy ensign Esteban José Martínez was sent to occupy Nootka in early 1789. He was also charged with enforcing Spain's sovereignty claims in the region by preventing Russian or British commercial activity without licensure from Spanish authorities. From May to July, he seized several vessels under British command. Under subsequent orders from Viceroy Manuel Antonio Flores (1787–1789), Martínez abandoned Nootka and returned to San Blas in December 1789 with James Colnett, who had been placed under arrest as the British captain of one of the seized vessels.

A few months later in April 1790, Spain reoccupied Nootka under the command of veteran Royal Navy officer Francisco de Eliza y Reventa. The commandant at San Blas, Francisco de la Bodega y Quadra, ordered Eliza to reestablish Spanish naval control at Nootka and create a settlement there. During 1790, expeditions of exploration and possession taking were undertaken in Prince William Sound under Salvador Fidalgo y Lopegarcía, who documented several Russian settlements in the region, and in the Strait of Juan de Fuca by Manuel Quimper Benítez del Pino. Quimper's voyage of reconnaissance was hurried, although his officers did enter the present-day Rosario and Haro Straits and concluded that a possible passage into the continent existed farther to the north and east. Spanish authorities were aware that British and American fur traders had previously entered the Strait of Juan de Fuca, but Quimper's exploration was the first to document the extent of the strait's interior.

Quimper arrived in San Blas in October 1790, the same month that the First Nootka Convention was signed. His report made the implications of the convention more significant, as geopolitical and geographical possibilities became linked in the strategizing of authorities in Madrid and Mexico City. News of the convention reached Mexico in mid-November 1790, followed by the royal order of December 25. The rapidity with which this order was rendered suggests the level of attention that King Carlos IV and his ministers attached to the high-stakes gamesmanship in the Pacific Northwest. Because the British could no longer be prevented from occupying or trading in the waters around Nootka, perhaps they could at least be contained.

Based on the royal order, Viceroy Revillagigedo constructed a map with demarcation lines drawn at the entrance to the Strait of Juan de Fuca at latitude ~48° north, above which both Britain and Spain would enjoy equal rights of occupancy and trade but below which Spain would retain sole sovereignty. Revillagigedo envisioned establishing a northern base of operations on the south side of the strait at or below latitude 48° north. In addition, he drew a line from approximately the southwest corner of present-day Vancouver Island, from

48° to 60°. East of the line would be considered Spanish sovereign territory. The British and any other foreign power would be restricted from the interior of the continent at these latitudes and thus be discouraged from accessing territory in Spanish-controlled New Mexico and the interior provinces of New Spain, which included much of both present-day Texas and the American Southwest and northern Mexico. In order to affect this strategy, however, it was important that the geography of the interior of the Strait of Juan de Fuca be explored and charted. In short order these plans were formulated, and the Galiano-Valdés expedition of 1792 soon followed.

89. Search for the Northwest Passage, 1792

INTRODUCTION

Near the end of the 18th century, the search for a passage from the Northwest Coast across North America was a principal motivator for the Spanish to finance a last major exploration to the Pacific Northwest. The Galiano-Valdés expedition of 1792 was important enough that men at all levels of Spanish authority, including the king, participated in its planning. In a rare departure from long-standing policy of withholding exploratory information, an account of the voyage was published in Madrid in 1802 as Relación del viage hecho por las goletas Sutil y Mexicana en el año de 1792 para reconocer el Estrecho de Fuca. *Fewer than 500 copies were printed, however, so circulation was limited. These excerpts are from the first published English translation of the account by Cecil Jane in 1930.*

The 1802 work was written anonymously, and historians have ascribed authorship to several contemporaries of Galiano's. In a recent study of the origins of the Relación del viage, *John Kendrick (1985) questions whether its author distorted some of the results of the voyage. Nevertheless, the* Relación del viage *is valuable as a document because of its description of some of the motivations for Galiano's and Valdés's voyage.*

Relación del Viage

Down to the year 1789, the only information which we possessed of Juan de Fuca Strait was the confused account of its discovery in 1592, left by the Greek pilot who gave it his name. Don Estéban Martínez . . . related that in 1774 . . . he sighted what appeared to be a very wide sound at 48° 20′ north . . . thinking that this might be the entry discovered by Fuca. . . .

In the following year [1791], [Don Francisco] Eliza was ordered by the viceroy of New Spain to complete the investigation [of the Strait of Juan de Fuca] which he had begun [in 1790] and by which the curiosity of geographers had been aroused. . . . [He] entered the strait on 27 May [1791]. He remained in it until 7 August, when he was forced to leave it owing to an outbreak of scurvy among his crews and to lack of suitable food for them. . . .

On his return to Nootka [on Vancouver Island], he wrote an account of his voyage to the viceroy of New Spain, and in the course of the comments which he made, remarked: "Your Excellency may rest assured that the passage to the ocean, for which foreign nations have sought with so much zeal, lies, if it exists at all, nowhere save in this great channel."

These reports were transmitted to His Majesty, and as he was always anxious to advance scientific knowledge and to extend hydrographical information, he at once sent orders to the viceroy of New Spain that appropriate means should be taken to effect such an exploration of Fuca Strait as to resolve all doubts concerning its extent. The Count of Revillagigedo immediately appointed Don Francisco Maurell, lieutenant of frigates, to execute this commission, employing for the purpose the schooner *Sutil* and a frigate's launch.

At this moment [October 1791] there arrived at Acapulco His Majesty's corvettes *Descubierta* and *Atrevida* [of the Malaspina Expedition], engaged in making a voyage round the world, the preliminary of which was to be a search for the supposed passage to the Atlantic, of which the western entrance, according to the well-known account given by Captain Lorenzo Ferrer Maldonado, lay at 59°30′ north on the north-west coast of North America. . . .

[O]bstacles of a different kind eventually precluded him [Malaspina] from examining the entry visited in 1775 by Don Bruno de Eceta [the Columbia River] . . . or from finding the river of Martin del Aguilar [charted by Vizcaíno in 1603]. . . .

[T]he commander [Malaspina] proposed to the viceroy that this work [exploring Strait of Juan de Fuca and de Hezeta's entrance] should be undertaken by officers belonging to his command, who should be supplied with clocks and other instruments required for its adequate performance. . . .

This opinion was communicated to the viceroy, who signified his approval of their [schooners *Sutil* and *Mexicana*] selection and also of the appointment of the officers for the expedition, these officers being the captains of frigates, Don Dionisio Alcalá Galiano and Don Cayetano Valdés.

Source: Cecil Jane, trans., *A Spanish Voyage to Vancouver and the North-West Coast of America* (London: Argonaut, 1930), 3–5.

ANALYSIS

The years 1790 through 1792 were the peak of Spanish geopolitical and exploratory activity in the Pacific Northwest, which culminated in the Galiano-Valdés expedition from May to November 1792. The First Nootka Convention had driven authorities in Madrid and Mexico City to plan expeditions in the Strait of Juan de Fuca to acquire information needed to contain incursion from other imperial powers, especially Britain. As these documents illustrate, the preparations for these expeditions were motivated by many prior accounts of discovery, some apocryphal and some based on incomplete or misinterpreted observations.

One of the most enduring geographical beliefs was the apocryphal tale of a Greek pilot, Juan de Fuca, who claimed that in 1592 he had sailed into a broad inlet on the coast between 47° and 48° and had eventually reached the North Sea (Atlantic Ocean) far into the interior of the continent. The mariner's story was propagated first in text and subsequently on maps as early as 1752. Although de Fuca's voyage did not occur, the eventual European discovery of a broad strait at 48° sustained the idea that a major water passage into the North American continent lay beyond the limits of exploration up to mid-1791. In 1774 during the first Bucareli expedition along the Northwest Coast, second-in-command pilot Esteban José Martínez reportedly observed a possible "wide sound at 48°20′." In 1789 Martínez sent a small sloop to reconnoiter the entrance of the strait, and he recounted information obtained from Natives that the strait extended far to the south and southeast (probably today's Puget Sound). In 1790 Manuel Quimper explored the strait to the San Juan Islands, and in June and July 1791 Francisco Eliza explored much of the eastern interior, including Rosario and Haro Straits and the Gulf of Georgia. Eliza described numerous channels along the British

Columbia mainland that he did not have time to explore, although he was convinced that one or more of them would lead far into the interior of the continent.

Spanish authorities were also anxious about the reports of British and American fur traders who had entered the Strait of Juan de Fuca in 1789–1791. The most troubling was the 1790 publication of Englishman John Meares, which included several maps. On "A Chart of the Interior Part of North America Demonstrating the Very Great Probability of an Inland Navigation from Hudson's Bay to the West Coast," Meares depicted a River Oregon, which emptied by a dotted line into John de Fuca's Straits. He also drew a river farther north that emptied into Cook's R. at ~62°, and he incorporated a legend in the interior between latitudes 48° and 55° illustrating a "Sketch of the Track of the American Sloop Washington in the Autumn of 1789." Although there was no documentation that these rivers or waterways actually existed, Meares's map suggested that the British may have gathered important geographical knowledge. Other British and French maps also depicted a "River of the West" that extended from the interior to the entrance of the Strait of Juan de Fuca.

Additional apocryphal information influenced Spanish authorities. In April 1791, Alejandro Malaspina in Acapulco received orders from Madrid to examine the Northwest Coast at 60° to find a passage to the Atlantic described in an 1609 account by navigator Lorenzo Ferrer Maldonado. Maldonado had reported to the king of Spain that in 1588 he had discovered a navigable water route north across the continent to the Pacific. In November 1790 the astronomer Philippe Buache presented Maldonado's narrative to the French Academy of Sciences, and his account was also published earlier in Spain. Malaspina did not discover a passage at 59°, but he named a bay in the region Ferrer Maldonado.

Two additional river systems farther to the south commanded the attention of the planners of the Galiano-Valdés expedition. Both the entry visited in 1775 by Don Bruno Hezeta (Columbia River) and the river of Martín del Aguilar had been observed during earlier Spanish voyages. Hezeta's Entrada appeared on all Spanish manuscript maps of the area after 1775, but it was not observed by any other Europeans until May 1792. Many European maps of the period showed a conceptualized "River of the West" that extended southwest from the interior to the coast at an Entrance discovered by Martín del Aguilar in 1603 at ~43°. In planning for the expedition, Spanish authorities wanted Galiano and Valdés to examine both rivers.

The expedition was motivated by a number of factors, including a plethora of both recent and centuries-old accounts whose information was observed, conceptualized, or simply apocryphal. By the turn of the 18th century few armchair geographers believed in the possibility of a navigable water passage into the continent, and within a few months after orders were issued to Galiano and Valdés, all prior accounts had been either verified or discarded.

90. Malaspina's and Revillagigedo's Orders to Galiano, 1791, 1792

INTRODUCTION

These documents are translations of instructions to 31-year-old Dionisio Galiano and 24-year-old Cayetano Valdés y Flores in preparation for their voyage in 1792. The first set was issued by

Alejandro Malaspina on December 27, 1791, after he had returned to Acapulco from exploring the coast nearly to latitude 60° north in search of the hypothetical passage described by Lorenzo Ferrer Maldonado. The instructions began with a long testimony of the potential for glory for his officers if they successfully carried out a voyage of exploration based on 21 objectives. Only those objectives that pertain to the Pacific Northwest are included here. Viceroy Revillagigedo wrote his own set of instructions from Mexico City in January 1792; they were delivered on February 7, with an accompanying letter. Revillagigedo duplicated many of Mapaspina's objectives, but several, such as number six, were more expansive.

Instructions of Malaspina and Revillagigedo

Instructions of Malaspina

1st The principal and almost the only objective of your commission (entirely subordinated to His Excellency [Revillagigedo]) is that of exploring all the internal channels of the strait of Fuca, giving priority to the maximum penetration of the sea or navigable rivers towards the east to decide once and for all the excessively confused and complicated questions of the communication or proximity of the Pacific Ocean and the Atlantic in this parallel [of latitude]. Afterwards, a secondary objective is to look into the veracity of the words of the English Captain Meares in relation to the discoveries of the Lady Wasington and the Princesa Real and finally to ascertain for the true utility of Geography what are the true limits of the continent and how far to the east the archipelago extends, which so far has been explored between 48° and 56° of latitude. . . .

6th I have indicated to His Excellency the Viceroy that it would now be appropriate to substitute for the method thus far followed of taking possession for the Crown of Castile according to the formalities of the Laws of the Indias, the less equivocal and more universal method of an exact hydrographic determination, and of one coin or another buried in a bottle in which the day, month, and year of discovery is stated, and the vessels which have made it, adding to this precaution that of buying from the natives with the greatest possible solemnity one or another of the places which appear most appropriate for a mercantile establishment, secure, healthy, and with a sheltered harbour, with easy communication with the sea, and an abundant trade in peltries, with fertile wooded soil suitable for seeding. I believe that our rights will be much less exposed to continual disputes with rival European Powers, and for us this would be of real and true advantage.

8th . . . The best method of tying together your explorations and ours will be to continue the coastal voyage [to the south of the Strait] . . . and above all not omitting the examination of the entry of Ezeta [Columbia River], whose depth of only 25 fathoms at the entrance makes me believe that it does not extend far inland. A good harbour in these latitudes would be a most useful discovery for the fur trade and for the National interest.

Instructions of Revillagigedo

6th If the results of this inspection [of navigable river passages] disclose any communication with the Atlantic by way of the Bay of Hudson, Baffin, et cetera, its entrance shall be fixed in the utmost detail, and the commanders of the goletas will set their course from it to Europe if permitted by the situation and their food supply, trying [to avoid] by every means touching at any foreign port, but in case they are obliged to do so through necessity, they will endeavour by every imaginable means [to ensure] that their place of departure and duties shall not be known, or at any rate the location of the connection that has been found. . . .

8th If, as is probable, there is no such connection, the one to the Southern Sea [Pacific Ocean] treated by the English voyager Meares shall be sought, and from this they will proceed south by the route inside the Islas de la Reyna Carlota, and attempting to explore the Archipelago de Princesa Real, of which Meares also spoke.

9th A clarification will be attempted as to whether there is a connection between this part and Fuca . . . sailing into the Boca de Ezeta as far as supplies and the time permit, determining the true position of the Rio de Martín de Aguilar, and entering ultimately the port of San Francisco.

Source: John Kendrick, trans., *The Voyage of Sutil and Mexicana, 1792* (Spokane, WA: Arthur H. Clark, 1991), 40–43, 50–51. Reproduced with permission from University of Oklahoma Press.

ANALYSIS

In 1791 while Francisco Eliza was still in the interior of the Strait of Juan de Fuca, Viceroy Revillagigedo planned another voyage to explore an extensive area from Alaska to Monterey, including the interior of the strait. A single ship, the schooner *Mexicana,* under the command of seasoned pilot Francisco Antonio Mourelle, was to depart by December 1. By then, however, Mourelle was ill, the voyage was delayed, and Malaspina had returned to Acapulco. At the time the results of Eliza's exploration were known, including the tantalizing possibility of a passage from one or more channels into the interior of the continent.

Malaspina convinced Revillagigedo of an alternative proposal: to send two schooners with the capacity for coastal and narrow channel reconnaissance to be commanded by two of his own officers. Valdés had accompanied Malaspina during all segments of his voyage, while Galiano had stayed in Mexico City to work on drafts of the scientific observations during Malaspina's recent foray to the Northwest Coast. Malaspina apparently developed his proposal over a very short period of time, and the viceroy rapidly accepted it. In less than a month, both sets of instructions were written. Although the two vessels, the *Sutil* and the *Mexicana,* required repairs and did not sail from Acapulco until early March 1792, the haste of planning for the voyage suggested how vital officials believed it was to gather information on the interior of the Strait of Juan de Fuca.

The instructions to Galiano and Valdés stipulated that they should focus almost entirely on obtaining geographic information. Possession taking, which had been used with elaborate ceremony on almost all previous Spanish voyages, was relegated to burying a coin and notes in a bottle. In fact, during the subsequent voyage the commanders eliminated possession-taking ceremonies. Although Malaspina encouraged his officers to consider purchasing land for a "mercantile establishment" (perhaps to ensure occupancy as proof of sovereignty), he did not encourage his crews to engage in trading for furs. Elsewhere in the instructions, Malaspina even stated that "all other objectives of botany, zoology, and lithology you will regard as fortunate accessories which cost neither the slightest risk or sacrifice of your own safety, nor the slightest loss of time on the part of the commission" (Kendrick 1991, 41). Finally, in an optimistic directive, Revillagigedo instructed Galiano and Valdés to proceed directly and surreptitiously to Europe if they discovered an ocean-to-ocean passage, an objective of European voyages of discovery for 300 years.

91. Tetacus, Native of Vancouver Island, June 1792

INTRODUCTION

European explorers' accounts of encounters with indigenous people in the Pacific Northwest often told readers as much about the observers as the ones being observed. The primary objectives of the Galiano-Valdés expedition did not include gathering and recording information about Native people, and with the exception of José Cardero, the ship's manifest did not include scientists or artists. The relatively brief time that the Spanish officers spent at any one location during their circumnavigation of Vancouver Island and the physical requirements of their survey left little opportunity to engage in ethnographic study, even among people who may have had little or no prior contact with Europeans. Nevertheless, Galiano and Valdés wrote about and reflected extensively on a Native chief they traveled with on June 6–10, 1792, in the Strait of Juan de Fuca. Tetacus, as he was referred to in their diary and journals, was known to other European explorers and writers as Tatoosh, Tatooch, Te-Ta-Koos, Tutuzi, Tetaku, Tatooth, Tootooch, and perhaps Tataciin.

From these documents, Tetacus emerges as a person with notable prestige and uncommon character. Contemporary and later readers of the 1802 report of the voyage, Relación del viage hecho por las goletas Sutil y Mexicana en el año de 1792, *may have been familiar with Tetacus from an engraved portrait by José Cordero that appeared in the atlas accompanying the* Relación del viage.

Diary of the Campaign Executed under the Command of Commanders Don Dionisio Galiano and Don Cayetano Valdés in the Estrecho or Canal de Juan de Fuca, 1792

The conduct of this Indian [Tetacus] during the crossing and the stay in Cordova [Esquimalt Harbor] . . . caused us to draw very different inferences about these Indians from what up to the present time voyagers have said about them. What they call ferocious treachery only seemed to us to be bold manliness. . . .

Tetacus, one of the principal chiefs of the entrance and who, according to reports which were borne out by his conduct, was among those most attached to the Spaniards, also came alongside and most politely asked leave to see the ships. It was granted and he examined them with a curiosity which we had not yet observed among these Indians. Afterwards he said in the *Sutil* that it was his wife who was alongside in the canoe. They called her Maria, a name which would have seemed a corruption of defect of pronunciation if we had not listened carefully to the way Tetacus pronounced it. Upon learning this we pressed her to come on board but she excused herself with an air of doubt or irresolution. We gave Tetacus to understand that we had no other object in urging his wife to come on board than to be polite to her, and that . . . she would not receive the slightest insult or want of respect. Her husband then ordered her to come on board, and she obeyed and held out her hand for us to assist her. . . . They departed leaving us well pleased with the thought that this chief had sufficient confidence to remain with his favorite wife, alone and unarmed, in a vessel he had only just seen. . . .

We found much to admire in the conduct of this chief. . . . He observed and asked questions with a curiosity which showed that his chief object in accepting the passage offered to him was to get information about our internal administration and the management of the vessels. He watched the work of the ship attentively, inquired about the use of the ropes, asked their names, and allowed few things to escape his examination, endeavoring not to be troublesome . . . and giving some information about the usages of his country and the names of various things which he thought might interest us. . . .

His intercourse gave continual proof of his rapid comprehension. He recognized on the chart the configuration of the strait and islands that had been discovered, and told us the names he gave to them. . . . He ate with curiosity whatever was given him, and imitated in everything our actions which he always carefully observed. He remembered the names of all the English and Spanish captains who have visited the coast of the mainland and Archipelago . . . and informed us that there were two large vessels inside the strait [the *Discovery* and the *Chatham* of George Vancouver].

Source: Henry R. Wagner, *Spanish Explorations in the Strait of Juan de Fuca* (Santa Ana, CA: Fine Arts Press, 1933), 210, 235–238, 240. Reprinted with permission.

ANALYSIS

By the time Galiano and Valdés met Tetacus at Nuñez Gaona (present-day Neah Bay) on June 6, Tatoosh, as he was commonly known, had already had extensive experience with European commercial and naval explorations near the entrance to the Strait of Juan de Fuca. In June 1788, for example, John Meares wrote that Tatoosh was "so surly and forbidding a character we had not yet seen," and Meares reportedly named present-day Tatoosh Island off Cape Flattery. In March 1789, Robert Haswell wrote in his log of the American sloop *Washington* that "Tatooth" had purchased the furs of the Natives in the area and controlled the price (Howay 1990, 72). In June 1791, John Hoskins noted in his journal of the voyage of the *Columbia* that "Tootooch" controlled the trade in sea otter skins and effectively prevented their sale to the Americans by other Natives (Howay 1990, 197). In 1790, Manuel Quimper encountered "Tutusi" and described him and the "rest of his tribe" as "warlike, intrepid, thievish and treacherous" (Wagner 2001, 132). Tutusi was also mentioned in the diary of an officer of the 1791 expedition to the strait commanded by Francisco Eliza (Wagner 2001, 189).

Undoubtedly, Galiano and Valdés were influenced by these reports to expect the worst, but in a letter to the viceroy they interpreted Tetacus's conduct as "bold manliness" instead of "treacherous" (Wagner 2001, 210). They were impressed by his concern for the safety of his wives, who paddled canoes from Neah Bay to Esquimalt Harbor while he sailed aboard the *Mexicana.* Galiano and Valdés were particularly affected when the chief addressed one of his wives as Maria, a name that might be expected to appeal to Spanish officers. Furthermore, not only did Tetacus show uncommon interest in the workings of the ship, but he also was able to recount the names of British, American, and Spanish officers he had encountered. Finally, Galiano and Valdés wrote that Tetacus appeared to understand how to read and interpret a map (presumably a copy of one prepared by Quimper or Eliza) and that he added Native names to islands in the strait.

Historians have speculated whether Tetacus was a Makah from the Cape Flattery area or a member of the Esquimalt people near present-day Victoria. Spanish officers identified his village at Puerto de Cordoba (Esquimalt), where he took leave of the *Mexicana* on June 10. Whatever his heritage, he certainly appeared to exert influence over and control of the fur trade in the region. At some point José Cardero sketched an impression of Tetacus, and he also painted a watercolor portrait of him wearing a woven coned hat with a whaling scene. Cardero's sketches were reworked to some extent by a commissioned artist in Madrid. While the engraving of Tetacus that appeared in the atlas accompanying the 1802 publication of the *Relación* appears to be quite similar to the watercolor, the portrait quality of the image brings into doubt the fidelity of the published print to Cardero's original sketch. This question of authenticity is even more relevant with Cardero's engraved image of Maria, Tetacus's wife, whose facial features appear distinctively European.

Sometimes overt, sometimes subtle, Eurocentric biases and perspectives were often embedded in accounts of early encounters with indigenous people. At best, Galiano and Valdés contributed a snapshot of Tetacus to add to the gallery of other European descriptions of this enterprising man.

92. Ship Manifests and Technology, 1792

INTRODUCTION

The account of the Galiano and Valdés expedition and the manifests of the Sutil *and the* Mexicana *clearly illustrate that the Spanish Royal Navy was practicing advanced navigational technology by 1792. A principal objective of the voyage was to prepare an accurate map of an extremely complex hydrography, and Alejandro Malaspina chose Galiano to command the expedition because of his extensive training and experience in conducting coastal surveys. Malaspina also provided two ships outfitted with the most advanced astronomical and scientific instruments available. Extreme physical conditions beyond the Strait of Juan de Fuca often precluded the mariners from making precise measurements of latitude and longitude, but when possible they made meticulous observations both on ship and in makeshift observatories onshore. These documents illuminate how the process of mapping resulted in charts of great accuracy and demonstrate that by the last quarter of the 18th century, navigators in the Spanish Royal Navy had acquired skills equivalent to those of the British.*

Ship Manifest of the *Sutil*

Astronomical and Physical Instruments

1 "Licot" pendulum
1 Dollon equatorial theodolite of 1½ feet with heliometer
1 two foot achromatic telescope
1 chronometer, Arnold No. 344
2 thermometers
1 barometer
2 hand telescopes
2 sextants, the personal property of the officers. . . .

The latitude of Nutca [Nootka Sound] had been established as 49° 35′ 16″, and in repeating this we found only 4″ of difference, more to the north. On the 28th the emergence of the second [satellite of Jupiter] was observed, with a difference of 19′ more to the west than the satellite observation of the 16th. The chronometers were judged to be satisfactory, the small one being preferable. The barometers could not be set up because the tubes were broken. . . . [T]he eudiometer gave the following results:

	1st experiment	*2nd experiment*
The air in the main room in the Commandant's house	60 parts	53 parts
The open air	54	52
The hospital ward	54	52 . . .

[June 14] We showed him [Lieutenant Broughton] our chronometer, Arnold 344, and told him we had chronometer 61. The officer said they carried an Arnold chronometer, and one of Kendal. . . .

[June 27] At eleven thirty we went to the southwest point of the Island to observe the latitude, accompanied by Mr. Vancouver, his second lieutenant Mr. Pujet and the same midshipman who accompanied him when we had the honour of their visit at the anchorage of the Punta de Lángara [Point Grey]. Our observations of the meridian altitude were within 20 seconds.

At three in the afternoon Mr. Vancouver came, bringing the artificial horizon, and Galiano and Vernacci [Spanish officer] embarked with the theodolite and the base of the achromatic [telescope] on which we mounted the theodolite in order to use it. We had already made [an observation] with our artificial horizons at the Cala del Descanso [Descanso Bay] and we wanted to compare their glasses. . . .

[August 22] The regulation of the chronometers being dealt with, Galiano left on the 22nd in the launch to follow the channel leading to the west, with the object of finding a good anchorage close to the exit to the sea so that the situation of the goletas could be improved and they could make use of the first occasion when the wind was northwest to proceed to Nutca. He ran along the length of the south shore of the Canal de Salida without finding an anchorage, as far as the Puerto de Gorostiza [Shushartie Bay], beyond which he observed latitude and east west hour angles from Punta de Sutil [Sutil Point], and very close to it he made use of an artificial horizon made of the crew's wine, since the mercury had been forgotten. In this way he determined the location of the mouth of the Canal de la Salida de las Goletas [Goletas Channel].

Source: John Kendrick, trans., *The Voyage of Sutil and Mexicana, 1792* (Spokane, WA: Arthur H. Clark, 1991), 250, 83, 113, 139–140, 190. Reproduced with permission from University of Oklahoma Press.

ANALYSIS

For centuries navigators used dead reckoning to determine their course at sea in the absence of visible landmarks. Using this method, a pilot charted his daily course by taking a compass bearing and judging the ship's speed to estimate the distance traveled. It was a practical method but prone to errors, which sometimes resulted in a tragic outcome. By the 18th century, a more precise determination of longitude was accomplished by measuring either lunar distances or the relationship of the four satellites of Jupiter. These methods required the use of extensive tables of predetermined data obtained from a known reference point, usually Greenwich, England. One set of tables, the *Nautical Almanac,* became widely available to European navigators within a few years of its publication in England in 1767.

The most reliable method of determining longitude—a dependable sea clock or chronometer—was first invented by the London-based watchmaker John Harrison in 1737. Eventually others also made chronometers, including Larcum Kendall and John Arnold in London and Ferdinand Berthoud in Paris. James Cook demonstrated the reliability of the chronometer during his second voyage in 1772, and the Spanish Admiralty maintained a person in London and Paris to acquire knowledge about the instrument and to purchase materials. In 1774 the Admiralty purchased four chronometers from Berthoud and two years later purchased four more from Arnold. The Admiralty was slow to use the instruments, however, and naval officers apparently navigated using dead reckoning during Spain's first three voyages from San Blas to the Pacific Northwest in 1774, 1775, and 1779.

After 1783, the Admiralty decided to adopt the most recent scientific information and equipment available, especially for purposes of navigation. Galiano and Valdés were thoroughly schooled in the latest methods of determining longitude, both by making celestial observations and using chronometers. The *Sutil*'s manifest lists both an Arnold Box chronometer (No. 61 in a marked series) and an Arnold pocket chronometer 344, while Vancouver had two chronometers, a Kendall No. 3 and an Arnold 82. Both parties carried editions of the annual *Nautical Almanac* and its French equivalent. Galiano probably had a copy of a Spanish equivalent of the almanac *Almanaque náutico y efemérides astronómicas, para el año bisiesto de 1792,* first published in 1791. This allowed him to calculate longitudes directly from Cádiz.

The Spanish also carried a "Dollon," an equatorial theodolite that allowed measurement of vertical and horizontal angles, probably built in the late 1780s by the British firm of John and Peter Dollond. In the late 1760s, John Dollond had perfected an achromatic telescope with a lens system that eliminated distracting colored light. Galiano used astronomical observations not only to confirm the accuracy of his chronometer but also to calculate latitude with considerable precision. Galiano and Vancouver had similar expertise in using instruments and performing calculations. Later, Galiano exercised ingenuity by successfully substituting wine for mercury when using an artificial horizon.

Galiano also used a eudiometer, an interesting device invented in England in 1775 to measure the oxygen content of the air, which in the understanding of the day equated to its purity. Malaspina, and thus Galiano, embraced the idea of regularly measuring air quality aboard ship or, as in Galiano's case, in the commandant's house and a hospital ward.

93. The Importance to the Viceroy of Place-Names on Maps, 1793

INTRODUCTION

The explorations of all maritime nations to the Pacific Northwest resulted in maps that conveyed both geographical and social knowledge. The mariners' seamanship and their use of advanced navigational technology and instruments produced information that cartographers plotted so that all Europeans could understand their maps. Native people, who had different mapmaking traditions, also understood the wayfinding function of European maps, and they often contributed information to mapmakers. At the same time, social information conveyed by placing names on maps reflected the values and beliefs of each imperial nation, usually at the expense of another nation and particularly of indigenous peoples.

This document illustrates the importance of these two forms of knowledge to Spanish officials. On October 23, 1793, Viceroy Revillagigedo wrote to Galiano after receiving the journal and maps that the mariner had spent nearly a year completing after his return from Nootka Sound in November 1792.

Letter from Revillagigedo to Galiano, October 23, 1793

Two nights before the day you started your journey from this capital [Mexico] to that port [Veracruz] you brought me the description and maps of the coasts explored by yourself and Frigate Captain Don Cayetano Valdés in the goletas Sutil and Mexicana from the 8th of March when you left Acapulco to the 23rd of the following November when they returned to San Blas. I have examined carefully and in detail these documents, which prove the worth of the hardships and anxieties with which they have been formed, showing with all possible accuracy the points on the part of the coast that could not be explored by the corvettes Descubierta and Atrevida [ships of Alejandro Malaspina] in the year 91.

The same applies with respect to the discoveries in the Strait of Juan de Fuca carried out in the three months of navigation from the entry of the goletas in latitude 48 deg. 20 m. until your exit in 50 deg. 10 m., resulting from which [it is known] that Nootka is an island whose territory occupies the space of more than 60 leagues in length and in some places 20 leagues of width.

The description appears to me adequate for the clearest understanding of the two maps, but in it I have noted various errors of the pen and defects of orthography, and in the maps some place names have been forgotten or omitted, namely the Puntas de Moreno de la Vega, the Isla de Bonilla, Isla and Canal de Güemes, the Canal de Pacheco, the Bocas de Winthuysen, and the Puntas de San Rafael and Lángara.

In consequence, I have ordered the correction of the material defects of the description and the addition of the omitted points for [the purpose of] sending them to Their Excellencies the Ministers of State and of the Navy, with the perfection demanded by the importance and the use of these documents.

My comments on them affirm the care with which I have examined them, in proof of the appreciation which is deserved by documents which ought to be of the greatest importance and which I anticipate are worthy of The Royal pleasure of His Majesty and his Sovereign approval.

Source: John Kendrick, trans., *The Voyage of Sutil and Mexicana, 1792* (Spokane, WA: Arthur H. Clark, 1991), 244. Reproduced with permission from University of Oklahoma Press.

ANALYSIS

Revillagigedo's comments reflect both the ways that Spanish officials thought about maps and his own desire to be recognized in the process. His statement that Galiano's journal and maps would "be of the greatest importance . . . and worthy of The Royal pleasure" was an expression of the king's intense interest in the Northwest Coast. Accurate geographical information about the region was critical, and Revillagigedo was almost effusive when he acknowledged that Galiano had constructed a map that depicted "with all possible accuracy . . . points . . . on the coast."

The viceroy also noted, however, "various errors of the pen" and listed eight "place names that have been forgotten or omitted." Spanish naval officers who had been in the Pacific Northwest in 1790–1791 had applied these names to various locations on Vancouver Island

and in the Strait of Juan de Fuca. Two—Güemes and Pacheco—were alternative names for Revillagigedo himself, and the viceroy's remark that they had been "forgotten or omitted" may have been a thinly disguised rebuke. He ended his letter with self-flattering references to his editing skills, presumably to impress the minister of state in Madrid, to whom he sent copies of his correspondence.

This emphasis on place-names derived from a long history of Spanish colonial policies directed at the Americas. Naming had been an important part of the ritual of claiming dominion on foreign land from the first encounters of Christopher Columbus in the Western Hemisphere, and advisers to Spanish monarchs believed that "no one could well lay claim to a nameless city, and that a province without a name was hardly a province at all" (Stewart 1982, 12). Those places occupied by indigenous people were not nameless, of course, and it is likely that an indigenous name was ignored for almost every European name bestowed on a geographical feature. Naming was part of the process of empire building, and scholar Stephen Greenblatt (1991) has written that the application of European place-names on the American landscape in the 16th century was also a form of "linguistic colonialism." Culturally specific names helped establish a culturally specific identity, and European names were included on maps read by Europeans and to be understood within a European consciousness.

The naming and renaming of places by rival imperial nations had implications. Places with Spanish names, for example, suggested Spanish ownership, and both indigenous peoples and competing European nations could be "mapped out" by removing certain names. While geographical features may be immutable, place-names are not. As we have seen, most Spanish names in and around Vancouver Island were supplanted by English names, and over time common usage of these names resulted in their use as wayfinding elements—for example, "I have to go to Gray Point on the Burrard Inlet near the Port of Vancouver." In this manner, place-names become part of the infrastructure of a map much like other requisites of an accurate map, such as a coordinate system of longitude and latitude.

Wallace Stegner wrote that "nothing is comprehended, much less possessed, until it has been given a name" (Stewart 1982, xxviii). Viceroy Revillagigedo certainly understood this. All societies with a West European mapping tradition have used names to legitimize one cultural group while delegitimizing another. By studying the place-names on maps from multiple maritime explorations in the Pacific Northwest, we can appreciate more of the rich cultural heritage of the region.

94. Perils in the Strait of Juan de Fuca, July 1792

INTRODUCTION

The primary objective of the Galiano-Valdés expedition was to produce a map of a part of the Northwest Coast that Europeans had not yet explored. The process of producing this map was multifocal and included the latest navigational and surveying technology, remarkable seamanship, arduous labor, and some of the most challenging sailing ever encountered by Spanish mariners.

Galiano and Valdés were capable men who had faced many challenging situations during their years at sea. These documents illuminate their attempt to inform the viceroy and the reading public about some of the hazards they experienced during four days in a particularly fearsome section of the passage between present-day Vancouver Island and mainland British Columbia. The translation by Henry R. Wagner of the 1802 Spanish publication of the voyage, Relación del viage *(Narrative of a Voyage), and another account translated by John Kendrick (1991) contain much the same descriptive information, as does an unpublished summary of events that the officers submitted in diary form to Viceroy Revillagigedo.*

Voyage of the *Sutil* and *Mexicana*, July 17–20, 1792

[July 17, 1792] The latitude was observed ashore and on board ship by double altitudes of the sun, and the variation by the theodolite's needle, and points of reference were made to continue our geometric work. . . .

[July 18] The extraordinary swiftness the waters acquired was a phenomenon worthy of the greatest attention. The current of the Angostura de la Esperanza in the Strait of Magellan runs 7½ miles an hour near the shore, and its velocity is much greater in mid-channel. Yet the difference apparent at first sight between the two currents is so considerable, that there is no exaggeration whatever in reckoning that of the Angostura de los Comandantes [Narrows of the Commandants, present-day Arran Rapids] at 12 miles. The aspect is a most strange and picturesque one. The waters looks like a race from a waterfall, and on them great numbers of fish are constantly jumping. . . .

The Indians received the commanders with the greatest friendship, and gave them to understand that they ought not to risk passing through the channel in the longboat for they would be swamped beyond help in the whirlpools, as happened to them in their canoes when they had the misfortune to be caught by the current. Galiano and Valdés thanked them for their warnings and applied themselves to ascertain how to make so dangerous a passage with the schooners. . . .

The Indians, utilizing the sun's path, indicated to us very clearly that when that body was near the top of a high mountain on the mainland, the favorable moment we desired would come. . . .

[T]hey allowed few moments to pass without trying to point out to us the dangers we were about to encounter, and the way and time to overcome them. They explained to us the method they pursued in this passage and the continual mishaps which nevertheless befell them, deducing that the size and resistance of our vessels would not promise us a more happy lot, but rather a more disastrous one than they had met with their canoes. For their humane and benevolent attitude we continued to call them "Good Indians" and we strove to give them whatever we knew might contribute to their satisfaction and comfort. . . .

[July 20] The schooners passed the narrows at an extraordinary rate steadied by the fore topstaysails as the wind blew fresh against them in an opposite direction to that of the channel. The *Sutil* having tacked near an island, went about on the opposite tack, and being caught in a strong whirlpool, turned round thrice with such violence that those on board were made giddy. She freed herself from this danger by rowing with full strength. . . .

The continual eddies and whirlpools, now favorable, now contrary, retarded one schooner and advanced the other, rendering it impossible to steer, and carrying us at their mercy, alternately flattering and frustrating our desires to make a little cove which was very near at hand. The *Sutil* succeeded, with the aid of the boat, in running out a cable to the east point, but at that very moment, caught by another whirlpool, she began to turn round again, and at the first turn tore the cable from the hands of those who were making it fast.

The wind increased later [at anchorage that evening], and we heard it whistling in the tops and among the trees in the forest. At the same time the violent rapidity of the waters in the channel caused a frightful noise and a most noticeable surf. This combination presented a terrifying situation, the more so as we had not explored the channel.

Source: Henry R. Wagner, *Spanish Explorations in the Strait of Juan de Fuca* (Santa Ana, CA: Fine Arts Press, 1933), 274–276, 278. Reprinted with permission.

ANALYSIS

On July 12, 1792, British reconnaissance found a passage to the northwest into open seas from the east side of present-day Vancouver Island (through Discovery Passage, Johnstone Strait, and Broughton Strait). George Vancouver resolved to follow this route in the *Discovery* (overall length 99 feet) and the *Chatham* (80 feet), and Galiano and Valdés decided to take advantage of the smaller size of their ships to continue reconnaissance along the channels farther to the east and north. On July 13, the two exploring expeditions parted company. Both the *Sutil* and the *Mexicana* were specially rigged schooners with a deck length of approximately 46 feet; each carried a complement of 20 men. From July 17 through July 20, these relatively small ships managed to explore one of the most treacherous sections of waterway from present-day Stuart Island to an anchorage probably between Dent Island and Henry Point in the present-day Cordero Channel between the Georgia and Johnstone Straits.

On July 17 they made precise determinations of latitude, including noting the magnetic variation of the compass needle. The "double" recording of the altitude of the sun enabled the officers to calculate latitude without waiting until noon. During most of the next three days, they were often at the mercy of currents, tides, whirlpools, and adverse winds. In order to progress north, they often had to use longboats to tow the schooners under power of oars. It must have been daunting to proceed under such conditions, moving ahead into unknown waters without the aid of tidal or navigational charts.

During this period many Native people likely encountered Europeans for the first time, and relations were peaceful. The Natives supplied the Spanish with food and information, including warnings about where not to proceed. At one point, Galiano and Valdés recorded an interesting method used by the Indians to use the altitude of the sun above a certain mountain to signal the most favorable time to proceed into the current.

On July 19, the officers observed a particularly treacherous inlet they called the Angostura de los Comandantes (Narrows of the Commandants) in present-day Cordero Channel. Today's published sailing directions for ships attempting to navigate through parts of the channel record the same violent currents, eddies, and whirlpools that Galiano and Valdés recounted on July 19, 1792, including velocities of current of up to nine knots (slightly more than nine miles an hour). On July 20 a little farther northwest at the entrance to the Angostura de Carvajal (north of present-day Sonora Island), the men had perhaps their most frightening experience when both ships were overcome by currents and whirlpools. The *Sutil* spun so completely out of control that the men became dizzy. The officers finally attained safe anchorage on the night of July 20 at Anchange del Refugio, but the crew spent a restless evening precariously moored between whistling treetop winds on one side and the "frightful noise" of the surf on the other.

95. Exchange of Information between Galiano and Vancouver, June–July 1792

INTRODUCTION

One of the most absorbing parts of the Galiano-Valdés expedition occurred during the joint exercises of the Spanish and the British, under the command of George Vancouver. On June 13, 1792, near present-day Point Roberts, Washington, in the Strait of Juan de Fuca, Spanish officers encountered Royal Navy lieutenant William Broughton, and on June 21 they met Vancouver near present-day Birch Bay. Until July 13 the two exploring parties worked together off and on to chart the complex array of archipelagoes and side channels as they sailed north between the east coast of Vancouver Island and mainland British Columbia. These documents describe the exchange of cartographic information between Spanish and British officers who had been entrusted with obtaining geographical information by the two greatest imperial rivals in the Pacific Northwest. The translated excerpts are from the narrative of the voyage published by John Kendrick in 1991.

The Voyage of *Sutil* and *Mexicana,* 1792

We told him [Broughton] what was known of navigation in the Strait up to the time of the explorations of the previous year [the voyage of Eliza], and that we knew of his anchorage, and the nearby lagoon. We told him the burthen of our ships, their draft, and the names of their commanders, and our objective, which was none other than to explore the inlets and plot exact maps of them. . . .

After seeing our map of the strait, Mr. Vancouver reverted to the undertaking of joining the two expeditions. The advantage which could result in geographical knowledge if all worked to a common end, and the intimate friendly reciprocation of our two courts did not leave us an instant's vacillation in accepting the proposition, which for all reasons ought to reflect accord with the generosity of our sovereigns, whose concerns are to acquire knowledge and present it to the view of all Europe for the advancement of navigation in general, for the advantage of commerce, and for the understanding of natural history, considering the human species as such an important part of it. For ourselves, we would draw the advantage of completing [our work] on this coast earlier, the exploration of the named entries having reduced our interest in them. We could then direct ourselves to the Entrada de Ezeta [Columbia River], and still have enough time to correct the chart to the southward of Fuca. . . .

[W]e parted [on July 13] after mutual demonstrations of the friendship and harmony which arose from passing a great part of the time together with the greatest cordiality and frankness, which were used to facilitate reciprocally not only a part of the exploration of the channels, but [to share] other geographical knowledge which we possessed, and methods of making use of this. We produced our copies of the maps of the parts explored up to the time of the separation, while the English gave to us a table of positions of the principal points of the coast of New California from forty five degrees to the entry of Fuca, and we gave to the English another table, of various points on the west coast of America, including that of South America. . . .

[At Nootka Sound, August 31] *Discovery* and *Chatham* had also arrived, having examined the coast as far as 52°30′ north latitude. The part extending from the Boca de Pinedo [present-day Tribune Channel] as far as Cabo Norte is copied from their discoveries, which they had communicated to us with the established freedom in the same way as we did, giving them copies of the parts they chose, to add to their maps. . . .

At twelve at night [on August 31] we set sail, having given to the English a copy of the part of our explorations which they selected, and having taken from them the portion of the north coast from Cabo Norte, with the adjacent islands, as well as the map of the shore of the entry of Fuca as far as 45° of latitude.

Source: John Kendrick, trans., *The Voyage of Sutil and Mexicana, 1792* (Spokane, WA: Arthur H. Clark, 1991), 113, 130–132, 149, 199, 206. Reproduced with permission from University of Oklahoma Press.

ANALYSIS

The chance encounter in the interior of the Strait of Juan de Fuca was the first recorded meeting at sea in the Pacific Northwest of Spanish and British naval officers. The primary objective of both exploring parties was to gather information to create accurate maps. From the perspective of Viceroy Revillagigedo, however, geographical information was linked to a broader geopolitical purpose: creating demarcation lines to constrain possible British encroachment on Spain's interior provinces.

In 1774, the first Spanish voyage of exploration to the Pacific Northwest since 1602 had been undertaken because officials in Madrid and Mexico City feared that both Russia and Britain would attempt to establish sovereignty in the region. One of the principal objectives of the third Bucareli expedition, from San Blas to the North Pacific in 1779, had been to interdict the expedition of James Cook. Ten years later, the arrest of British merchant captain James Colnett by Esteban José Martínez at Nootka Sound resulted in an intense, though bloodless, confrontation between Britain and Spain. Subsequently, British and Spanish diplomats signed the First Nootka Convention in 1790. In 1792, the British crown sent George Vancouver to the Pacific Northwest to negotiate details of the agreement and to conduct an extensive exploring expedition of the region. Because of the Spanish policy of keeping the results of prior explorations secret, Vancouver had little knowledge of Spanish activities on the Northwest Coast.

Not long before his first encounter with Galiano and Valdés, Vancouver performed his first act of possession on the Northwest Coast near present-day Everett, Washington, from ~48° south to ~39°. He subsequently applied the toponym New Albion between these latitudes on his maps, which were published in 1798. Viceroy Revillagigedo was concerned about British attempts to establish the northern limits of Spanish dominion at 39°. He hoped that if a naval base could be established on the southern shore of the Strait of Juan de Fuca and if the interior of the strait was accurately mapped, then he could make a strong argument that the demarcation line of Spanish dominion could be established at 48°. Because of Revillagigedo's distrust of the British, he ordered Bodega not to share Spanish maps with them at Nootka. It is unclear whether Galiano received similar orders.

In the interior of the Strait of Juan de Fuca in June and July 1792, however, the potential benefits from an exchange of cartographic information overcame mistrust and rancor. The exchange was all the more impressive because Vancouver enjoyed a considerable advantage in armament and men. This amicable first encounter may have tempered Vancouver's anxiety over his anticipated meeting in August with Bodega to negotiate the terms of the First Nootka Convention.

The commanders of the two exploring parties carried out their reconnaissance more or less jointly, although at times one examined an area that already had been visited by the other. Vancouver had the advantage with four boats while Galiano had only two, but the smaller size and shallower draft of the Spanish vessels allowed for a more detailed exploration of the inlets along the mainland of British Columbia. Vancouver's previous survey of present-day Admiralty Inlet and Puget Sound relieved Galiano of that task.

In the end, the information obtained by both groups contributed to maps that were strikingly similar except for place-names. Publication in 1798 of Vancouver's maps, which were largely populated with British names, meant that many readers and researchers could not appreciate the cartographic contributions made by Galiano and Valdés during the joint exploration in mid-1792.

96. The Placement and Replacement of Place-Names, 1795

INTRODUCTION

Galiano and Valdés produced an elegantly detailed map of their explorations between present-day Vancouver Island and mainland British Columbia, but several factors precluded the recognition of their achievement: delay in the Spanish publication of the map until 1802, the early and widespread dissemination of the maps of the Vancouver expedition in 1798, and the ramifications of the Spanish retreat and the ascendency of British hegemony in the Pacific Northwest following the Nootka Convention of 1790. Nevertheless, this map is important because it illustrates both the principal results of the surveys and the role of nationally derived place-names as markers of geographical discovery.

The translated title of the map, "Nautical Chart of Survey Made on the North Coast of America from the Part Starting with Narrow Channels at the Entrance of Juan de Fuca to the Exit of the Goletas Sutil *and* Mexicana *in the Year 1795," made clear that Spanish ships were performing their own surveys in the Strait of Juan de Fuca and made their own determination of the insularity of Vancouver Island. The complicated process of creating the map began with sketches and observations made during the surveys. On the expedition's return to San Blas in October 1792, Galiano and Valdés began compiling the map and their account of the voyage. They continued their work in Mexico City in February 1793 and then in October 1793 provided Viceroy Revillagigedo with the narrative and a manuscript map. The viceroy made some changes and in December forwarded the documents to the minister of state in Madrid. A manuscript copy of the original map, dated 1793, was made in Mexico (it is now in the U.S. Library of Congress). Spanish authorities acted quickly to promote the expedition's discoveries, possibly in anticipation of the publication of Vancouver's work. This map was printed in 1795 and was published sometime after that in the* Atlas Maritímo de España. *The atlas, however, did not receive the recognition and distribution accorded the Vancouver account and atlas, which were published in 1798.*

ANALYSIS

This 1795 printed map, which measured about 40 by 22½ inches, covered the same general region of the original 1793 Galiano map and incorporated both Galiano's and Valdés's

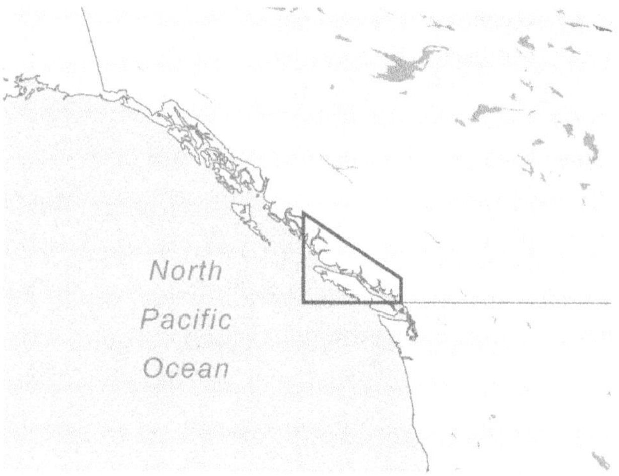

place-names as well as adding names on the west coast of Vancouver Island attributed to Malaspina. On the 1793 map longitudes were plotted from both Cádiz, Spain, and Tenerife in the Canary Islands, which cartographers had used as a prime meridian for more than two centuries. In the upper right portion of the printed map, the engraver included two profile views of part of the west coast of Vancouver Island, near Cape Frondoso and outside Nootka Sound (Nutka). These views were probably taken from artists traveling with Malaspina.

The map's principal focus is on the regions that Galiano and Valdés explored, extending from the northern reaches of Texada Island at the lower right edge of the full map from ~49°35′ to the northern part of Vancouver Island at ~50°50′ on the left. The lighter engraved coastal lines and island archipelagoes farther north were derived from shared information from Vancouver and others. Occasional vertical double crosses indicate potential navigational hazards.

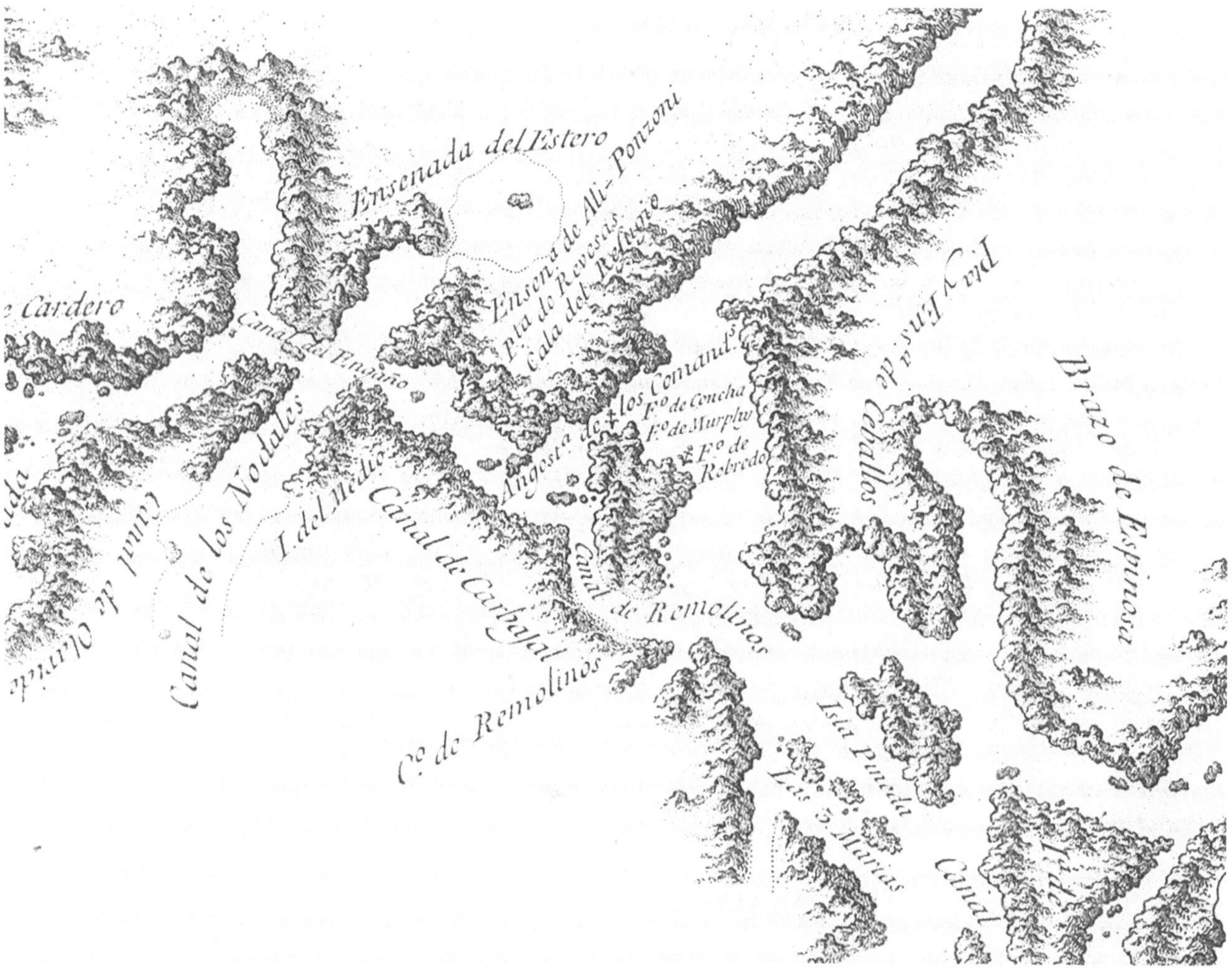

Detail of "Nautical Chart of Survey Made on the North Coast of America," 1795. (Courtesy of the John Carter Brown Library at Brown University)

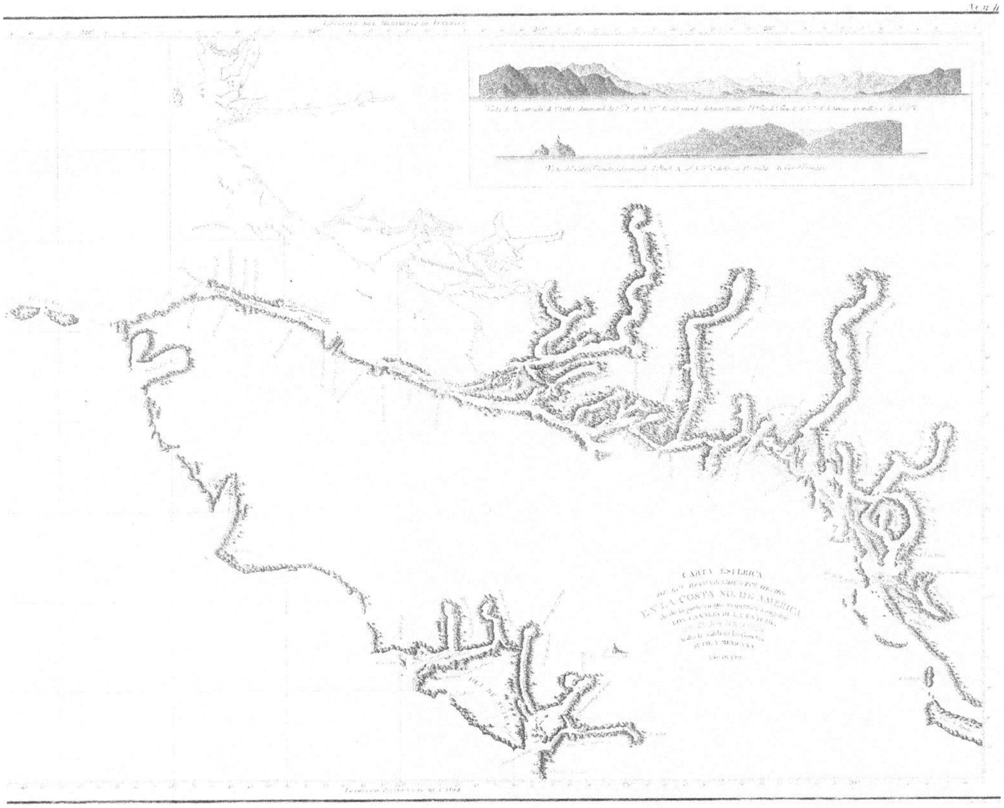

"Nautical Chart of Survey Made on the North Coast of America," 1795. (Courtesy of the John Carter Brown Library at Brown University)

By reading the account of the voyage and examining the map, it is possible to infer that the lower right edge of the map corresponded approximately to the ships' position on June 24, 1792, three days after the officers of the Spanish and British expeditions came together to conduct their surveys. Farther north on July 13 near the upper reaches of the Canal de Separación at the lower right corner (Canal) of the enlarged inset map, the two groups parted company, and the Spanish undertook much of the rest of the charting of tortuous channels and inlets. On August 26, Galiano's and Valdés's ships reached a point midway between Cape Scott and the Islas de Lanz to the northwest of Vancouver Island.

The map is replete with place-names of Spanish origin, but comparisons with modern charts of the region show that many of these names were subsequently eliminated in lieu of English names. The following is a compilation of several place-names and their present-day equivalent and origin, proceeding from the southeast to the northwest (Wagner 1968):

Spanish Name to Modern Name

Isla de Texada, now Texada Island, originally given by Spanish explorers in 1791, changed by Vancouver to Feveda and corrected later

Punta de Sarmiento, now Sarah Point, named by Vancouver after his sister Sarah

Isla de Quema, now Kinghorn Island, one of the Discovery Islands renamed in the 19th century

Brazo de Toba (Tova), now Toba Inlet, originally named by Valdés as Tabulas but later changed to Tova, one of the officers with Malaspina

Names on the Inset Map

Canal de Separación, now Lewis Channel, probably named in the 19th century after Captain Herbert Lewis of the Hudson's Bay Company

Canal de Remolinas, now Yuculta Rapids, officially named in the 20th century after a local Native tribe

Islas 3 (Tres) Marias, now Rendezvous Islands, part of the Discovery Islands; origin of name unknown

Isla Pineda, now Raza Island, named by Galiano after a lieutenant with the Malaspina expedition

Brazo de Espinosa, now Ramsey Arm, named after José Espinosa on the Malaspina expedition

Canal de Carbajal, now Cordero Channel, part of Cordero Channel, named after a member of Galiano's crew, José Cardero

Angost. de los Comand., now Arran Rapids, named by 19th-century British surveyors after an island in Scotland

Ensenada de Ali-Ponzoni, named after Fabio Aliponzoni, a marine on the Malaspina expedition

Canal del Engaño, now Cordero Channel, part of Cordero Channel

Canal de los Nodales, now Nodales Channel, named by Galiano after two Spanish brothers who explored the Strait of Magellan in 1618

Additional Names on the Full Map

Brazo de Salamanca, now Loughborough Inlet, named by Vancouver after Lord Chancellor Loughborough

Brazo de Vernacci, now Knight Inlet, named by Vancouver after Captain John Knight

Canal de Neuvo Remolinos, now Wellbore Channel, named in the 19th century after Royal Navy captain John Wellbore Spencer

Brazo de Retamal, now Call Inlet, named by Vancouver after Sir James Call

Canal de la Descubierta, now Johnstone Strait, named by Vancouver after James Johnstone

Salida de las Goletas, now Goletas Channel, named by Galiano after the two schnooers

Punta Sutil, now Sutil Point, named by Galiano, changed in 1860 to Cape Commerell; the original name was restored in 1906

Cabo Scot, now Cape Scott, probably named by fur trader James Strange in 1786 after his patron; later adopted by the Spanish

Islas de Lanz, now Scott Islands, named by James Strange in 1786

Although this list of names is incomplete, it is clear that some of the names bestowed by Galiano and Valdés were retained, some were changed and later restored, some were modified, and many were discarded by Vancouver or others who replaced them with British names. In this example, names of Native origin are rare. Although place-names on modern maps of the Pacific Northwest are an important part of the wayfinding process of navigation and travel, they are also a legacy of the multinational presence of explorers in the region. Historic maps such as this one fail to recognize the presence of Native peoples, who also had a rich culture of naming.

97. Coastal Map of Vizcaíno's Voyage, 1802

INTRODUCTION

In a departure from policy, Spanish authorities publicized the cartographic results of the Galiano-Valdés expedition. In 1795 Spain had two maps engraved and, probably within the year, published them as part of an extensive Atlas Maritímo de España. *Publication may have been related to Malaspina's arrest for criticizing the crown's colonial policies in November 1795. Authorities confiscated documents and other materials relating to his voyage, but after the results of George Vancouver's voyage were published (1798), they apparently decided to promote their own exploration of the Northwest Coast.*

The narrative written by Galiano and Valdés, who remained in royal favor, was published in 1802 in Madrid by the prestigious La Imprenta Real (Royal Press) as Relación del viage hecho por las Goleta Sutil y Mexican en al año de 1792 para reconocer el estrecho de Fuca *(Relation of the Voyage Made by the Schooners Sutil and Mexicana to Explore the Strait of de Fuca). Malaspina's name was almost entirely expunged from the text. An accompanying atlas,* Atlas para el Viage de las Goletas Sutil y Mexicana, *contained numerous copper-engraved plates and nine maps, including four maps of sections of the Pacific Coast from Acapulco to Alaska and the Aleutian Islands. This map (the second of four) was titled "Carta Esferica de los Reconocimientos hechos en la Costa N. O. de America en 1791 y 92 por las Goletas Sutil y Mexicana y otros Buques de S. M." (Nautical Chart of Surveys on the Coast of North America in 1791 and 1792 by the Schooners Sutil and Mexicana and other Vessels of Spain). "Other vessels" was a reference to Malaspina's ships, the* Descubierta *and the* Atrevida.

ANALYSIS

Carta Esferica is one of four maps in the atlas that illustrated Spanish knowledge of the Pacific Coast of North America by 1800. The maps were engraved with consummate skill and printed on high-quality, elegantly watermarked paper. This section, Numero 2, covers approximately the same area as the Vancouver map but on a slightly different scale, and longitudes were calculated from Cádiz, not Greenwich.

While the title referred to the surveys of Galiano and Valdés, information on the map also came from English and American sources. Naval officers and captains of fur-trading ships occasionally shared nautical information, including the results of recent surveys. Ships' officers and captains generally applied new names to topographical features, retained names from other sources of places they had not visited, and occasionally retained names from a rival imperial power. Native names were rarely used.

Galiano and Valdés never entered Canal de Hood (Hood Canal) or Canal del Almirantazgo (Admiralty Inlet), and they took the delineation of the Puget Sound area from Vancouver's survey. British cartographers retained the Spanish Canal de Haro in its present location, but the Canal del Rosario became the Strait of Georgia, with the former name transferred to the present-day canal east of the San Juan Islands. Seno de Gaston became Bellingham Bay (from Vancouver), and Punta de Martinez was changed to Cape Flattery (from Cook).

Even though Spanish explorers had not yet crossed the bar into the Columbia River in 1795, this map labels the river with the name of an American ship and the shape of a survey

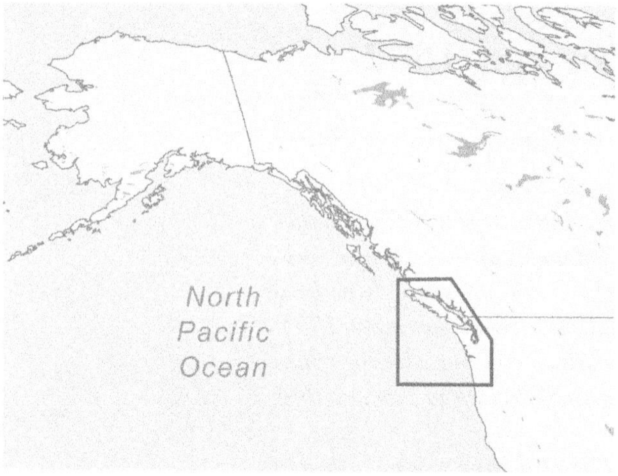

by an English navigator while carrying the legend that Bruno de Hezeta first recognized the river in 1775. To the north, the Spanish cartographer retained the name of Gray's Harbor (Puerto de Gray), which an officer with Vancouver originally bestowed in recognition of Robert Gray. To the south, James Cook's place-names—Cape Foulweather and Cape Perpetua—were retained by both Vancouver and Galiano.

Originally published in a limited edition of 500 copies, the *Relación* and *Atlas para el viage* were not reprinted in Spain until 1958. Unlike the journals of Cook, Vancouver, and La Pérouse, neither volume was reprinted by foreign presses. Perhaps more important, the British Admiralty quickly incorporated Vancouver's maps (and thus his toponyms) into charts, which were used by commercial mapmakers such as Aaron Arrowsmith in London. Thomas Jefferson, Meriwether Lewis, and William Clark used Arrowsmith's wall map of North America, updated to 1802, when planning their expedition from St. Louis, Missouri, to the Pacific Ocean. Some French mapmakers continued to favor Spanish in lieu of British names, but popular use of the latter eventually prevailed. A notable exception was geographer and explorer Alexander von Humboldt, working in Mexico in 1803–1804, who used Spanish charts (including this one) and navigational information in compiling his own maps.

It is fitting to end this book with this elegant map of 1802, which effectively summarizes the geographical and geopolitical information accumulated during maritime explorations in the 18th century. Representations of the Oregon and Washington coast, the tortuous complexity of Vancouver Island, and the inlets of mainland British Columbia are rendered with great accuracy, although some prominent features (such as the Fraser River) remained unrecognized. Perhaps more important is that the Native, English, Spanish, American, and French place-names on the map provide evidence of the multinational nature of this part of North America. In fact, other maps in the *Atlas para el Viage* incorporated Russian place-names in areas farther north and west.

The map was published at a time when the focus of mapping in the Pacific Northwest was shifting from maritime to overland exploration. Alexander McKenzie's account and map of his journey to the Pacific Ocean (1793–1794) was published in 1801, and in January 1803 Thomas Jefferson delivered his message to Congress requesting support for the Lewis and Clark expedition.

Further Reading

Barry, J. Neilson. "Columbia River Exploration, 1792." *Oregon Historical Quarterly* 33 (March 1932): 31–42.

Elliott, T. C., ed. "Remnant of Official Log of the Columbia." *Oregon Historical Quarterly* 22 (December 1921): 352–356.

Gibson, James R. *Otter Skins, Boston Ships, and China Goods: The Maritime Fur Trade of the Northwest Coast, 1785–1841.* Seattle: University of Washington Press, 1992.

Greenblatt, Stephen. *Marvelous Possessions: The Wonder of the New World.* Chicago: University of Chicago Press, 1991.

Hayes, Edmund. "Gray's Adventure Cove." *Oregon Historical Quarterly* 68 (June 1967): 101–110.

"*Num. 2 Carta Esférica de los Reconocimientos hechos en la Costa N.O. De America en 1791 y 1792, por las Goletas Sutil y Mexicàna y otros Buques de S.M.*" (Copyright © 2000 by Cartography Associates)

Howay, F. W., ed. "Captains Gray and Kendrick: The Barrell Letters." *Washington Historical Quarterly* 12 (October 1921): 243–271.

Howay, F. W., ed. "Letters Relating to the Second Voyage of the Columbia." *Oregon Historical Quarterly* 24 (June 1923): 132–152.

Howay, F. W., ed. "A Short Account of Robert Haswell." *Washington Historical Quarterly,* 24 (April 1933): 83–90.

Howay, F. W., ed. *Voyages of the Columbia to the Northwest Coast, 1787–1790 and 1790–1793.* Portland: Oregon Historical Society Press, 1990.

Howay, F. W., T. C. Elliott, and F. G. Young, eds. "John Boit's Log of the Columbia—1790–1793." *Oregon Historical Quarterly* 22 (December 1921): 257–351.

Jane, Cecil, trans. *A Spanish Voyage to Vancouver and the North-West Coast of America.* London: Argonaut, 1930.

Kendrick, John. *The Men with Wooden Feet: The Spanish Exploration of the Pacific Northwest.* Toronto, Ontario: NC Press, 1985.

Kendrick, John, trans. *The Voyage of* Sutil *and* Mexicana, *1792: The Last Spanish Exploration of the Northwest Coast of America.* Northwest Historical Series no. 16. Spokane: Arthur H. Clark, 1991.

Lamb, W. Kaye. *George Vancouver; A Voyage of Discovery to the North Pacific Ocean and Round the World, 1791–1795,* Vol. 2. London: Hakluyt Society, 1984.

Meany, Edmond S. "The Widow of Captain Robert Gray." *Washington Historical Quarterly* 20 (July 1929): 192–195.

Nokes, J. Richard. *Columbia's River: The Voyages of Robert Gray.* Tacoma: Washington State Historical Society, 1991.

Scofield, John. *Hail Columbia! Robert Gray, John Kendrick, and the Pacific Fur Trade.* Portland: Oregon Historical Society Press, 1990.

Stewart, George R. *Names on the Land: A Historical Account of Place-Naming in the United States.* 4th ed. San Francisco: Lexikos, 1982.

Tovell, Freeman M. *At the Far Reaches of Empire: The Life of Juan Francisco de la Bodega y Quadra.* Vancouver: University of British Columbia Press, 2008.

Wagner, Henry R. *The Cartography of the Northwest Coast of America to the Year 1800.* Amsterdam: N. Israel, 1968.

Wagner, Henry R. *Spanish Explorations in the Strait of Juan de Fuca.* Reprint ed. Mansfield Centre, CT: Martino Publishing, 2001.

Selected Bibliography

Anderson, Bern, ed. "The Vancouver Expedition: Peter Puget's Journal of the Exploration of Puget Sound, May 7–June 11, 1792." *Pacific Northwest Quarterly* (April 1939): 177–217.

Angus, T. S. "Captain Cook's Hygiene." *British Medical Journal* 2(3495) (December 31, 1927): 1233–1234.

Archer, Christon I. "The Transient Presence: A Re-appraisal of Spanish Attitudes towards the Northwest Coast in the Eighteenth Century." *BC Studies* 18 (Summer 1973): 3–32.

Ashley, Scott. "How Navigators Think: The Death of Captain Cook Revisited." *Past & Present* 194 (February 2007): 107–137.

Barrera-Osorio, Antonio. *Experiencing Nature: The Spanish American Empire and the Early Scientific Revolution.* Austin: University of Texas Press, 2006.

Barry, J. Neilson. "Who Discovered the Columbia River?" *Oregon Historical Quarterly* (June 1938): 152–161.

Bartholomew, Michael. "James Lind and Scurvy: A Reevaluation." *Journal of Maritime Research* (January 2002): 1–14.

Beaglehole, J. C. "Cook the Navigator." *Proceedings of the Royal Society of London.* Series A: *Mathematical and Physical Sciences* 114(1516) (December 16, 1969): 27–38.

Beaglehole, J. C. *The Exploration of the Pacific.* Palo Alto, CA: Stanford University Press, 1966.

Beaglehole, J. C. *The Life of Captain James Cook.* Palo Alto, CA: Stanford University Press, 1974.

Beaglehole, J. C. "On the Character of Captain James Cook." *Geographical Journal* (December 1956): 417–429.

Beaglehole, J. C., ed. *The Journals of Captain James Cook on His Voyages of Discovery.* Vol. 3: *The Voyage of the* Resolution *and* Discovery, *1776–1780.* 3 vols. Cambridge: Cambridge University Press, 1967.

Beals, Herbert K., et al., eds. *Four Travel Journals: The Americas, Antarctica and Africa, 1775–1874.* London: Hakluyt Society, 2007.

Beals, Herbert K., trans. *For Honor and Country: The Diary of Bruno de Hezeta.* Portland: Oregon Historical Society Press, 1985.

Beals, Herbert K., trans. *Juan Perez on the Northwest Coast.* Portland: Oregon Historical Society Press, 1989.

Bentley, Anne E. "The Columbia-Washington Medal." *Proceedings of the Massachusetts Historical Society,* 3rd Series, 101 (1989): 120–127.

Beresford, William. *A Voyage Round the World; but More Particularly to the North-West Coast of America: Performed in 1785, 1786, 1787 and 1788, in the King George and Queen Charlottes, Captains Portlock and Dixon.* Internet Archive, https://archive.org/stream/cihm_50493#page/n425/mode/2up.

Black, J. L. *G.-F. Müller and the Imperial Russian Academy.* Kingston and Montreal: McGill-Queen's University Press, 1986.

Blue, Verne. "French Interest in Pacific America in the Eighteenth Century." *Pacific Historical Review* (September 1935): 246–266.

Blumenthal, Richard. *With Vancouver in Inland Waters: Journals of 12 Crewmen, April–June 1792.* Jefferson, NC: McFarland, 2007.

Bolton, Herbert E. *Spanish Exploration in the Southwest, 1542–1706.* New York: Scribner, 1916.

Breschini, Gary S. "Sebastián's Vizcaíno's Expedition of Monterey in 1602–1603." Monterey County Historical Society, http://www.mchsmuseum.com/vizcaino.html.

Brown, Stephen R. *Madness, Betrayal, and the Lash: The Epic Voyage of Captain George Vancouver.* Vancouver: Douglas and McIntyre, 2008.

Caldwell, Francis E. *Land of the Ocean Mists: The Wild Ocean Coast West of Glacier Bay.* Culver City, CA: ProStar Publications, 2003.

Carpenter, Kenneth J. *The History of Scurvy and Vitamin C.* Cambridge: Cambridge University Press, 1986.

Carpenter, Kenneth J. "Sebastián Vizcaíno: Exploration of California." *Southwestern Historical Quarterly* 23(4) (April 1920): 285–301.

Clayton, Daniel W. *Islands of Truth: The Imperial Fashioning of Vancouver Island.* Vancouver and Toronto: University of British Columbia Press, 2000.

Coclanis, Peter A. "Pacific Overtures: The Spanish Lake and the Global Economy, 1500–1800." Common-Place 5(2) (January 2005), http://www.common-place-archives.org/vol-05/no-02/coclanis/index.shtml.

Cook, James. "The Method Taken for Preserving the Health of the Crew of His Majesty's Ship the Resolution during Her Late Voyage Round the World: By Captain James Cook, F. R. S. Addressed to Sir John Pringle, Bart. P. R. S." *Philosophical Transactions of the Royal Society of London* 66 (1776): 402–406.

Cook, Warren L. *Flood Tide of Empire: Spain and the Pacific Northwest, 1543–1819.* New Haven, CT: Yale University Press, 1973.

Crawford, Virginia. "Northwest Coast Indian Art." *Bulletin of the Cleveland Museum of Art* 65(9) (November 1978): 299–309.

Cruikshank, Julie. *Do Glaciers Listen? Local Knowledge, Colonial Encounters, and Social Imagination.* Vancouver: University of British Columbia Press, 2005.

Cutter, Donald C. *The California Coast: A Bilingual Edition of Documents from the Sutro Collection.* Norman: University of Oklahoma Press, 1969.

Cutter, Donald C. *Malaspina & Galiano Spanish Voyages to the Northwest Coast, 1791 & 1792.* Vancouver: Douglas and McIntyre, 1991.

Cutter, Donald C., ed. *Journal of Tomas de Suria of His Voyage with Malaspina to the Northwest Coast of America in 1791.* Fairfield, WA: Ye Galleon, 1980.

Dalrymple, Alexander. *Plan for Promoting the Fur-trade, and Securing It to This Country, by Uniting the Operations of the East-India and Hudson's-Bay Companys.* London: George Bigg, 1789.

David, Andrew. "An Eighteenth Century Joint Survey on the North-West Coast of America by British and Spanish Surveyors." *International Hydrographic Review,* no. 3 (May 2010), Ihttp://www.iho.int/mtg_docs/IHReview/2010/IHB_Review_final_10_05_10.pdf.

David, Andrew. "John Sherriff on the Columbia, 1792: An Account of William Broughton's Exploration of the Columbia River." *Pacific Northwest Quarterly* (April 1992): 53–59.

David, Andrew, et al., eds. *The Malaspina Expedition, 1789–1794: Journal of the Voyage by Alejandro Malaspina.* 3 vols. London: Hakluyt Society, 2001–2004.

Deacon, G. E. R., and Margaret Deacon. "Captain Cook as a Navigator." *Notes and Records of the Royal Society of London* 24(1) (June 1969): 33–42.

De Laguna, Federica. "Some Dynamic Forces in Tlingit Society." *Southwestern Journal of Anthropology* (Spring 1952): 1–12.

De Laguna, Federica. *The Story of a Tlingit Community.* Smithsonian Institution, Bureau of Ethnology Bulletin 172. Washington, DC: U.S. Government Printing Office, 1960.

De Laguna, Federica. *Under Mount Saint Elias: The History and Culture of the Yakutat Tlingit.* Washington, DC: Smithsonian Institution, 1972.

Delesseps, Jean-Marie Barthelemy. *Travels in Kamtschatka during the Years 1787 and 1788.* London, 1790.

Dening, Greg. *Mr Bligh's Bad Language: Passion, Power and Theatre on the Bounty.* Cambridge: Cambridge University Press, 1992.

Dmytryshyn, Basil, E. A. P Crownhart-Vaughan, and Thomas Vaughan, eds. *Russian Penetration of the North Pacific Ocean, 1700–1797,* Vol. 2, *A Documentary Record.* Portland: Oregon Historical Society Press, 1988.

Dunmore, John. *Pacific Explorer: The Life of Jean-François de La Pérouse, 1741–1788.* Annapolis, MD: Naval Institute Press, 1986.

Dunmore, John. *Who's Who in Pacific Navigation.* Honolulu: University of Hawaii Press, 1991.

Dunmore, John, ed. *The Journal of Jean-François de Galaup de La Pérouse, 1785–1788.* 2 vols. London: Hakluyt Society, 1994, 1995.

Efimov, A. V. *Atlas of Geographical Discoveries in Siberia and Northwestern American in the 17th–18th Centuries.* Moscow: Nauka, 1964.

Elliott, T. C. "Captain Robert Gray's First Visit to Oregon." *Oregon Historical Quarterly* (June 1928): 162–188.

Emmons, G. T. "Native Account of the Meeting between La Pérouse and the Tlingit." *American Anthropologist* (April–June 1911): 294–298.

Engstrand, Iris H. W. *Spanish Scientists in the New World: The Eighteenth-Century Expeditions.* Seattle: University of Washington Press, 1981.

Espinosa y Tello, José. *Relación del viage hecho por las goletas Sutil y Mexicana en el año de 1792.* Madrid: Imprenta Real, 1802.

Etulain, Richard W., ed. *Western Lives: A Biographical History of the American West.* Albuquerque: University of New Mexico Press, 2004.

Fireman, Janet R. "The Seduction of George Vancouver: A Nootka Affair." *Pacific Historical Review* (August 1987): 427–443.

Fisher, Raymond H. *Bering's Voyages: Whither and Why.* Seattle: University of Washington Press, 1977.

Fisher, Robin. *Contact and Conflict: Indian-European Relations in British Columbia, 1774–1890.* 2nd ed. Vancouver: University of British Columbia Press, 1992.

Fradkin, Philip L. *Wildest Alaska: Journeys of Great Peril in Lituya Bay.* Berkeley: University of California Press, 2001.

Frost, Alan, et al. *Pacific Empires: Essays in Honour of Glyndwr Williams.* Vancouver: University of British Columbia Press, 1999.
Frost, Orcutt. *Bering: The Russian Discovery of America.* New Haven, CT: Yale University Press, 2003.
Fry, Howard T. *Alexander Dalrymple and the Expansion of British Trade.* London: Frank Cass, 1970.
Galvin, John, ed. *The First Spanish Entry into San Francisco Bay, 1775.* San Francisco: John Howell Books, 1971.
Gerhard, Peter. *A Guide to the Historical Geography of New Spain.* Norman: University of Oklahoma Press, 1993.
Gibson, James R. *Otter Skins, Boston Ships, and China Goods: The Maritime Fur Trade of the Northwest Coast, 1785–1841.* Seattle: University of Washington Press, 1992.
Giesecke, E. W. "The Ship Columbia and the Sloop Washington, a Little-Known Lithograph by George C. Wales." *Oregon Historical Quarterly* (Summer 1982): 204–205.
Golder, F. A. *Russian Expansion on the Pacific, 1641–1850.* Gloucester, UK: Peter Smith, 1960.
Gough, Barry M. *Juan de Fuca's Strait Voyages in the Waterway of Forgotten Dreams.* Vancouver, British Columbia: Harbour Publishing, 2012.
Gough, Barry M. *The Northwest Coast: British Navigation, Trade, and Discoveries to 1812.* Vancouver: University of British Columbia Press, 1992.
Greenblatt, Stephen. *Marvelous Possessions: The Wonder of the New World.* Chicago: University of Chicago Press, 1991.
Grinev, Andrei Val'terovich. *The Tlingit Indians in Russian America, 1741–1867.* Lincoln: University of Nebraska Press, 2005.
Hansen, Dagny B. "Captain James Cook's First Stop on the Northwest Coast: By Chance or by Chart?" *Pacific Historical Review* 62(4) (November 1993): 475–484.
Harlow, Vincent T. *The Founding of the Second British Empire, 1763–1793.* 2 vols. London: Longmans, Green, 1964.
Haycox, Stephen, James K. Barnett, and Caedmon A. Liburd. *Enlightenment and Exploration in the North Pacific, 1741–1805.* Seattle: University of Washington Press, 1997.
Hayes, Derek. *Historical Atlas of the North Pacific Ocean: Maps of Discovery and Scientific Exploration, 1500–2000.* Seattle: Sasquatch Books, 2001.
Hayes, Derek. *Historical Atlas of the Pacific Northwest: Maps of Exploration and Discovery.* Seattle: Sasquatch Books, 1999.
Hayes, Edmund. "Gray's Adventure Cove." *Oregon Historical Quarterly* (June 1967): 101–110.
Hebda, Richard J., and Rolf W. Mathewes. "Holocene History of Cedar and Native Indian Cultures of the North American Pacific Coast." *Science* 225(4663) (August 17, 1984): 711–713.
Howay, F. W. "Captains Gray and Kendrick: The Barrell Letters." *Washington Historical Quarterly* (October 1921): 243–271.
Howay, F. W. "Captain Simon Metcalfe and the Brig 'Eleanor.'" *Washington Historical Quarterly* (April 1925): 114–121.
Howay, F. W. "Early Followers of Captain Gray." *Washington Historical Quarterly* (January 1927): 11–20.
Howay, F. W. "Four Letters from Richard Cadman Etches to Sir Joseph Banks, 1788–1792." *British Columbia Historical Quarterly* 6(2) (April 1942): 125–140.
Howay, F. W. "Indian Attacks upon Maritime Traders of the North-West Coast, 1785–1805." *Canadian Historical Review* 6 (1925): 287–309.

Howay, F. W. "Letters Relating to the Second Voyage of the Columbia." *Oregon Historical Quarterly* (June 1923): 132–152.
Howay, F. W. "A List of Trading Vessels in Maritime Fur Trade, 1785–1794." *Transactions of the Royal Society of Canada* 24(2) (1930): 111–134.
Howay, F. W. "The *Resolution* on the Oregon Coast, 1793–94." *Oregon Historical Quarterly* (September 1933): 207–215.
Howay, F. W. "A Short Account of Robert Haswell." *Washington Historical Quarterly* (April 1933): 83–90.
Howay, F. W. "Some Notes on Cook's and Vancouver's Ships, 1776–80, 1791–95." *Washington Historical Quarterly* 21(4) (October 1930): 268–270.
Howay, F. W. "The Widow of Captain Robert Gray." *Washington Historical Quarterly* (July 1929): 192–195.
Howay, F. W., ed. *The Dixon-Meares Controversy.* Bibliotheca Australiana Extra Series. Reprint ed. New York: Da Capo, 1969.
Howay, F. W., ed. *The Journal of Captain James Colnett aboard the* Argonaut. Toronto: Champlain Society, 1940.
Howay, F. W., ed. *Voyages of the Columbia to the Northwest Coast, 1787–1790 and 1790–1793.* Portland: Oregon Historical Society Press, 1990.
Howay, F. W., T. C. Elliott, and F. G. Young, eds. "John Boit's Log of the Columbia 1790–1793." *Oregon Historical Quarterly* (December 1921): 257–351.
Howay, F. W., and Albert Matthews. "Some Notes upon Captain Robert Gray." *Washington Historical Quarterly* (January 1930): 8–12.
Inglis, Robin, ed. *Spain and the North Pacific Coast: Essays in Recognition of the Bicentennial of the Malaspina Expedition, 1791–1792.* Vancouver, British Columbia: Vancouver Maritime Museum, 1992.
Jackson, Donald, ed. *Letters of the Lewis and Clark Expedition, with Related Documents, 1783–1854.* 2nd ed. 2 vols. Urbana: University of Illinois Press, 1978.
Jane, Cecil, trans. *A Spanish Voyage to Vancouver and the North-West Coast of America.* London: Argonaut, 1930.
Joppling, Carol F. *The Coppers of the Northwest Coast Indians: Their Origin, Development, and Possible Antecedents.* Philadelphia: American Philosophical Society, 1989.
The Journal of Captain James Colnett aboard the Argonaut. Toronto: Champlain Society, 1940.
Kendrick, John. *Alejandro Malaspina: Portrait of a Visionary.* Montreal: McGill-Queen's University Press, 1999.
Kendrick, John. *The Men with Wooden Feet: The Spanish Exploration of the Pacific Northwest.* Toronto, Ontario: NC Press Limited, 1985.
Kendrick, John. "Seas No Mariner Has Ever Sailed." *Northern Mariner* (July 1994): 1–17.
Kendrick, John, trans. *The Voyage of* Sutil *and* Mexicana, *1792: The Last Spanish Exploration of the Northwest Coast of America.* Northwest Historical Series no. 16. Spokane, WA: Arthur H. Clark, 1991.
Knight, T. M. "Cook the Cartographer." *Cartography* 7 (June 1971): 110–118.
Kodicek, Egon H., and Frank G. Young. "Captain Cook and Scurvy." *Notes and Records of the Royal Society of London* 24(1) (June 1969): 43–63.
Krasheninnikov, Stepan P. *Explorations of Kamchatka, 1735–1741.* Portland: Oregon Historical Society Press, 1972.
Lamb, W. Kaye, ed. *George Vancouver; Voyage of Discovery to the North Pacific Ocean and Round the World, 1791–1795.* 4 vols. London: Hakluyt Society, 1984.

Lang, William L. "George Vancouver, Lewis and Clark, and David Thompson: Exploring the Pacific Northwest by Tide and Trail." In *Western Lives: A Biographical History of the American West,* edited by Richard W. Etulain, 89–116. Albuquerque: University of New Mexico Press, 2004.

Ledyard, John. *Travels and Adventures of John Ledyard.* 2nd ed. London: Henry Colburn, 1834.

Malaquais, Dominique. "Eighteenth-Century French Voyages to the Northwest Coast of North America." *Anthropology and Aesthetics* (1990–1991): 211–233.

The Malaspina Expedition "In the Pursuit of Knowledge." Santa Fe: Museum of New Mexico Press, 1977.

Manby Diary. MS 2389 Oregon Historical Society, Portland, Oregon.

Marchand, Etienne. *A Voyage Round the World Performed during the Years 1790, 1791, and 1792.* London: Longman, 1801.

Martin, James C., and Robert Sidney Martin. *Maps of Texas and the Southwest, 1513–1900.* Austin: Texas State Historical Association, 1999.

Mathes, W. Michael. "Sebastián Vizcaíno and San Diego Bay." *San Diego Historical Society Quarterly* 1 (Spring 1972): 2.

Mathes, W. Michael. *Spanish Approaches to the Island of California, 1628–1632.* San Francisco: Book Club of California, 1975.

Mathes, W. Michael. *Vizcaíno and Spanish Expansion in the Pacific Ocean, 1580–1630.* San Francisco: California Historical Society, 1968.

McClellan, James, and Francois Regourd. "The Colonial Machine: French Science and Colonization in the Ancien Regime." *Osiris,* 2nd Series, 15 (2000): 31–50.

McCord, Carey P. "Scurvy as an Occupational Disease." *Journal of Occupational Medicine* 13(9) (September 1971): 441–447.

Meany, Edmond S., ed. "New Log of the Columbia." *Washington Historical Quarterly* (January 1921): 3–50.

Meany, Edmond S., ed. *A New Vancouver Journal on the Discovery of Puget Sound by a Member of the Chatham's Crew.* Seattle: n.p., 1915.

Meares, John. *Voyages Made in the Years 1788 and 1789 from China to the N.W. Coast of America.* London, 1790.

Morse, William Inglis, ed. *Letters of D. Alexandro Malaspina (1790–1791).* Translated by Christopher Dawson. Boston: McIver-Johnson, 1944.

Müller, Gerhard F. *Voyages from Asia to America.* Bibliotheca Australiana No. 26. Amsterdam: N. Israel, 1967.

Munford, James Kenneth, ed. *John Ledyard's Journal of Captain Cook's Last Voyage.* Corvallis: Oregon State University Press, 1963.

Naish, John M. *The Interwoven Lives of George Vancouver, Archibald Menzies, Joseph Whidby, and Peter Puget: Exploring the Northwest Coast.* Lewiston, NY: Mellen, 1996.

Newcombe, C. F., MD., ed. *Menzies' Journal of Vancouver's Voyage, April to October, 1792.* Victoria: British Columbia Legislative Assembly, 1923.

Nokes, J. Richard. *Almost a Hero: The Voyages of John Meares, R.N., to China, Hawaii and the Northwest Coast.* Pullman: Washington State University Press, 1998.

Norris, John. "The Strait of Anian and British Northwest America: Cook's Third Voyage in Perspective." *BC Studies,* no. 36 (Winter 1977–1978): 1–22.

Oberg, Kalervo. *The Social Economy of the Tlingit Indians.* Seattle: University of Washington Press, 1973.

Obeyesekere, Gananath. *The Apotheosis of Captain Cook: European Mythmaking in the Pacific.* Princeton, NJ: Princeton University Press, 1992.

Petchick, Derek. *First Approaches to the Northwest Coast.* Seattle: University of Washington Press, 1979.
Postnikov, Alexei V. "The Mapping of Russian America: A History of Russian-American Contacts in Cartography." American Geographical Society Collection Special Publication No. 4, 1995.
Sahlins, Marshall David. *How "Natives" Think: About Captain Cook, for Example.* Chicago: University of Chicago Press, 1995.
Salmond, Anne. *The Trial of the Cannibal Dog: The Remarkable Story of Captain Cook's Encounters in the South Seas.* New Haven, CT: Yale University Press, 2003.
Samwell, David. "Some Account of a Voyage to South Seas." 1778–1779, MSS EG 2591. British Museum, London. Accessible in UW Microform A1687, Suzzallo Library, University of Washington, Seattle.
Schurz, William Lytle. "Acapulco and the Manila Galleon." *Southwestern Historical Quarterly* 22(1) (1918): 18–37.
Schurz, William Lytle. *The Manila Galleon.* 1939; reprint, New York: E. P. Dutton, 1959.
Servin, Manuel P., trans. "The Instructions of Viceroy Bucareli to Ensign Juan Pérez." *California Historical Society Quarterly* 40 (1961): 237–248.
Simsarian, James. "The Acquisition of Legal Title to Terra Nullius." *Political Science Quarterly* 53(1) (March 1938): 111–128.
Skelton, R. A., and R. V. Tooley. "The Marine Surveys of James Cook in North America, 1758–1768." *Map Collectors' Series* 4(37) (1967).
Stewart, George R. *Names of the Land: A Historical Account of Place-Naming in the United States.* San Francisco: Lexikos, 1982.
Stoddart, David R. "Captain Cook and How We Understand Him." *Geographical Review* 87(4) (October 1997): 537–541.
Svet, Yakov M., and Svetlana G. Federova. "Captain Cook and the Russians." *Pacific Studies* 2(1) (Fall 1978): 1–19.
Swanton, John R. "Social Condition, Beliefs, and Linguistic Relationship of the Tlingit Indians." 26th Annual Report, Bureau of American Ethnology. Washington, DC, 1908, 391–485.
Thornton, Thomas F. "Know Your Place: The Organization of Tlingit Geographic Knowledge." *Ethnology* (Autumn 1997): 295–307.
Tovell, Freeman M. *At the Far Reaches of Empire: The Life of Juan Francisco de la Bodega y Quadra.* Vancouver: University of British Columbia Press, 2008.
Tovell, Freeman M. "The Hezeta-Bodega Voyage of 1775: Its Significance for Spain's Presence in the Pacific Northwest." *Terrae Incognitae* 27 (1995): 57–65.
Tovell, Freeman M. "The Other Side of the Coin: The Viceroy, Bodega y Quadra, Vancouver, and the Nootka Crisis." *BC Studies* (Spring 1992): 3–29.
Tovell, Freeman M., et al., eds. *Voyage to the Northwest Coast of America, 1792: Juan Francisco de la Bodega and the Nootka Sound Controversy.* Norman, OK: Arthur H. Clark, 2012.
Vancouver, George. *A Voyage of Discovery to the North Pacific Ocean and Round the World.* 3 vols. London: G. G. and J. Robinson, and J. Edwards, 1798.
Vaughan, Thomas, and E. A. P. Crownhart-Vaughan. *Voyages of Enlightenment: Malaspina on the Northwest Coast, 1791/1792.* Portland: Oregon Historical Society, 1977.
Venegas, Miguel. *A Natural and Civil History of California Containing an Accurate Description of That Country.* London: Printed for James Rivington and James Fletcher, 1759.
Wagner, Henry R. "Apocryphal Voyages to the Northwest Coast of America." *Proceedings of the American Antiquarian Society* (1932): 179–234.

Wagner, Henry R. *The Cartography of the Northwest Coast of America to the Year 1800.* Amsterdam: N. Israel, 1968.

Wagner, Henry R. "Creation of Rights of Sovereignty through Symbolic Acts." *Pacific Historical Review* 7(4) (December 1938): 297–326.

Wagner, Henry R. "Journal of Tomás de Suría of His Voyage with Malaspina to the Northwest Coast of America in 1791." *Pacific Historical Review* (September 1936): 234–276.

Wagner, Henry R. "Some Imaginary California Geography." *Proceedings of the American Antiquarian Society* (April 1926): 22–49.

Wagner, Henry R. *Spanish Explorations in the Strait of Juan de Fuca.* Reprint ed. Mansfield Centre, CT: Martino Publishing, 2002.

Wagner, Henry R. *Spanish Voyages to the Northwest Coast of America in the Sixteenth Century.* San Francisco: California Historical Society, 1929.

Walker, Alexander. *An Account of a Voyage to the North West Coast of America in 1785 & 1786.* Edited by Robin Fisher and J. M. Bumsted. Seattle: University of Washington Press, 1982.

Waters, D. W. "The Cook Bicentenary: Reflexions upon Some of the Effects of the Three Voyages of Captain James Cook to the Pacific between 1768 and 1780." *British Journal for the History of Science* 4(2) (December 1968): 160–165.

Williams, Glyn. *The British Search for the Northwest Passage in the Eighteenth Century.* Primary source ed. London: Longmans, 1962.

Williams, Glyn. "An Eighteenth-Century Spanish Investigation into the Apocryphal Voyage of Admiral Fonte." *Pacific Historical Review* (November 1961): 319–327.

Williams, Glyn. *Voyages of Delusion: The Quest for the Northwest Passage.* New Haven, CT: Yale University Press, 2003.

Witthoff, John, and Frances Eyman. "Metallurgy of Tlingit, Dene, and Eskimo." *Expedition* (1969): 12–23.

Woollen, William Watson. *The Inside Passage to Alaska, 1792–1920.* Cleveland, OH: Arthur H. Clark, 1924.

Index

About the Authors

William L. Lang is professor emeritus of history at Portland State University. The author or editor of seven books on Pacific Northwest history, Lang was director of the Center for Columbia River History (1990–2003) and founding editor of *The Oregon Encyclopedia.*

James V. Walker is a retired physician and map collector living in Eugene, Oregon, with research interests focusing on the Pacific Northwest and early 19th-century Trans-Mississippi West material. He is a member of the steering committee of the Philip Lee Phillips Map Society of the Library of Congress. Walker attended Amherst College and the Case Western Reserve University School of Medicine and practiced nephrology and internal medicine for 26 years.

www.ingramcontent.com/pod-product-compliance
Lightning Source LLC
Chambersburg PA
CBHW081416220926
51613CB00011B/261